Genetic Analysis of Complex Diseases

Genetic Analysis of Complex Diseases

Third Edition

Edited by William K. Scott and Marylyn D. Ritchie

WILEY Blackwell

This third edition first published 2022

Edition History
2nd edition © 2006 by John Wiley & Sons, Inc. All rights reserved.

Registered Office
John Wiley & Sons, Inc., 111 River Street, Hoboken, NJ 07030, USA

Editorial Office
9600 Garsington Road, Oxford, OX4 2DQ, UK
For details of our global editorial offices, customer services, and more information about Wiley products visit us at www.wiley.com.

Wiley also publishes its books in a variety of electronic formats and by print-on-demand. Some content that appears in standard print versions of this book may not be available in other formats.

Library of Congress Cataloging-in-Publication Data

Names: Scott, William K., 1970– editor. | Ritchie, Marylyn DeRiggi, 1977– editor.
Title: Genetic analysis of complex diseases / edited by William K. Scott and Marylyn D. Ritchie.
Description: Third edition. | Hoboken, NJ : Wiley-Blackwell, 2022. | Preceded by Genetic analysis of complex diseases / [edited by] Jonathan L. Haines, Margaret Pericak-Vance. 2nd ed. c2006. | Includes bibliographical references and index.
Identifiers: LCCN 2021009896 (print) | LCCN 2021009897 (ebook) | ISBN 9781118123911 (paperback) | ISBN 9781119104087 (adobe pdf) | ISBN 9781119104070 (epub)
Subjects: MESH: Genetic Diseases, Inborn–genetics | Disease–genetics | Chromosome Mapping–methods | Genetic Predisposition to Disease | Research Design | Genetic Research
Classification: LCC RB155 (print) | LCC RB155 (ebook) | NLM QZ 50 | DDC 616/.042–dc23
LC record available at https://lccn.loc.gov/2021009896
LC ebook record available at https://lccn.loc.gov/2021009897

Cover Design: Wiley
Cover Image: © ESB Professional/Shutterstock

Set in 9.5/12.5pt STIXTwoText by Straive, Pondicherry, India

SKY10031113_110221

Contents

List of Contributors

Susan H. Blanton
Dr. John T. Macdonald Foundation
Department of Human Genetics
University of Miami Miller School of Medicine
Miami, FL, USA

Adam Buchanan
Genomic Medicine Institute
Geisinger
Danville, PA, USA

William S. Bush
Department of Population and Quantitative Health Sciences
Case Western Reserve University
Cleveland, OH, USA

Ren-Hua Chung
Institute of Population Health Sciences
Division of Biostatistics and Bioinformatics
National Health Research Institutes (Taiwan)
Hsinchu, Taiwan

Dana C. Crawford
Department of Population and Quantitative Health Sciences
Case Western Reserve University
Cleveland, OH, USA

Abigail Deppen
InformedDNA
St Petersburg, FL, USA

Logan Dumitrescu
Department of Neurology
Vanderbilt University
Nashville, TN, USA

Kayla Fourzali
University of Miami Miller School of Medicine
Miami, FL, USA

Susan Estabrooks Hahn
Genomic Services
Quest Diagnostics
North Andover, MA, USA

Jonathan L. Haines
Department of Population and Quantitative Health Sciences
Case Western Reserve University
Cleveland, OH, USA

Dale J. Hedges
Center for Applied Bioinformatics
St. Jude Children's Research Hospital
Memphis, TN, USA

Elizabeth Heise
Clinical Genetics Program
GeneDX, Inc
Gaithersburg, MD, USA

Allison Ashley Koch
Duke Molecular Physiology Institute
Duke University Medical Center
Durham, NC, USA

Eden R. Martin
Dr. John T. Macdonald Foundation
Department of Human Genetics
University of Miami Miller School of Medicine
Miami, FL, USA

Jacob L. McCauley
Dr. John T. Macdonald Foundation
Department of Human Genetics
University of Miami Miller School of Medicine
Miami, FL, USA

Sarah A. Pendergrass
Human Genetics
Genentech
San Francisco, CA, USA

Margaret A. Pericak-Vance
Dr. John T. Macdonald Foundation
Department of Human Genetics
University of Miami Miller School of Medicine
Miami, FL, USA

Evadnie Rampersaud
Center for Applied Bioinformatics
St. Jude Children's Research Hospital
Memphis, TN, USA

Marylyn D. Ritchie
Department of Genetics
Perelman School of Medicine at the
University of Pennsylvania
Philadelphia, PA, USA

William K. Scott
Dr. John T. Macdonald Foundation
Department of Human Genetics
University of Miami Miller School of Medicine
Miami, FL, USA

Stephen D. Turner
Signature Science
LLC
Charlottesville, VA, USA

Shefali Setia Verma
Department of Pathology and Laboratory
Medicine
Perelman School of Medicine at the
University of Pennsylvania
Philadelphia, PA, USA

Yogasudha Veturi
Department of Genetics
Perelman School of Medicine at the
University of Pennsylvania
Philadelphia, PA, USA

Chantelle Wolpert
Physician Assistant Program
Thomas Jefferson University
Philadelphia, PA, USA

Foreword

This book grew from our four-day NIH-sponsored course, which, for 20 years, was focused on providing an overview and guide to the design and execution of human genetic mapping studies for these common (and genetically complex) diseases, melding the genomic technology with the statistical rigor needed to apply and interpret the results. When we developed the concept for the first edition of this book in 1996, the Human Genome Project was just reaching full speed, combining continual breakthroughs in DNA gene mapping and sequencing technology with emerging applications to human disease to shed the first light on the organization of the human genome and the variations that cause disease. The first applications of the Human Genome Project data were to find the location, and ultimately the causative mutations, for rare Mendelian inherited diseases. It was dogma then that the genetic architecture of common diseases was beyond our reach, based on the naïve belief that Mendelian disease represented how genetic variation impacted disease. However, we soon demonstrated, with the discovery that multiple apolipoprotein E (*APOE*) alleles had differing and strong effects on the risk of Alzheimer disease, that these technologies and approaches could be adapted to illuminate the genetic underpinnings of common diseases.

The rapid advances in both DNA technology and statistical methodology demanded that a significant update to the book was needed, with the second edition of the book in 2006. By this point the blood and protein markers of the 1970s had been surpassed by the restriction fragment length polymorphisms (RFLPs) of the 1980s, the microsatellite repeats of the 1990s, and the single nucleotide polymorphisms (SNPs, of which RFLPs are a subset) for the past 20 years. Naturally, the analyses of these data also advanced from early mainframe applications of genetic linkage analysis in small numbers of families, to PC-powered analyses of thousands of cases and controls for association.

In the past 15 years since that second edition, increasingly dense SNP arrays and whole exome or whole genome sequencing have created new horizons for dissecting complex diseases. In addition, the explosion of other "omics" data, particularly gene expression data, provide biological context for the discovered DNA variations, adding biological interpretation as a critical element of genetic studies.

With all these advances, it became apparent that a new edition of this book was warranted, and new and fresh perspectives were needed. Thus, we turned over the editing of this new edition to two of our brilliant younger colleagues, who have been active in both developing and applying methods at the forefront of genetics and genomics. While the inclusion of genome-wide association studies, integration of genomic data, and data mining are new, the breadth of the book in describing the overall process of designing and executing successful projects remains.

Finally, we fondly acknowledge the continuing impact of our mentor, Dr. P. Michael Conneally, who inspired both of us to inquire, question, investigate, and solve, the often difficult, constantly emerging human genetic puzzles. He encouraged us to help educate researchers, physician-scientists, and physicians in the complex nature of genetic studies. He wrote the forward for the first two editions, and although he passed away in 2017, his legacy remains in our work and the work of our trainees and collaborators.

We are immensely grateful to Bill and Marylyn for taking on this important task and developing this excellent third edition of the book.

Jonathan L. Haines, PhD
Margaret A. Pericak-Vance, PhD

1

Designing a Study for Identifying Genes in Complex Traits

William K. Scott[1], Marylyn D. Ritchie[2], Jonathan L. Haines[3], and Margaret A. Pericak-Vance[1]

[1] *Dr. John T. Macdonald Foundation Department of Human Genetics, University of Miami Miller School of Medicine, Miami, FL, USA*
[2] *Department of Genetics, Perelman School of Medicine at the University of Pennsylvania, Philadelphia, PA, USA*
[3] *Department of Population and Quantitative Health Sciences, Case Western Reserve University, Cleveland, OH, USA*

Introduction

Disease gene discovery in humans has a long history, predating even the identification of DNA as the genetic molecule (Watson and Crick 1953) and the determination of the number of human chromosomes (Ford and Hamerton 1956; Tjio and Levan 1956). In fact, as early as the 1930s some simple statistical methods for the analysis of genetic data had been developed (Bernstein 1931; Fisher 1935a,b). However, these methods were severely limited in their application (more on basic concepts of genetics in Chapter 2). Not only were genetic markers lacking (the ABO blood type was one of the few that had been described), but these methods were restricted to small, two to three generation pedigrees. Any calculations were performed by hand, of course, making analysis laborious.

There were two hurdles to overcome before human disease gene discovery would become routine. First, appropriate statistical methods were lacking, as were ways of automating the calculations. Second, sufficient genetic markers to cover the human genome needed to be identified. Morton (1955), building on the work of Haldane and Smith (1947) and Wald (1947), described the use of maximum likelihood approaches in a sequential test for linkage between two loci. He used the term "LOD score" (for logarithm of the odds of linkage) for his test. This score is the basis for most modern genetic linkage analyses and represents a milestone in human disease gene discovery. However, the complex calculations had to be done by hand, severely limiting the use of this approach. Elston and Stewart (1971) described a general approach for calculating the likelihood of any non-consanguineous pedigree. This algorithm was extended by Lange and Elston (1975) to include pedigrees of arbitrary complexity. Soon thereafter, the first general-purpose computer program for linkage in humans, LIPED (Ott 1974), was described. Thus, the first of the two major hurdles was overcome.

By the mid-1970s there were 40–50 red cell antigen and serum protein polymorphisms available as genetic markers. A few markers could be arranged into initial linkage groups, but these markers covered only approximately 5–15% of the human genome. In addition to this limited coverage, genotyping these polymorphisms was labor intensive, time consuming, and often quite technically

Genetic Analysis of Complex Diseases, Third Edition. Edited by William K. Scott and Marylyn D. Ritchie.

demanding. This remaining hurdle was crossed with the description of restriction fragment length polymorphisms (RFLPs) by Botstein et al. (1980). Not only were these markers easier to genotype in a standard manner, but they were frequent in the genome, covering the remaining 85–95% of the genome for the first time.

With these tools in place, the field of human disease gene discovery blossomed. The first successful disease gene linkage using RFLPs was reported (Gusella et al. 1983), localizing the Huntington disease gene to chromosome 4p. This discovery marked the beginning of disease gene identification through the *positional cloning* approach. Early successes using positional cloning were for diseases inherited in Mendelian fashion: autosomal dominant, autosomal recessive, or X-linked. Although confounding factors such as genetic heterogeneity, variable penetrance, and phenocopies might exist for single-gene or Mendelian traits, it is generally possible with a known genetic model to determine the best and most efficient approach to identifying the responsible gene. The success of these tools is apparent since by mid-2017 over 3350 single-gene disorders had at least one causative genetic variant identified (OMIM, accessed May 2017 at http://omim.org).

However, the inheritance patterns for traits such as the common form of Alzheimer's disease, multiple sclerosis, and non-insulin-dependent diabetes (to name a few) do not fit any simple genetic explanation, making it far more difficult to determine the best approach to identifying the unknown underlying effect. In addition to the confounding factors involved in single-gene disorders, such as genetic heterogeneity and phenocopies, gene–gene and gene–environment interactions must be considered when a complex trait is dissected. However, the tools that enabled efficient mapping of Mendelian trait loci through positional cloning were not as effective in dissecting these more complex traits. New statistical tools, study designs, and genotyping technologies were needed to perform large-scale analysis of genetic factors underlying these complex traits. As these technologies were developed, a new approach to complex disease gene identification via genome-wide association studies (GWAS) was enabled. The shift to this approach was predicted by a seminal perspective published by Risch and Merikangas (1996), in which they showed that large-scale case–control analyses of complex traits would be a powerful and efficient method of identifying alleles underlying complex traits, once genotyping technology allowed the cost-effective determination of a dense map of genetic markers. The first GWAS was published in 2005 (Klein et al. 2005), identifying the association of variation in the *CFH* gene with age-related macular degeneration. This was simultaneously confirmed using alternate study designs (Edwards et al. 2005; Haines et al. 2005) proving that GWAS worked, allowing this new era of complex disease genetics to begin in earnest.

With the dawn of the GWAS era, a corresponding shift in the prevailing hypotheses for these studies occurred. No longer were studies solely searching for one or a few rare mutations in a single gene that cause a rare and devastating disease. Studies of common complex diseases were searching for multiple alterations in one or more genes acting alone or in concert to increase or decrease the risk of developing a trait. Early GWAS tended to test the "common disease-common variant" (CDCV) hypothesis: the risk for common diseases, across ethnic groups, arises from evolutionarily old variants that have had substantial time to spread throughout the human population. Many studies successfully identified thousands of variants associated with the risk of complex diseases. An interactive catalog of these variants is maintained by the National Human Genome Research Institute and the European Molecular Biology Laboratory at http://www.ebi.ac.uk/gwas. Despite these successes, many studies testing the CDCV hypothesis failed to explain all the heritable variation in the risk of the complex traits under study – a phenomenon termed "missing heritability" (Manolio et al. 2009). One explanation for this was that the effect of rare variants was not well studied by early GWAS – an alternative hypothesis termed the "common disease-rare variant"

(CDRV) hypothesis. This hypothesis suggests that risk of common complex diseases arises from a larger number of rare variants in one or more genes, perhaps occurring more recently.

As was the case with common variants and the exploration of the CDCV hypothesis being enabled by GWAS approaches and high-throughput genotyping technology, exploration of the CDRV hypothesis was enabled by advances in high-throughput sequencing technology and accompanying statistical analysis methods. Initial screens of coding-sequence variants in Mendelian traits via whole-exome sequencing (WES) were published by Ng et al. (2009, 2010) and Choi et al. (2009), demonstrating that in some cases, disease gene mapping could skip the positional cloning strategy and proceed directly to evaluating segregation of mutations in families. This proof of principle has been used to justify this approach for testing the CDRV hypothesis in complex traits but has been met with mixed success. A successful example is the recent analysis of 50 000 individuals in the MyCode Community Health Initiative successfully identified rare variants underlying cardiovascular traits and lipid levels (Dewey et al. 2016). The rapid and continuing decrease in whole-genome sequencing (WGS) costs suggests that within a few years, it will be possible (and perhaps commonplace) to test the CDRV hypothesis using WGS in large sample sizes – essentially performing genome-wide association for common and rare variants with direct genotype determination via sequencing.

Study design, laboratory methods, and analytic approaches differ by trait type (Mendelian or complex) and hypothesis being tested (rare disease-rare variant, Mendelian positional cloning; CDCV [GWAS]; CDRV [WES or WGS and individual variant or set-based association]). These approaches are described in the following sections.

Components of a Disease Gene Discovery Study

Each genetically complex trait has its own peculiarities that require special attention. However, a guiding paradigm can be applied to most conditions. Originally, the general approach that was used for Mendelian single-gene disorders was *positional cloning*. With the completion of the human genome reference sequence, cloning was no longer a necessary step – and therefore this general approach is better described as *disease gene discovery*. The classical approach (Figure 1.1) follows a generally linear series of events: defining the phenotype, identifying multi-case families, collecting blood samples, genotyping markers, analyzing data for initial disease gene localization, refining the initial localization to define the minimum candidate region, and then sequencing genes within this region to find the causative mutation(s).

In contrast to the classical approach, the current approaches to finding genes for common and genetically complex traits are not linear, and many steps are works in progress, subject to further defining, refining, or replacement by subsequent steps. Figure 1.2 illustrates the stepwise and recursive nature of the components of a complex trait study. Each step has its own key factors that must be considered, and for complex traits, the order and emphasis of these steps on the approach will vary from study to study. This fact is underappreciated and contrasts strongly with the classical

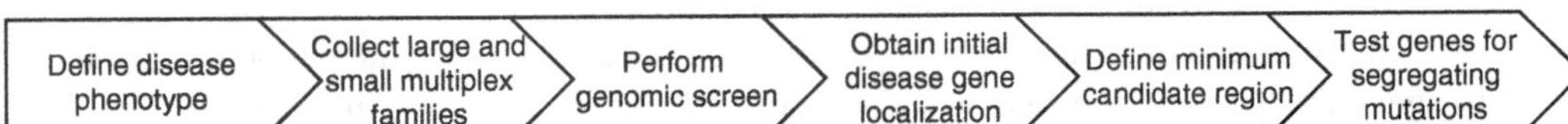

Figure 1.1 Steps in a Mendelian disease gene discovery (positional cloning) study.

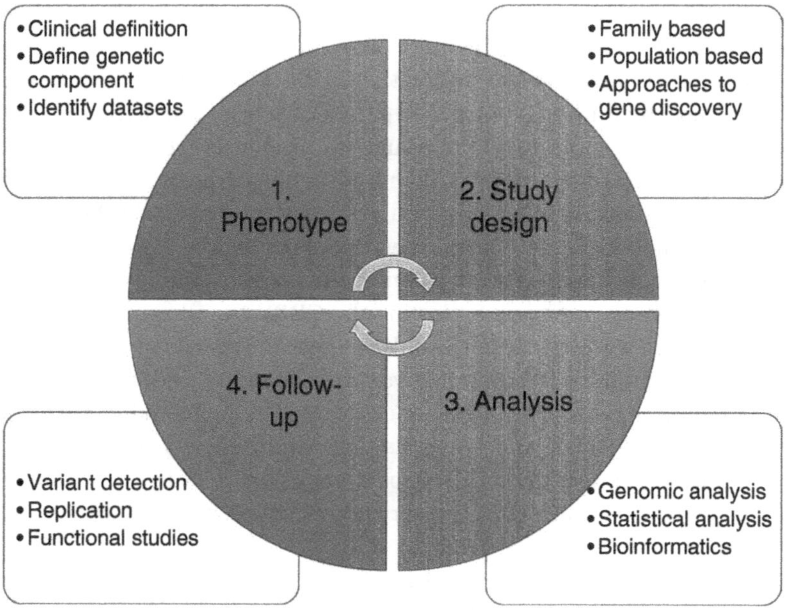

Figure 1.2 Study cycle for a complex trait gene identification study.

disease gene discovery approach. Indeed, many of the difficulties reconciling discordant studies of the same complex trait arise from study-specific decisions made in the approach.

This section discusses the steps in Figure 1.2, providing an overview of each component and a guide to the chapter(s) providing more detail on these points.

Define Disease Phenotype

The first step in any disease gene discovery process is to know what phenotype is being studied. This may sound obvious, but specifying the exact measures that will be used to reliably and validly determine the phenotype is often overlooked in the rush to move forward. There are three aspects that need to be considered: clinical definition, determining that a trait has a genetic component, and identification of datasets that can be studied.

Clinical Definition

It is not enough to define a trait in binary terms, such as the presence or absence of Huntington's disease or diabetes. In Huntington's disease, for example, there can be wide variation in the symptoms, with some only psychological or very mild motor disturbances detectable by expert examination, and the age at which these symptoms begin is similarly variable. In diabetes, there are distinct subtypes (insulin-dependent diabetes mellitus and non-insulin-dependent diabetes mellitus) as well as variable age at onset. Additionally, blood glucose levels (a quantitative trait) are strongly associated with diabetes (a qualitative trait) and could be used as a surrogate measure or endophenotype. One critical role of the clinician in study design is to assess the various diagnostic procedures and tools and determine which ones best define a consistent phenotype. Additionally, dissecting genetically complex diseases usually requires large datasets to supply enough power to unravel genetic effects. For this reason, participant ascertainment often extends to multiple sites. It is critical for multi-site studies to establish consensus diagnostic procedures and criteria and apply them consistently across sites. For example, the establishment of a consensus diagnostic scheme (McKhann et al. 1984) played an important role in a successful complex disease linkage study in late-onset familial Alzheimer's disease (Pericak-Vance et al. 1991) and subsequent identification of the association of Alzheimer's disease and common variation in the *APOE* gene (Corder et al. 1993; Corder et al. 1994).

The phenotype assignment must be done in a rigorously consistent fashion. Even a small rate of phenotype error might alter analytic results – in some cases leading to false-positive results and in others to false-negative results. Thus, which data will be used to assign the trait status must be carefully determined. Must detailed clinical records of an examination specifically addressing the phenotype be obtained and reviewed for consistency on every participant? Is the self-report of a participant or a participant's relative sufficient? Is a note documenting a diagnosis (but no examination findings) from a medical record adequate? Or is direct examination of every participant using a standardized research protocol required? Additionally, investigators must consider whether to collect additional biomarker data (e.g. antibody titers, protein assays) or clinical tests (e.g. electroencephalogram, electrocardiogram, magnetic resonance imaging) that might correlate with the trait of interest. The goal of the phenotyping protocol is to standardize procedures, minimize error in determining the phenotype, and maximize the power of the dataset to detect genes underlying the trait.

Determining that a Trait Has a Genetic Component

It is critical that as much as possible be known about the genetic basis of a complex trait prior to determine the most appropriate study design for gene identification. That a trait "runs in families" is insufficient evidence, since this phenomenon can occur for several reasons other than shared genetic susceptibility, including shared environmental exposure and biased ascertainment. As outlined in Chapter 3, there are numerous lines of evidence that can be examined, including family studies, segregation analysis, twin studies, adoption studies, heritability studies, and population-based risks to relatives of probands (the initially identified individual with disease). For most traits being contemplated, some such data already exist in the literature. A thorough review of this literature may provide most of the necessary information and point out any missing data. The data may not only indicate the strength of the genetic effect on the trait but also give some indication of the underlying genetic model. For example, there may be obvious evidence of a single "major" gene, such as in Huntington's disease, or multiple genes interacting in complex ways, such as in multiple sclerosis (Sadovnick et al. 1996).

Identification of Datasets

It is helpful to identify early on what potential datasets exist or can be collected. Do large families exist or are most cases apparently sporadic? Are large cohort or case–control studies available? Are there repositories of multiplex families with associated clinical data available? Are there existing clinical networks or large specialty clinics available? Is the necessary phenotype data available in a biobank linked to an existing electronic health record? The answers to these questions determine what study designs are feasible for the trait under study, as discussed in Chapters 3 and 4.

Develop Study Design

Developing your study design and delineating the phenotype are not independent steps. Review of the available data may indicate that a trait as originally defined has little or no evidence of a genetic component. However, there may be strong evidence that a subset of the trait is strongly genetic. For example, there had for many years been debate about the role of genetics in Alzheimer's disease. Over time it became increasingly clear that a subset of individuals with the onset of Alzheimer's disease before age 65 existed and strongly clustered in families with apparent autosomal dominant inheritance. Within each of these families, Alzheimer's disease appeared to be caused by a single gene. By restricting sample collection and genetic analysis to these types of families, three genes (*APP*, *PSEN1*, and *PSEN2*) were identified with mutations causing early-onset Alzheimer's disease (Goate et al. 1991; Levy-Lahad et al. 1995; Rogaev et al. 1995; Sherrington et al. 1995).

The exact approach to the disease gene discovery process should be outlined as completely as possible before the project gets underway. With the clinical phenotype in hand, it is possible to determine the best strategy for defining what type of dataset to collect. Participant recruitment is perhaps the longest and most labor-intensive step in the entire process. It is imperative that the enrollment of participants (particularly if studying multiple members of the same family) proceeds with careful consideration of the wishes and norms of the participating individuals, families, and communities. The rights of individuals to participate or refuse participation should receive careful consideration, and the informed consent process should provide adequate explanation of the study and answer any questions, and, critically, confidentiality must be carefully protected. These issues are outlined in detail in Chapter 5.

Determination of the study design (case–control, cohort, case series, family-based) is based on the characteristics of the phenotype, the estimated genetic model, and the research objective. For example, the existence of large families with apparent Mendelian segregation suggests that a single major gene could be detected, and a family-based study would be appropriate. A phenotype with weaker estimated heritability, a pattern of recurrence risks suggesting many genes of small effect, and little familial aggregation would suggest that a case–control study design is most feasible. The process of selecting a study design to answer a research question is reviewed in Chapter 4.

It is also important to have some sense of the sample size required to identify the genes being sought. When pedigree structures are already available in family-based studies of single-gene disorders, power is easily calculated with high confidence for specific genetic models using computer simulation programs. For complex traits, however, genetic models are not as easily specified in advance, and computer simulations often must consider a range of parameter values for the genetic model to describe the power across several competing alternatives. Chapter 12 provides an overview of the available approaches and tools for sample size, power estimation, and genetic simulations.

Family-Based Studies

Family-based studies include large extended families, smaller multi-case families (often affected sibpair or other affected relative pairs), and discordant sibpair studies. Depending on family structure and number of individuals collected, these families may be used in linkage analyses (as discussed in Chapter 6) or association studies (Chapter 8). Depending on the genetic architecture of the trait and the frequency of the disease-associated alleles being sought, this design may offer increased power over population-based designs.

Population-Based Studies

Several types of observational designs may be considered for population-based studies, including case-series, case–control, and longitudinal cohort designs. The possible sampling frames for these types of studies include simple random samples of a defined geographical area, clinic- or hospital-based samples, convenience samples such as voluntary registries or biobanks, or hybrids of these (e.g. health-system-based biobanks linked to longitudinal electronic health records). These designs became much more frequent with the advent of high-throughput genotyping technologies, which enabled the efficient study of very large samples of unrelated individuals through GWAS (Chapter 9), an approach with substantially greater power than a similarly sized family-based study.

Approaches for Gene Discovery

There are two general, but not mutually exclusive, ways to approach gene discovery for complex traits. The first is to take a genome-wide screening approach. Genomic screening can aim to identify areas of *genetic linkage* in family-based designs (Chapter 6) or areas of *association* in either family- or population-based designs (Chapters 8 and 9). A good genomic screen will attempt to cover the entire human genome using markers evenly spaced across the genome. Current high-throughput genotyping technologies enable genotyping of hundreds of thousands to millions of single nucleotide polymorphisms in a rapid, inexpensive manner for use in linkage or association studies. More recently, high-throughput sequencing technology has been used to screen the entire coding sequence of the genome (WES) or the entire genome (WGS) for trait-associated variants, without first conducting genome-wide linkage or association studies. As sequencing costs continue to decline, a shift to "genotyping by sequencing" is likely, in which results from WGS might be used to conduct a genome-wide screen and follow-up in a single molecular experiment. These same high-throughput genotyping and sequencing technologies allow large-scale examination of gene expression (through gene expression microarrays or RNA-Seq) and epigenetic changes (through methylation arrays or Methyl-Seq) in trait-relevant tissues. The results of such experiments are often used in conjunction with genome-wide screens to identify high-priority candidate genes for follow-up studies. These technologies and their application to genomic studies are discussed in Chapter 10.

In contrast to the genomic screening approach, a directed screening approach may be used. This approach, sometimes termed a "candidate-gene" approach, focuses on an area of the genome selected for examination based on prior information. The additional information could come from many sources, including results from a previous genome-wide screen, results from gene expression studies, genes suggested by pathophysiology, or candidate genes identified in model systems. For example, multiple sclerosis is an autoimmune disease in which the myelin sheaths around nerves are attacked and often destroyed. This information suggests that certain genes, such as the human leukocyte antigen genes, T-cell receptor genes, and the myelin basic protein gene, are prime candidates for analysis. The strength and weakness of this approach arise from the confidence in the role of these genes. If the evidence is strong that a direct role is played, only a few such genes may need to be tested to find a trait-associated variant. If the evidence is more circumstantial, then many genes may have equal justification for being studied, and not much is gained over conducting a genome-wide screen. Such studies are now most often conducted as follow-up of prior genomic screens or other hypothesis-generating experiments.

Analysis

Genomic Analysis

Generally, genome-wide genotyping or sequencing is the first analytic step. Such studies may use newly collected blood samples or stored blood samples (or extracted DNA or RNA) made available by a biorepository. Depending on the goal of the study and its design, genome-wide genotyping, sequencing, gene expression, or epigenetic analysis may be performed on these samples. Some studies may be able to re-use stored genotype or sequence data available from public repositories (such as dbGaP [https://www.ncbi.nlm.nih.gov/gap] or the European Genome-phenome Archive [https://www.ebi.ac.uk/ega/home]) or from prior studies of the sample being used. The technologies and approaches to these molecular experiments are covered in Chapter 10. In each case, it is important to formulate a quality control plan to detect potential laboratory errors such as sample switches, failed genotyping probes, sequencing errors, and batch effects. When possible, coordinating laboratory analysis with initial analytic quality control is optimal for finding and correcting such errors. If archived genomic data are being used, careful review of the initial quality control protocols and further checks (when possible) in the subsequent analysis is recommended.

Statistical Analysis

The analysis of genetic and phenotypic data for a complex trait is multifaceted and depends on the research question, study design, genomic data available, and phenotypic characteristics. Methods to analyze these data are under constant development, and new approaches are continuously being released. Therefore, the analytic strategy for a genomic study must be reviewed periodically and revised if necessary to take advantage of newly developed approaches. Depending on the study design, the analytic plan may include linkage analysis (Chapter 6) in families or association studies in families or population samples (Chapters 8 and 9). These approaches are not mutually exclusive – a design may start with a linkage analysis of large families followed by association analysis within regions of linkage. Similarly, other multi-stage studies conduct a GWAS of individual SNPs (Chapter 9) and then incorporate gene–gene and gene–environment interactions to identify additional genetic loci. Additionally, "data mining" approaches may be applied to these datasets to extract even more genetic information using data reduction techniques, set-based tests, and pathway analyses. These more complex analyses are discussed in detail in Chapter 11.

Bioinformatics

The large amount of information generated by any genomic study of a complex trait requires careful attention to quality control, efficient and secure storage, and compliance with data-sharing requirements and privacy protections. These activities require a well-designed and secure database system. Such systems have evolved over time from text files to relational databases, to large-scale "data warehouses." Such datasets also require large-scale processing power with ample attached storage to facilitate linkage and association studies. High-throughput sequencing in particular requires a large amount of storage and computational power for genome alignment (or assembly) and base calling. For multi-site studies, these resources may need to be accessible from multiple locations, requiring levels of access and security depending on the role on the study and need to access other sites' information. In addition to maintaining local resources for a study, a bioinformatics team also must be familiar with many different public sources of genomic data (e.g. UCSC and Ensembl browsers, ENCODE databases, sequence repositories, dbGaP) and be able to submit results to public repositories for sharing with the wider research community. These issues are discussed in more detail in Chapter 7.

Follow-up

Variant Detection

Once a single gene (or region) is implicated by a screen (linkage or association), it is necessary to examine it for potentially functional variations that might explain the linkage or association signal. For positional cloning efforts, this generally consisted of sequencing the minimum candidate region and identifying mutations that segregated with the trait in families. For complex traits, this effort is more difficult, and the variant being sought may be a more common, yet functional, polymorphism. Several strategies, including haplotype analysis, conditional analysis, and exhaustive sequencing, may be used in this case. The analyses required for such efforts are discussed in Chapters 8 and 9. However, statistical analysis of a single dataset only goes so far to establish a trait-associated variant. Additional studies, including replication in independent datasets and functional studies in cellular and animal models, may be required to ultimately determine if a variant influences the biology underlying the complex trait.

Replication

The literature on most complex traits is at this point littered with initial reports of allelic or genotypic associations that cannot be replicated at all (or are replicated in a small minority of studies). Reproducibility of findings in independent samples is a critical characteristic most investigators seek when weighing the evidence for a trait-associated variant. Because of this, most studies (particularly those seeking government or foundation funding) now include a plan for replication of findings in a second dataset. These replication datasets should be independent of the initial finding (e.g. do not overlap with the discovery dataset) and be assessed in similar fashion (e.g. phenotype definitions agree, ascertainment is similar, genetic analysis is comparable). This does not mean that the datasets must be from the same population – indeed, demonstrating replication across populations (e.g. European, Asian, and African) for a common complex trait locus may add strength to the study. However, for rare variants, cross-population replication might be more difficult (due to population-specific alleles); for such studies, replication in a second sample from the sample population would be desirable.

Functional Studies

While most disease gene discovery efforts have claimed success based on finding variants that segregate with traits in pedigrees or polymorphisms significantly associated with the trait in population samples, this is, strictly, not sufficient evidence. More conclusive is evidence arising from biological systems (e.g. cultured cells, animal models, or human blood and tissue samples) that the trait can be either induced by introduction of the allele or ameliorated by blocking the action of the allele. In genetically complex traits, where the responsible variation may be a common polymorphism, it is even more critical that such evidence be found before success is declared.

Tests in biological systems can be of several types. Perhaps the most common is to test the action of the gene in a model organism, such as mouse, zebrafish, or fruit fly. With transgenic models, the proposed trait-associated variant is introduced into the germline of the organism and the resulting offspring are examined for evidence of the abnormal phenotype. With knockout models, the action of the gene in question is eliminated and the offspring are examined for evidence of an abnormal phenotype. Similar experiments can be performed in cultured cells, where the introduction of the variant (or gene knockout) is easier. However, finding the appropriate cell line and determining the appropriate cellular phenotype corresponding to the trait may be difficult. Recent advances in generating relevant cellular models have utilized inducible pluripotent stem cell (iPSC) technology, by which cells (blood, fibroblast) from an individual with a phenotype and genotype of interest can be reprogrammed and differentiated to a cell type of interest (such as neuron or retinal pigment epithelium). Such cells might be closer to the affected tissue type and have more recognizable phenotypes due to the genetic variant under study. A further advance incorporates gene editing technology (e.g. CRISPR/Cas9) into the approach, whereby an established iPSC line can be edited to introduce (or correct) a variant of interest. Such an approach eliminates the need to draw a sample from a person known to carry a variant of interest and allows examination of isogenic cell lines with and without the variant for phenotypic changes. These approaches are rapidly evolving, and frequently revised sources, such as *Current Protocols in Human Genetics*, should be consulted for the latest details on functional studies using these approaches.

Keys to a Successful Study

Foster Interaction of Necessary Expertise

To appropriately carry out any disease gene discovery study, one must use techniques from five different areas of expertise (Figure 1.3). These areas are clinical evaluation, molecular genetics, statistical genetics, bioinformatics, and epidemiology. The first provides the necessary diagnostic and participant recruitment skills needed to define the phenotype and help collect samples and data. The second provides genotyping, sequencing, and functional analysis skills necessary to help locate and identify the genes and variants of interest and evaluate their functional consequences. The third provides the statistical and analytical framework for the proper design of the study and the analysis of the generated data. The fourth provides computational and algorithmic expertise for the processing, storage, and dissemination of large-scale datasets. And the fifth provides expertise to incorporate environmental variables and apply results at the population level.

The initial focus of gene discovery on single-gene disorders resulted in a linear approach (Figure 1.1) that could be implemented by a single investigator with expertise in one of these areas, with periodic consultation with colleagues from other disciplines as needed. Complex traits require a multidisciplinary approach that is not easily implemented by a single investigator, and given differences in genetic architecture, available samples, and research questions, different approaches (and thus different teams) may need to be formed for each trait. Thus, experts in each of these

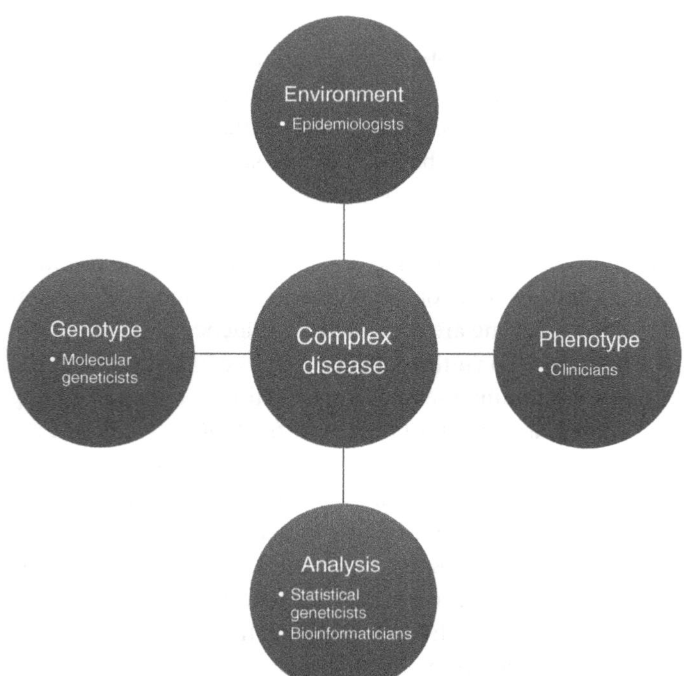

Figure 1.3 Components of a complex disease study and expertise needed to contribute.

fields must be intimately involved in all aspects of the study. Even with all this expertise in place, it is essential that the study not be divided into separate components with little interaction. For example, statistical geneticists and epidemiologists should be involved in the discussion of the clinical phenotype to determine the effect of potential changes to the phenotype definition on the genetic study design, screening approach, and statistical power.

Develop Careful Study Design

It may seem self-evident that a careful study design is necessary for a successful study. However, it is not enough to decide on a general design of "collect cases and controls, genotype on a genome-wide chip, do GWAS analysis." Each step in the study requires substantial thought, and the decisions made at one step will have implications for each of the others. Much as a team of engineers and architects must project unintended side effects from a change in a structural design, lest a catastrophic failure ensue, researchers must consider carefully all aspects of the experimental design lest they doom themselves to making inappropriate conclusions based on inadequately obtained and interpreted results.

References

Bernstein, F. (1931). Zur grundlegung der chromosomentheorie der vererbung beim menschen. *Z. Abst. Vererb.* 57: 113–138.

Botstein, D., White, R.L., Skolnick, M., and Davis, R.W. (1980). Construction of a genetic linkage map in man using restriction fragment length polymorphisms. *Am. J. Hum. Genet.* 32 (3): 314–331.

Choi, M., Scholl, U.I., Ji, W. et al. (2009). Genetic diagnosis by whole exome capture and massively parallel DNA sequencing. *Proc. Natl Acad. Sci. USA* 106 (45): 19096–19101.

Corder, E.H., Saunders, A.M., Strittmatter, W.J. et al. (1993). Gene dose of apolipoprotein E type 4 allele and the risk of Alzheimer's disease in late onset families. *Science* 261 (5123): 921–923.

Corder, E.H., Saunders, A.M., Risch, N.J. et al. (1994). Protective effect of apolipoprotein E type 2 allele for late onset Alzheimer disease. *Nat. Genet.* 7: 180–184.

Dewey, F.E., Murray, M.F., Overton, J.D. et al. (2016). Distribution and clinical impact of functional variants in 50,726 whole-exome sequences from the DiscovEHR study. *Science* 354 (6319): aaf6814.

Edwards, A.O., Ritter, R. 3rd, Abel, K.J. et al. (2005). Complement factor H polymosphism and age-related macular degeneration. *Science* 308 (5720): 421–424.

Elston, R.C. and Stewart, J. (1971). A general model for the genetic analysis of pedigree data. *Hum. Hered.* 21: 523–542.

Fisher, R.A. (1935a). The detection of linkage with dominant abnormalities. *Ann. Eugenics* 6: 187–201.

Fisher, R.A. (1935b). The detection of linkage with recessive abnormalities. *Ann. Eugenics* 6: 339–351.

Ford, C.E. and Hamerton, J.L. (1956). The chromosomes of man. *Nature* 178 (4541): 1020–1023.

Goate, A., Chartier-Harlin, M.C., Mullan, M. et al. (1991). Segregation of a missense mutation in the amyloid precursor protein gene with familial Alzheimer's disease. *Nature* 33: 53–56.

Gusella, J.F., Wexler, N.S., Conneally, M.P. et al. (1983). A polymorphic DNA marker genetically linked to Huntington's disease. *Nature* 306 (5940): 234–238.

Haines, J.L., Hauser, M.A., Schmidt, S. et al. (2005). Complement factor H variant increases the risk of age-related macular degeneration. *Science* 308 (5720): 419–421.

Haldane, J.B.S. and Smith, C.A.B. (1947). A new estimate of the linkage between the genes for color blindness and hemophilia in man. *Ann. Eugenics* 14: 10–31.

Klein, R.J., Zeiss, C., Chew, E.Y. et al. (2005). Complement factor H polymorphism in age-related macular degeneration. *Science* 308 (5720): 385–389.

Lange, K. and Elston, R.C. (1975). Extension to pedigree analysis. I. Likelihood calculations for simple and complex pedigrees. *Hum. Hered.* 25: 95–105.

Levy-Lahad, E., Wasco, W., Poorkaj, P. et al. (1995). Candidate gene for the chromosome 1 familial Alzheimer's disease locus. *Science* 269: 973–977.

Manolio, T.A., Collins, F.S., Cox, N.J. et al. (2009). Finding the missing heritability of complex diseases. *Nature* 461 (7265): 747–753.

McKhann, G., Drachman, G., and Folstein, M. (1984). Clinical diagnosis of Alzheimer's disease: report of the NINCDS-ADRDA Work Group under the auspices of the department of health and human services task force on Alzheimer's disease. *Neurology* 34: 939–944.

Morton, N.E. (1955). Sequential tests for the detection of linkage. *Am. J. Hum. Genet.* 7: 277–318.

Ng, S.B., Turner, E.H., Robertson, P.D. et al. (2009). Targeted capture and massively parallel sequencing of 12 human exomes. *Nature* 461 (7261): 272–276.

Ng, S.B., Buckingham, K.J., Lee, C. et al. (2010). Exome sequencing identifies the cause of a Mendelian disorder. *Nat. Genet.* 42 (1): 30–35.

Ott, J. (1974). Estimation of the recombination fraction in human pedigrees: efficient computation of the likelihood for human linkage studies. *Am. J. Hum. Genet.* 26: 588–597.

Pericak-Vance, M.A., Bebout, J.L., Gaskell, P.C. et al. (1991). Linkage studies in familial Alzheimer disease: evidence for chromosome 19 linkage. *Am. J. Hum. Genet.* 48 (6): 1034–1050.

Risch, N. and Merikangas, K. (1996). The future of genetic studies of complex human disorders. *Science* 273 (5281): 1516–1517.

Rogaev, E.I., Sherrington, R., Rogaeva, E.A. et al. (1995). Familial Alzheimer's disease in kindreds with missense mutations in a gene on chromosome 1 related to the Alzheimer's disease type 3 gene. *Nature* 376 (6543): 775–778.

Sadovnick, A.D., Ebers, G.C., Dyment, D.A., and Risch, N.J. (1996). Evidence for genetic basis of multiple sclerosis. *Lancet* 347 (1728): 1730.

Sherrington, R., Rogaev, E.I., Liang, Y. et al. (1995). Cloning of a gene bearing missense mutations in early-onset familial Alzheimer's disease. *Nature* 375: 754–760.

Tjio, J.H. and Levan, A. (1956). The chromosome number of man. *Hereditas* 42: 1–6.

Wald, A. (1947). *Sequential Analysis*. New York: Wiley.

Watson, J.D. and Crick, F.H. (1953). Molecular structure of nucleic acids; a structure for deoxyribose nucleic acid. *Nature* 171 (4356): 737–738.

2

Basic Concepts in Genetics

Kayla Fourzali[1], Abigail Deppen[2], and Elizabeth Heise[3]

[1] *University of Miami Miller School of Medicine, Miami, FL, USA*
[2] *InformedDNA, St Petersburg, FL, USA*
[3] *Clinical Genetics Program, GeneDX, Inc, Gaithersburg, MD, USA*

Introduction

For centuries, the hereditary basis of human disease has fascinated both scientists and the general public. The Talmud gives behavioral proscriptions regarding circumcision in sons born after a male sibling who died of a bleeding disorder, suggesting that the ancient Hebrews knew of hemophilia; nursing students in Britain in the 1600s tracked the recurrence of spina bifida in families; and questions as to whether Abraham Lincoln and certain celebrity sports figures had Marfan syndrome sometimes arise in casual dinner conversation.

In many respects, the study of the genetic factors in disease today remains, as it has for centuries, dependent on careful description of human pedigree data in which patterns of transmission from parent to offspring are characterized. Gregor Mendel provided the groundwork for the study of human genetics by carefully constructing quantified observations of the frequency of variable characteristics in the pea plant. The importance of detailed pedigree analysis was later exemplified in the delineation of patterns of transmission of the fragile X syndrome: the most common inherited intellectual disability (Turner et al. 1996).

This chapter explores the underpinnings for observational and experimental genetics. Concepts ranging from laws of Mendelian inheritance through molecular and chromosomal aspects of deoxyribonucleic acid (DNA) structure and function are defined. The chapter concludes with clinical examples of the various types of DNA mutation and their implications for human disease.

Historical Contributions

Segregation and Linkage Analysis

In 1865, Gregor Mendel, an Austrian monk, published his findings on the inheritance of a series of traits in the pea plant, including seed texture (round or wrinkled), seed color (yellow or green), and plant height (tall or short). He described three properties of heritable factors that explained his quantified observations of these traits. The first property was the unit inheritance, which is now

Genetic Analysis of Complex Diseases, Third Edition. Edited by William K. Scott and Marylyn D. Ritchie.

considered the basis for defining the gene. He hypothesized that a factor was transmitted from parent to offspring in an unchanged form. Such a factor produced an observable trait. This idea represented a radical departure from the scientific thinking at the time, which suggested parental characteristics were blended in the offspring.

Mendel also described the transmission of factors which were controlling observable traits, such as flower color or plant height as a single unit. He proposed that these factors were transmitted independently and with equal frequency to germ cells (egg and sperm) and this observation is referred to as Mendel's first law, the law of segregation. In his experiments, Mendel crossbred the offspring (the F1 generation) of two phenotypically different, pure-breeding parental strains of peas with one another. The offspring of these matings (the F2 generation) expressed the grandparental traits in a 3 : 1 ratio (Figure 2.1). Serendipitously, some of the traits Mendel had chosen to study were simple dominant traits, such that the presence of one factor was sufficient to express the trait, with the other trait being recessive (expressed only in the absence of the dominant factor). Later work showed that the factor defining a characteristic need not be dominant to the other. For instance, if each factor contributes to the trait equally, as in codominant systems, then three different classes of offspring from the same cross described above are possible: the two parental traits and a third intermediate trait. These classes occur in proportions parental to intermediate to parental of 1 : 2 : 1.

Mendel extended his observations from the transmission of a single trait from parent to offspring to the interaction of two traits. Mendel's law of independent assortment, also referred to as Mendel's second law, predicts that factors controlling different traits will segregate to offspring independently from one another. For instance, seed texture will segregate to offspring independently of seed color. In one experiment, Mendel crossed pure-breeding round, green seed plants to pure-breeding wrinkled, yellow seed plants. Since round is dominant to wrinkled and green is dominant to yellow, the resulting seeds (F1 generation) yielded entirely round, green seed plants. These F1 plants were then crossed with one another, confirming the predictions from his theory of independent assortment: The seed types (offspring) in the F2 generation were nine round/green, three round/yellow, three wrinkled/green, and one wrinkled/yellow, a ratio of 9:3:3:1 (Figure 2.2). Any observed departure from these expected ratios using identical parental crossing strategies suggests the two traits fail to segregate independently. As described later in the chapter, this may be a result of linkage, in which the genes for those traits are physically located near one another at a genetic region. We will describe later how the failure of two traits to segregate independently can be exploited to find genes and diagnose genetic disorders.

Mendel's observations remained largely obscure until the early 1900s, when they were independently rediscovered by plant geneticists and by Sir Archibald Garrod, who was studying the human hereditary disorder alkaptonuria (Garrod 1902). Garrod's work provided the basis of our understanding of alleles and genetic linkage, described later. Mendel's observations remain one of the most important contributions of critical descriptive science in the history of genetics.

Hardy–Weinberg Equilibrium

Another historical landmark in genetics occurred in the early 1900s as evolutionary biologists attempted to explain why the frequency of a dominant trait or disease in the population did not increase until everyone in the population was affected. The answer to this question was provided independently by Hardy (1908) and Weinberg (1908), who predicted the behavior of genetic traits in a population using the binomial theorem. Their proof, now called the Hardy–Weinberg theorem, shows that in a large, randomly mating population, the frequency of an

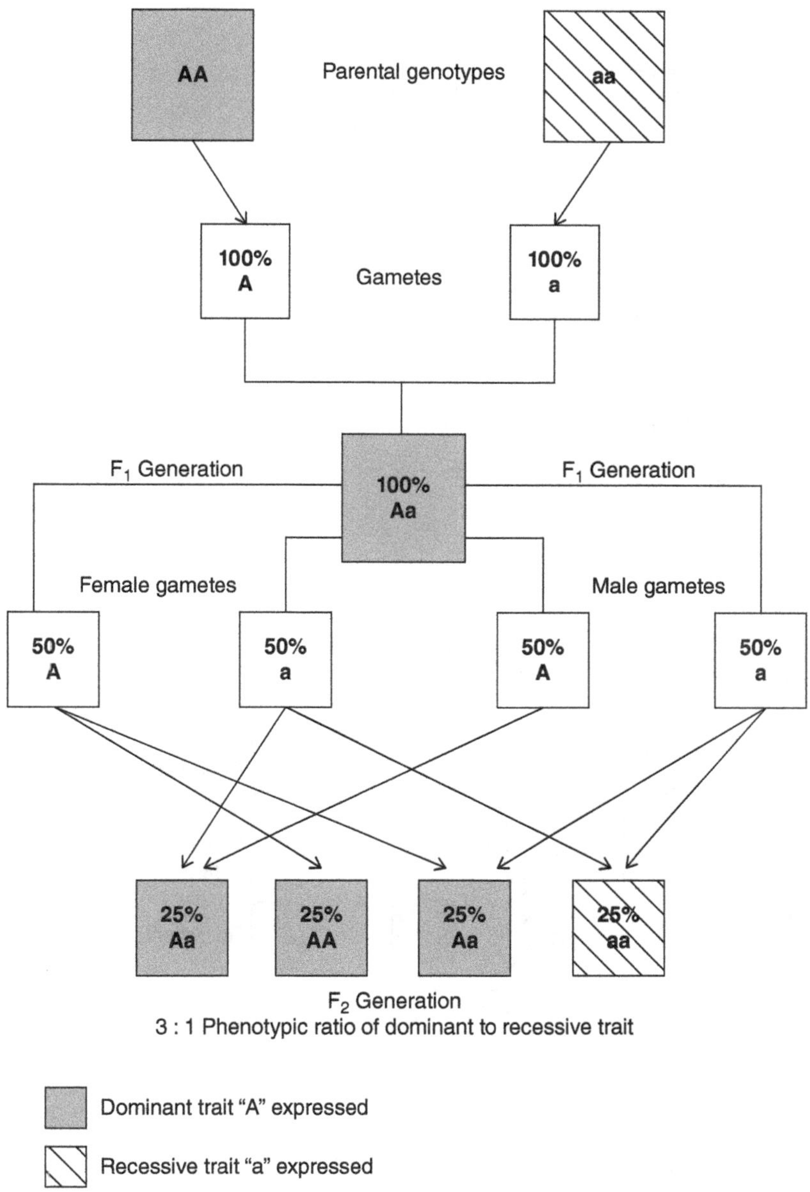

Figure 2.1 Principles of Mendel's first law of segregation of heritable characters for a dominant trait.

autosomal genetic trait (a trait that is not dependent on sex) will achieve and remain in a state of equilibrium after one generation. Several evolutionary forces can alter the equilibrium frequencies, including selection for or against a phenotype, migration into or out of a population, new mutations, or genetic drift, a phenomenon in which allele frequencies fluctuate in a small population due to chance.

In a two-allele autosomal system with alleles "A" and "a" (having frequencies p and q, respectively), $p + q = 1$ and $p^2 + 2pq + q^2 = 1$. The Hardy–Weinberg theorem predicts that the frequencies of genotypes "AA", "Aa," and "aa" are p^2, $2pq$, and q^2, respectively. Various manipulations of these

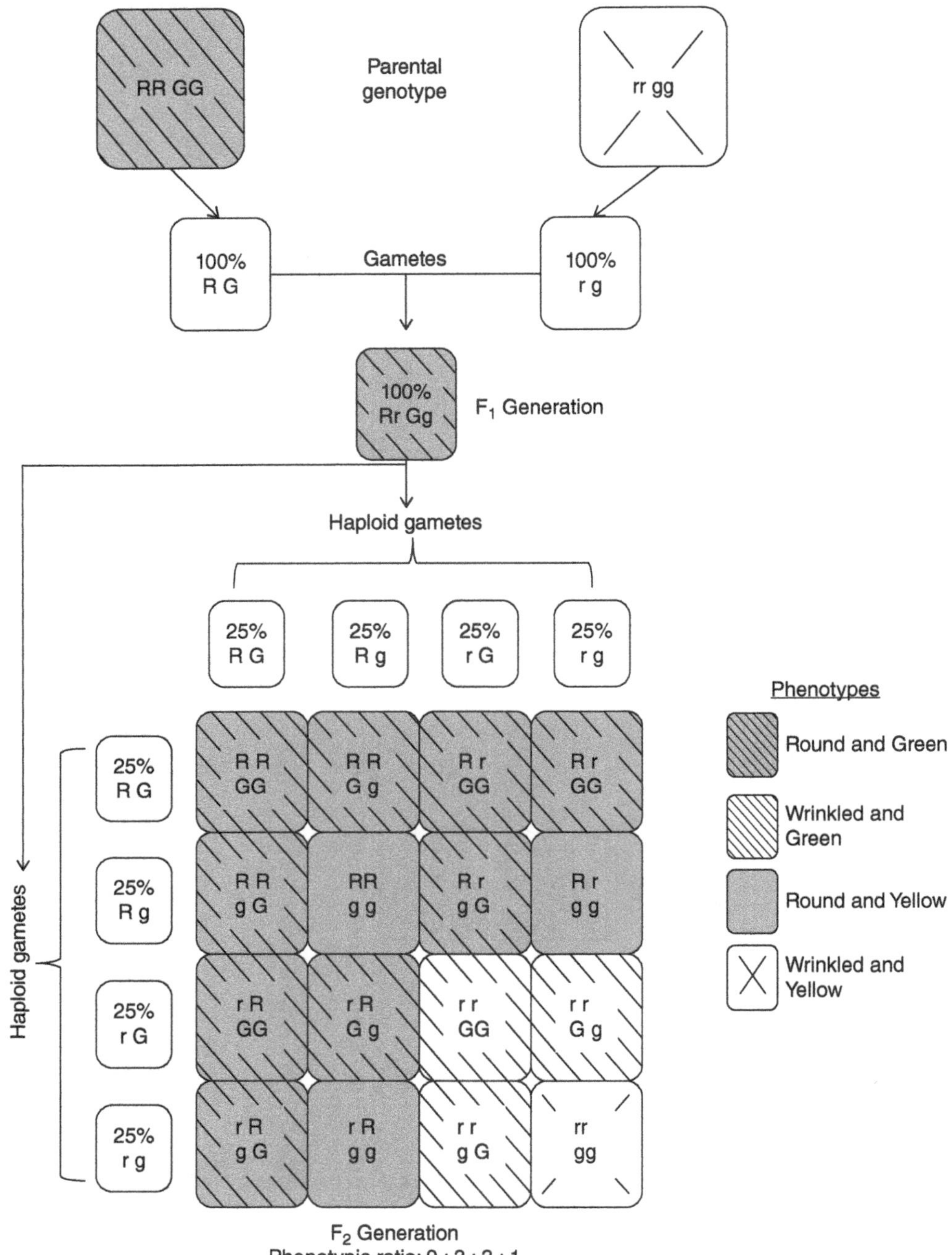

Figure 2.2 Principles of Mendel's second law of independent assortment with a dominant trait.

algebraic formulas allow many useful calculations, such as carrier frequencies of diseases, disease prevalence, and gross estimates of penetrance, the proportion of individuals with a specific genotype that exhibit a particular phenotype. Examples of applications of the Hardy–Weinberg theorem are shown in Table 2.1.

Table 2.1 Useful applications of Hardy–Weinberg theory.

Recall that $p+q=1$ and $p^2+2pq+q^2=1$

Example 1. Cystic fibrosis (CF), an autosomal recessive disease, has an incidence of 1 in 3200. What is the frequency of CF carriers in the general population?

- The population frequency of the disease (1 in 3200) is represented by q^2
- In order to calculate the frequency of the carrier state (2pq), one must first determine q
  - $q = \sqrt{(1/3200)} = 1/57$
  - Since $p+q=1$, $p=56/57$
- The frequency of CF carriers is calculated as $2pq = 2(1/57)(56/57) = 1/29$, or 0.0344.

Example 2. The frequency of the allele (q) for an autosomal dominant disorder in 1/100. What is the frequency of the disease itself in the population?

- Since the frequency of the disease allele in 1/100, the frequency of the normal allele $(p) = 1-1/100 = 99/100$.
- Since the disease is dominant, both heterozygous carriers and homozygous individuals are affected with the disease:
  - $2pq+q^2 = 2(99/100)(1/100)+(1/100)^2 = 0.0199$

Example 3. An autosomal dominant disorder with incomplete penetrance (f) has a population prevalence of 16/1000. If the allele frequency for the normal allele (p) is 0.99, what is the estimated penetrance of the disease allele?

- Since $p=0.99$, then $q=0.01$
- As in Example 2, both heterozygous and homozygous gene carriers are affected (assuming no difference in penetrance) between homozygotes and heterozygotes. Therefore,
  - $f(q^2)+f(2pq) = 0.016$
  - $f(q^2+2pq) = 0.016$
  - $f((0.01)^2+2(0.99)(0.01)) = 0.016$
  - $f(0.0199) = 0.016$
  - $f = 0.804$

DNA, Genes, and Chromosomes

Structure of DNA

When Mendel described the unit of inheritance, he did not know the underlying biological factor. It was 90 years later when the actual genetic molecule was identified. The fundamental unit of inheritance that Mendel's work uncovered was later termed "the gene." A gene contains the information for synthesizing proteins necessary for human development, cellular and organ structure, and biological function. DNA is the molecule that comprises the gene and encodes information for synthesizing both proteins and ribonucleic acid (RNA). DNA is present in the nucleus of virtually every cell in the body. It is made up of three components: a sugar, a phosphate, and a base. In DNA, the sugar is deoxyribose, whereas in RNA, the sugar is ribose. The four bases in DNA are the pyrimidines adenine (A) and guanine (G) and the purines cytosine (C) and thymine (T). A DNA sequence is often described as an ordered list of bases, each represented by the first letter of its name (e.g. ACTGAAACTTGATT). A nucleoside is a molecule made of a base and a sugar; a nucleotide is made by adding a phosphate to a nucleoside.

A single strand of DNA is a polynucleotide, consisting of nucleotides bonded together. A single strand of DNA is, however, unstable. The double-helical nature of DNA, which confers stability to the molecule, was hypothesized in 1953 by J. D. Watson and F. H.C. Crick. Their cohesive theory of the structure of DNA accounted for some of the previously identified properties of DNA (Watson and Crick 1953).

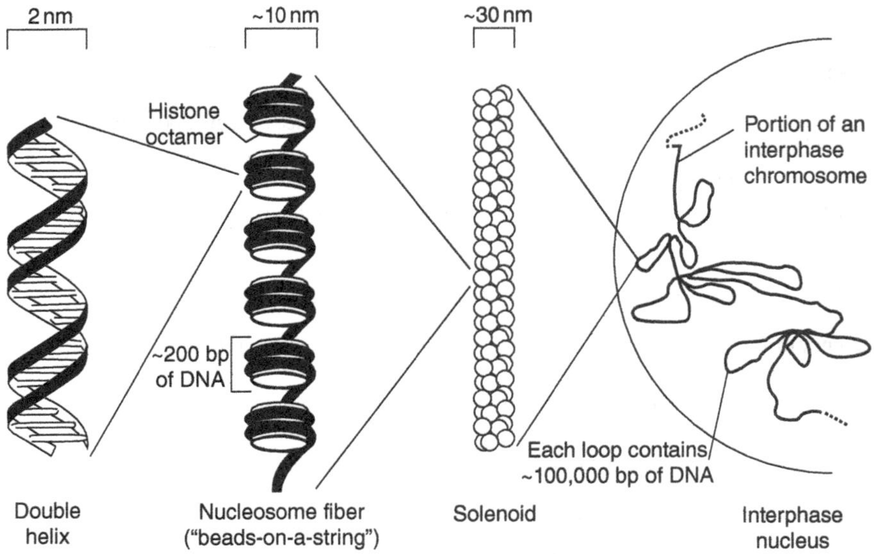

Figure 2.3 The DNA double helix is packaged and condensed in several different forms. (*Source:* Reprinted by permission from Thompson et al. (1991).)

Specifically, Watson and Crick postulated that DNA is a double-stranded structure and that the two strands of DNA are arranged in an antiparallel orientation (Watson 1968). In the central portion of the molecule, hydrogen bonds link a base with its complement, such that a purine always bonds with a pyrimidine (e.g. adenine always bonds with thymine and guanine always bonds with cytosine). The conformation of the resultant molecule is the double helix, which undergoes several levels of compacting to fit within the cell (Figure 2.3).

The sequence of DNA bases represents a code for synthesizing proteins. The fundamental unit of this genetic code is termed a codon, which consists of three nucleotides. Since there are four different nucleotides (one made with each of the four bases) and a codon is made of three nucleotides, there are 4^3, or 64, different codons. However, these 64 codons specify only 20 different amino acids, which are the building blocks of proteins. Thus, the genetic code is degenerated, meaning that different codons may code for the same amino acid (Figure 2.4). In addition, some codons act as punctuation. For instance, one specific codon signals the start for the reading of the DNA and several other codons signal the reading process to cease. These reading signals are called start and stop codons, respectively (see Figure 2.4).

While some of the DNA in a cell codes for a protein product, the vast majority of the DNA sequence does not carry information for the formation of a protein. Within a gene, exons are the portions utilized to make proteins. Introns are the sequences between exons that do not code for the final protein product. The size and number of introns and exons vary dramatically between genes.

The central dogma of genetics is that the utilization of DNA is unidirectional such that DNA → RNA → protein (Figure 2.5). Specifically, genes are encoded by the DNA. Then, in the nucleus, messenger RNA (mRNA) is transcribed (produced) through reading of the codons of DNA. The mRNA produced during transcription is a single-stranded complement to the DNA but with the base uracil instead of thymine. Subsequently, mRNA undergoes a series of posttranscriptional modifications: The introns are spliced out, a cap is added at the 5′ end of the molecule, and a string of adenylate residues (poly-A tail) is added to the 3′ end. The mRNA is then transported out of the nucleus into the cytoplasm, where it is translated into protein by means of cellular machinery called the ribosomes. Many excellent resources describe the very complicated process of transcription and translation (e.g. Strachan and Read 1996).

Second Base

First base (5′)	U	C	A	G	Third base (3′)
U	UUU Phe	UCU Ser	UAU Tyr	UGU Cys	U
U	UUC Phe	UCC Ser	UAC Tyr	UGC Cys	C
U	UUA Leu	UCA Ser	UAA *stop*	UGA *stop*	A
U	UUG Leu	UCG Ser	UAG *stop*	UGG Trp	G
C	CUU Leu	CCU Pro	CAU His	CGU Arg	U
C	CUC Leu	CCC Pro	CAC His	CGC Arg	C
C	CUA Leu	CCA Pro	CAA Gln	CGA Arg	A
C	CUG Leu	CCG Pro	CAG Gln	CGG Arg	G
A	AUU Ile	ACU Thr	AAU Asn	AGU Ser	U
A	AUC Ile	ACC Thr	AAC Asn	AGC Ser	C
A	AUA Ile	ACA Thr	AAA Lys	AGA Arg	A
A	AUG *Start* or Met	ACG Thr	AAG Lys	AGG Arg	G
G	GUU Val	GCU Ala	GAU Asp	GGU Gly	U
G	GUC Val	GCC Ala	GAC Asp	GGC Gly	C
G	GUA Val	GCA Ala	GAA Glu	GGA Gly	A
G	GUG Val	GCG Ala	GAG Glu	GGG Gly	G

Amino acid abbreviations

Alanine (Ala)	Glycine (Gly)	Proline (Pro)
Arginine (Arg)	Histidine (His)	Serine (Ser)
Asparagine (Asn)	Isoleucine (Ile)	Threonine (Thr)
Aspartate (Asp)	Leucine (Leu)	Tryptophan (Trp)
Cysteine (Cys)	Lysine (Lys)	Tyrosine (Tyr)
Glutamine (Gln)	Methionine (Met)	Valine (Val)
Glutamate (Glu)	Phenylalanine (Phe)	

Figure 2.4 The genetic code and abbreviations for amino acids.

Genes and Alleles

The physical site or location of a gene is called its locus. At any particular gene locus, there exist different forms of the gene, called alleles. These alleles are unique from each other due to variations of the nucleotides or structure in the gene, which may result in altered function of the gene and/or gene product. These alleles are analogous to the factors of inheritance and variation identified in the 1800s by Mendel. Except on the sex chromosomes of males, an individual has two alleles at each locus. In some instances, a gene may exhibit homozygosity in its two alleles, meaning that both alleles are indistinguishable from each other. Such an example is the presence of two specific disease-causing alleles of the HBB gene that together cause sickle cell anemia. In the heterozygous state, the two alleles can be distinguished from each other. Males with a normal chromosome complement represent an exception to this convention as they are "hemizygous" for loci on the X chromosome for which they do not have a Y chromosome complement.

The difference between two alleles may be as subtle as a single base-pair (bp) change, such as the adenine-to-thymine substitution in the HBB gene that alters the B chain of hemoglobin A from its wild type to its hemoglobin sickle cell state. Alternatively, allelic differences can be as extensive as large, multicodon deletions, such as those observed in Duchenne muscular dystrophy. Some bp changes have no deleterious effect on the function of the gene; nevertheless, these functionally neutral changes in the DNA still represent different forms of a gene. Rare changes in the genetic code that cause functional change of the gene or protein are often termed "mutations," while functionally neutral changes that are more common in the population (>1%) are given the name

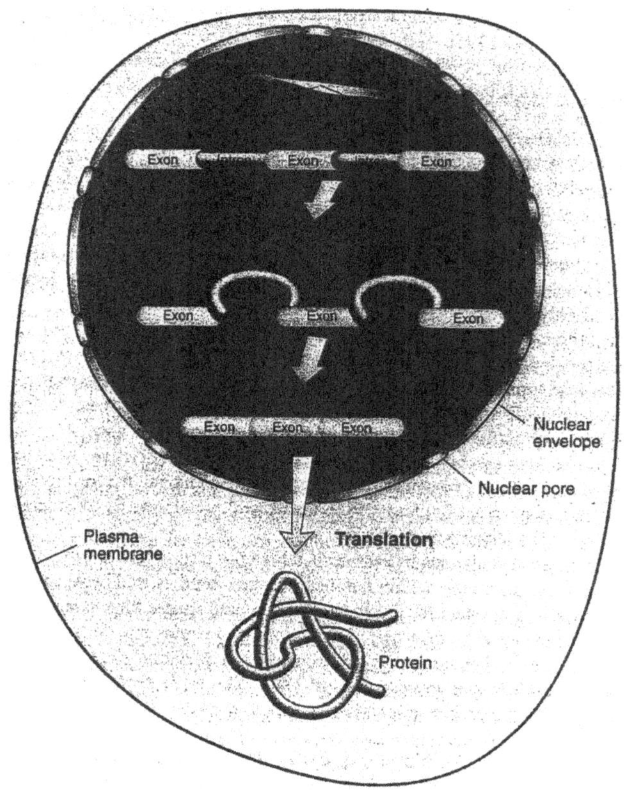

Figure 2.5 Central dogma of genetics: DNA → RNA → protein. (*Source:* Reprinted by permission from Jorde et al. (1995).)

"polymorphism." The term "pathogenic variant" is also commonly used to refer to rare genetic changes that cause functional change in the gene or protein. "Mutation" and "pathogenic variant" may be used interchangeably throughout the chapter.

Differences in alleles can be detected via laboratory testing. The ability to detect allele differences accurately within families, between families, and between laboratories is critically important for tracking the alleles that may be involved in Mendelian and genetically complex common disorders through linkage or association analysis. Allele detection strategies may be as simple as the presence (+) or absence (−) of a deletion or point mutation or as complicated as assessing the allele size in bp of DNA. The latter application is common when highly polymorphic microsatellite repeat markers are used in linkage analysis.

Genes and Chromosomes

Genes are organized as linear structures called chromosomes, with many thousands of genes on each chromosome. Each chromosome has distinguishable sites, known as centromeres that aid in cell division and in the maintenance of chromosome integrity. The centromere is visualized as the central constriction on a chromosome, and it separates the p (short) and the q (long) arms from one another. The centromere enables correct segregation of the duplicated chromosomal material during meiosis and mitosis. Telomeres are present at both ends of the chromosome and are required for stability of the chromosomal unit.

Using appropriate staining techniques, the chromosomes in a cell can be analyzed under the microscope following cell culture and the arrest of cell division at metaphase (when the chromosomes have duplicated and condensed). At this stage of the cell cycle, a chromosome has two

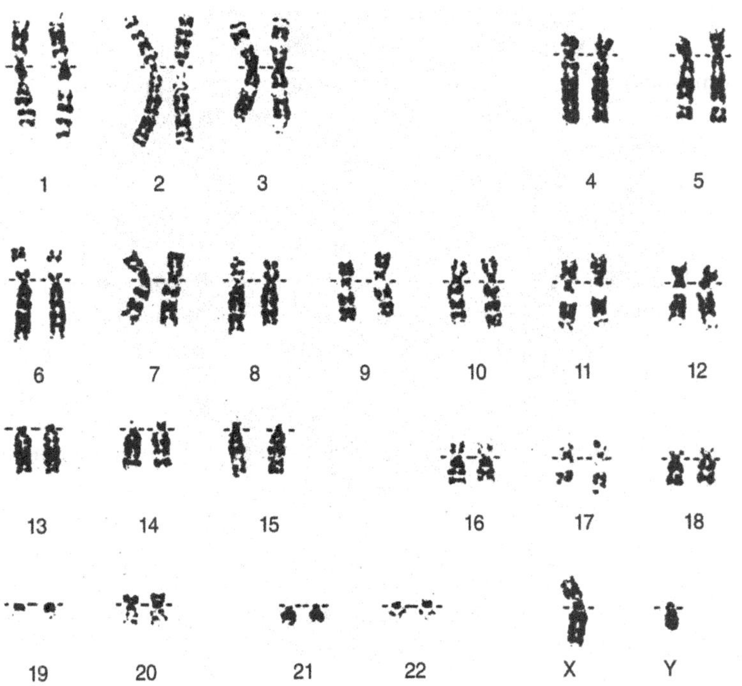

Figure 2.6 A G-banded human male karyotype. (*Source:* Courtesy of Mazin Qumsiyeh, Duke University Medical Center, Durham, NC.)

double-stranded DNA molecules. Together, the strands are called sister chromatids. The sister chromatids are held together by the centromere. Photographs are magnified, and the chromosomes are arranged into a karyotype. The normal human chromosome complement consists of 46 chromosomes arranged in 23 pairs, with one member of each pair inherited from each parent (Figure 2.6). The first 22 pairs, called autosomes, are arranged according to size and are the same in males and females. The pair of sex chromosomes generally predicts an individual's biological sex. Most females have two X chromosomes, whereas males have one X chromosome inherited from the mother and one Y chromosome inherited from the father. Therefore, the sex of an individual is primarily determined by the contribution of the father.

Because two copies of each chromosome are present in a normal somatic (body) cell, the human organism is diploid. In contrast, egg and sperm cells have haploid chromosomal complements, consisting of a single member of each chromosome pair. The correct number of chromosomes in the normal human cell was finally established in 1956, three years after the double-helical structure of DNA was described, when Tjio and Levan (1956) demonstrated unequivocally that the chromosomal complement is 46.

Regions of chromosomes are defined by patterns of alternating light and dark regions called bands, which become apparent after a chemical treatment has been applied. One of the most common types of banding process, called Giemsa or G banding, involves digesting the chromosomes with trypsin and then staining with a Giemsa dye. G banding identifies late-replicating, heterochromatic regions of DNA; these are the dark bands. Other chemical processes will produce different banding patterns and identify unique types of DNA.

A specific genetic locus can then be defined quite precisely along a chromosome, such as the gene FRAXA (fragile X syndrome), which is located on the X chromosome at band q27.3.

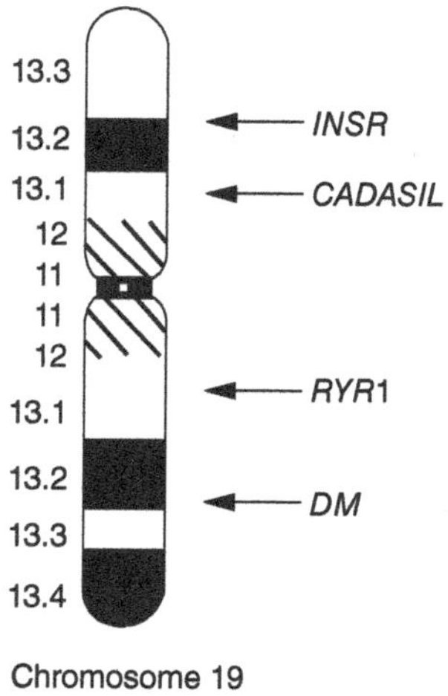

Figure 2.7 The myotonic dystrophy (DM) and insulin receptor (INSR) genes are distal (telomeric) to the ryanodine receptor 1 and CADASIL, respectively; RYR1 and CADASIL are proximal (centromeric) to DM and INSR, respectively.

Alternatively, its localization may be specified as an interval flanked by two genetic markers. Any two loci that occur on the same chromosome are considered to be syntenic or physically linked. The location of two loci on the same arm of the chromosome is specified by their positions relative to each other and to the centromere. The gene closer to the centromere is termed centromeric or proximal to the other; similarly, the gene further from the centromere is distal or telomeric to the other (Figure 2.7).

The X and Y chromosomes vary in their genetic composition except for an area at the distal end of the p arm of each, termed the pseudoautosomal region. The pseudoautosomal region behaves similarly to the autosomes during meiosis by allowing for recombination of the sex chromosomes. Just proximal to the pseudoautosomal region on the Y chromosome are the sex-determining region on the Y and testes-determining factor genes, which are critical for the normal development of male reproductive organs. When a recombination event extends past the boundary of the pseudoautosomal region and includes one or both of these genes, sexual development will most likely be adversely affected. For instance, the rare occurrences of chromosomally XX males and XY females are due to such aberrant recombination.

Genes, Mitosis, and Meiosis

A cell's ability to reproduce itself is critical to the survival of an organism. This cell duplication process, utilized by somatic cells, is called mitosis. Similarly, an organism's ability to reproduce itself is critical to the survival of the species. In sexual organisms, the reproductive process involves the union of gametes (sperm and egg cells), which are haploid. Meiosis is the process by which these haploid gametes are formed from a diploid cell and is the biological basis of linkage analysis.

Meiosis consists of two parts: meiosis I and meiosis II. In meiosis I, which is called the reduction division stage, each chromosome in a cell is replicated to yield two sets of duplicated homologous chromosomes. During meiosis I, physical contact between chromatids may occur, resulting in the formation of chiasmata. Chiasmata are thought to represent the process of crossing over or recombination, in which an exchange of DNA between two (of the four) chromatids occurs (Figure 2.8). A chiasma occurs at least once per chromosome pair; thus, each chromosome pair undergoes at least one recombination event per meiotic division. Despite being physically linked, or syntenic, loci on the same chromosome may segregate independently from each other. When two loci are unlinked to one another, the recombination fraction (θ) between them is 0.50. The upper limit for observed recombination between two unlinked loci is set at 50% because the frequency with which odd numbers of recombination events between a pair of loci occur should equal the frequency with which even numbers of recombination events occur; when an even number of recombination events occurs between two loci, the resultant gametes appear to be nonrecombinant and hence these recombination events are unobserved. However, two loci that are located closely on the same chromosome have a low likelihood of experiencing a recombination event between them and nearly always segregate together. These loci are considered to be genetically linked (Figure 2.9).

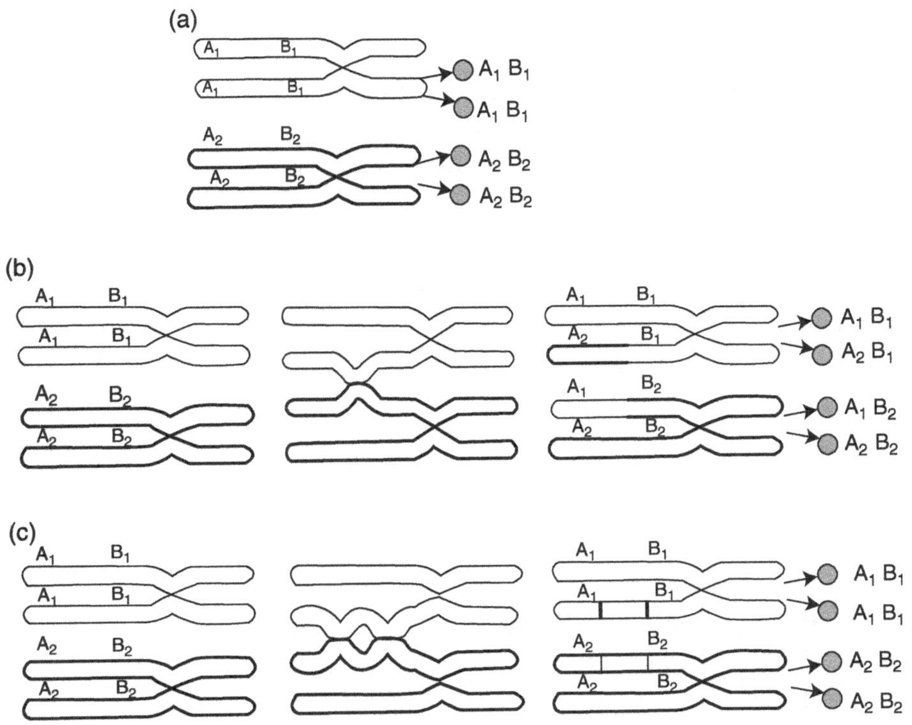

Figure 2.8 Genetic results of crossing over: (a) no crossover: A and B remain together after meiosis; (b) crossover between A and B results in a recombination (A and B are inherited together on a chromosome, and A and B are inherited together on another chromosome); (c) double crossover between A and B results in no recombination of alleles. (*Source:* Reprinted by permission from Jorde et al. (1995).)

Linkage analysis (Chapter 6) is a method of determining whether two loci are genetically linked when passed on from one generation to the next through measuring the recombination fraction (θ) between loci.

Following recombination, at least two of the four chromatids become unique, unlike those of the parent. The cellular division process that occurs ensures that one paternal homologue and one maternal homologue are transmitted to each of two diploid daughter cells. This cell division marks the end of meiosis I. The process of genetic recombination helps to preserve genetic variability within a species by allowing for virtually limitless combinations of alleles in the transmission from parent to offspring. Estimates of genetic recombination can also predict distance between two loci: The closer two loci are to one another, the less chance for recombination between them. The frequency of recombination is not uniform through the genome. Some areas of some chromosomes have increased rates of recombination (hot spots), while others have reduced rates of recombination (cold spots). For instance, recombination frequencies may vary between sexes or may vary depending on whether the loci are at the telomere or centromere of the chromosome.

The second phase of meiosis is identical to a mitotic (somatic cell) division, in which genetic material is transmitted equally, identically, and without recombination to daughter cells. However, in contrast to a mitotic division, which yields two identical diploid daughter cells, the end result of the entire meiotic process in sperm cells is four haploid daughter cells with chromosomal haplotypes different from those originally present in the parent; in egg cells, the final outcome is a single haploid daughter cell, with the remainder of the genetic material lost because of the formation of nonviable polar bodies.

The fundamental differences between meiosis and mitosis are summarized in Table 2.2.

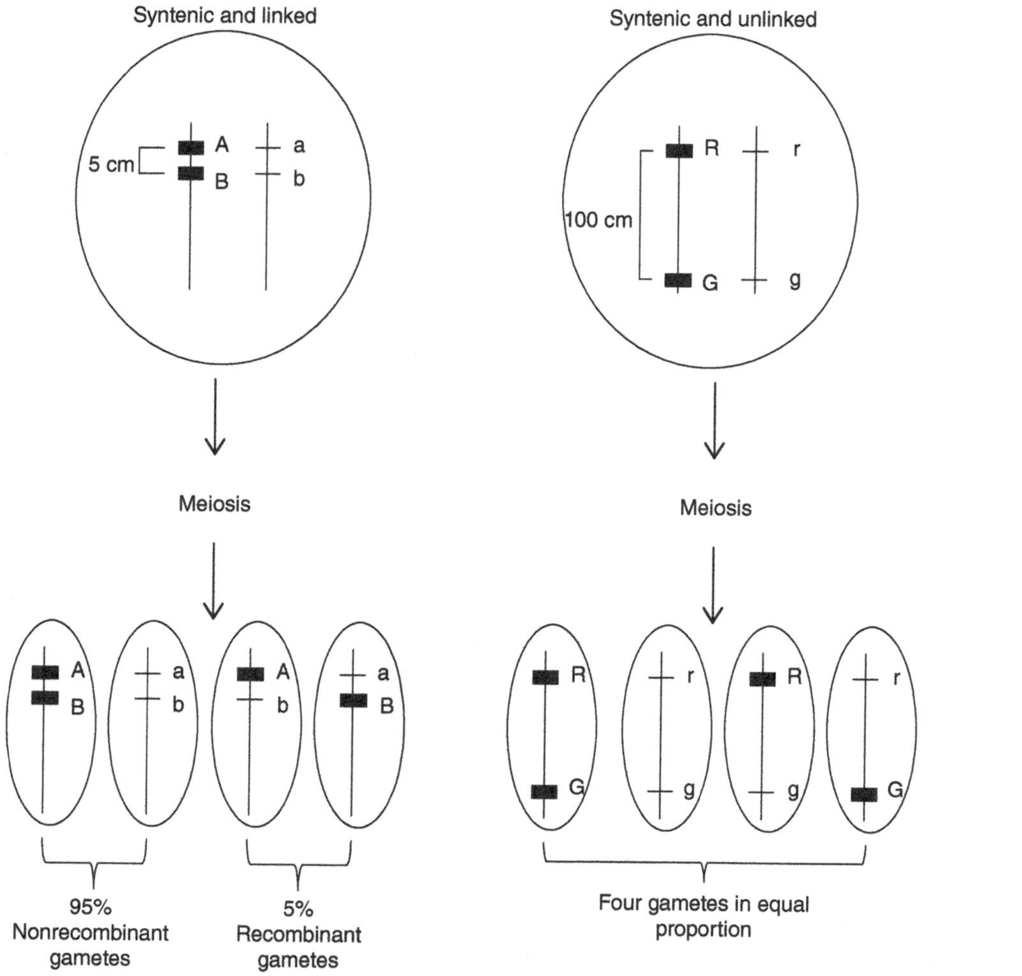

Figure 2.9 Genes that are on the same chromosome (syntenic) may be unlinked or linked. Loci A/a and B/b are syntenic and linked because their physical distance is short enough that recombination is less likely to occur between them, resulting in a lower recombinant to nonrecombinant gamete ratio. Loci R/r and G/g are syntenic and unlinked because their physical distance is great enough that a recombination event will occur between them during meiosis resulting in equal proportions of recombinant and nonrecombinant gametes.

Table 2.2 Differences between meiosis and mitosis.

	Meiosis	**Mitosis**
Purpose	Produce gametes; ensure or produce genetic variability through recombination	Replace somatic cells
Location	Gonads	Body cells
Number of cell divisions per cycle	Two: meiosis I and meiosis II (latter identical to mitosis)	One
Chromosome number in daughter cells	Halved from parental complement to produce gametes; resultant cells are haploid	Identical to parental complement; resultant cells are diploid
Recombination	Occurs in diplotene of prophase I of meiosis I	Occurs rarely, usually result of abnormality

When Genes and Chromosomes Segregate Abnormally

Failure of meiosis at either phase (meiosis I or II) is termed nondisjunction and leads to aneuploidy, or abnormal chromosomal complements. The most well-known aneuploidy is Down syndrome. Most cases of Down syndrome are caused by an extra copy of chromosome 21, but ~5% of cases are caused by a translocation such that the trisomy is not of the whole chromosome. Down syndrome is often called trisomy 21 because most individuals with Down syndrome have a total of 47 chromosomes, with three copies of chromosome 21. A monosomy, or the absence of a second member of a chromosome pair, is rarely viable. A noted exception is Turner syndrome, in which a female has a total of 45 chromosomes, including a single X chromosome. Nondisjunction events that lead to abnormal chromosome complements may occur in the egg or the sperm. However, in contrast to male sperm, which are produced throughout the lifespan, female egg cells are suspended in meiosis II since shortly after that female's conception. These eggs are subject to aging over the female lifespan, and errors of nondisjunction are more likely to occur.

Triploidy and tetraploidy are the terms for the presence of one or two entire extra sets of chromosomes, leading to a total of 69 or 92 chromosomes, respectively. These anomalies, which usually are not viable in humans, are due to errors in fertilization such as dispermy (two sperm fertilizing an ovum) or failure of the ovum's polar body to separate.

Inheritance Patterns in Mendelian Disease

Many of the traits Mendel studied in his peas were inherited in a simple dominant or recessive fashion, a manner in which many human traits and disease are also inherited. Characteristics of each of the different Mendelian inheritance patterns are summarized in Table 2.3. Pedigrees representing these well-known patterns of inheritance are shown in Figure 2.10.

Autosomal Recessive

A condition described as autosomal recessive maintains that the causative gene is located on an autosome and that both copies of the gene must harbor a causative allele. In most cases of recessive conditions, it is correctly assumed that each parent of an offspring with the trait carries a single causative allele. Carriers for recessive conditions do not typically show phenotypic features. Recessive conditions occur more commonly in offspring of consanguineous unions or in populations with high frequency of carriers for a particular condition.

Autosomal Dominant

In a dominant condition, a genetic mutation in a single copy of a gene is sufficient to cause a disease or trait. Dominant conditions may be inherited through an affected parent or alternatively may have occurred as the result of a de novo genetic mutation. Some de novo dominant mutations have been shown to be associated with increased paternal age (Friedman 1981). Some dominant conditions may exhibit reduced penetrance, in which a person carries a causative gene but does not exhibit any manifestations of the condition. On the other hand, variable expressivity refers to the range of features that may be observed in individuals with the same condition. Codominance, in which a trait is expressed from both alleles, has also been observed in certain traits, for example, in AB blood type.

Table 2.3 Hallmarks of Mendelian inheritance patterns of different types.

Inheritance pattern	Examples	Transmission features	Recurrence risk	Prevalence in population	Other critical features
Autosomal dominant	Marfan syndrome; neurofibromatosis; myotonic dystrophy	Transmitted from affected parent to affected offspring (vertical transmission) male-to-male possible transmission; de novo mutations may occur	For each offspring of affected parent, risk to child to inherit disease gene is 50%	$p^2 + 2pq$	Reduced penetrance may be observed
Autosomal recessive	Sickle cell anemia; cystic fibrosis	Carrier parents generally unaffected	For carrier parents, risk for each subsequent child is 25%	q^2	Consanguinity considered
X-linked	Duchenne muscular dystrophy; fragile X syndrome; hemophilia	No male-to-male transmission; de novo mutations may rarely occur	50% of offspring of carrier female have trait (if male, affected, if female carrier); all female offspring of affected male are carriers	Females: q^2; males: q	Females may show sub-clinical, atypical, or fully penetrant features of the condition. Non-random X inactivation may contribute to more severe female phenotype.
Y linked	Genes SRY and TDF, important in sex determination, are on the Y chromosome; no known diseases are located on Y	Exclusively male-to-male transmission	All sons of affected males are affected; no daughters of affected males are affected	Females: 0; males: q	Male-determining genes are located just proximal to pseudoautosomal region on Y chromosome; faulty recombination in pseudoautosomal region can lead to errors in sex determination

Autosomal codominant	MN blood group; microsatellite repeat markers	Each allele confers measurable component to phenotype	Varies according to mating type	Genotypes expected to occur in Hardy–Weinberg proportions of p^2, $2pq$, and q^2	
Mitochondrial	Leber's optic atrophy; KSS (Kearns–Sayre syndrome); MELAS (mitochondrial myopathy, encephalopathy, lactic acidosis and stroke-like episodes);	Exclusively maternal transmission through maternal mitochondria	All offspring of affected females are at risk to inherit mutation (may be affected or carrier). Proportion of affected offspring is variable based on maternal heteroplasmy. Offspring of affected male not at risk to inherit mutation.		Heteroplasmy may determine phenotypic severity. Majority of mitochondrial diseases are due to mutations in the nuclear genome rather than the mitochondrial genome and follow autosomal recessive inheritance pattern.

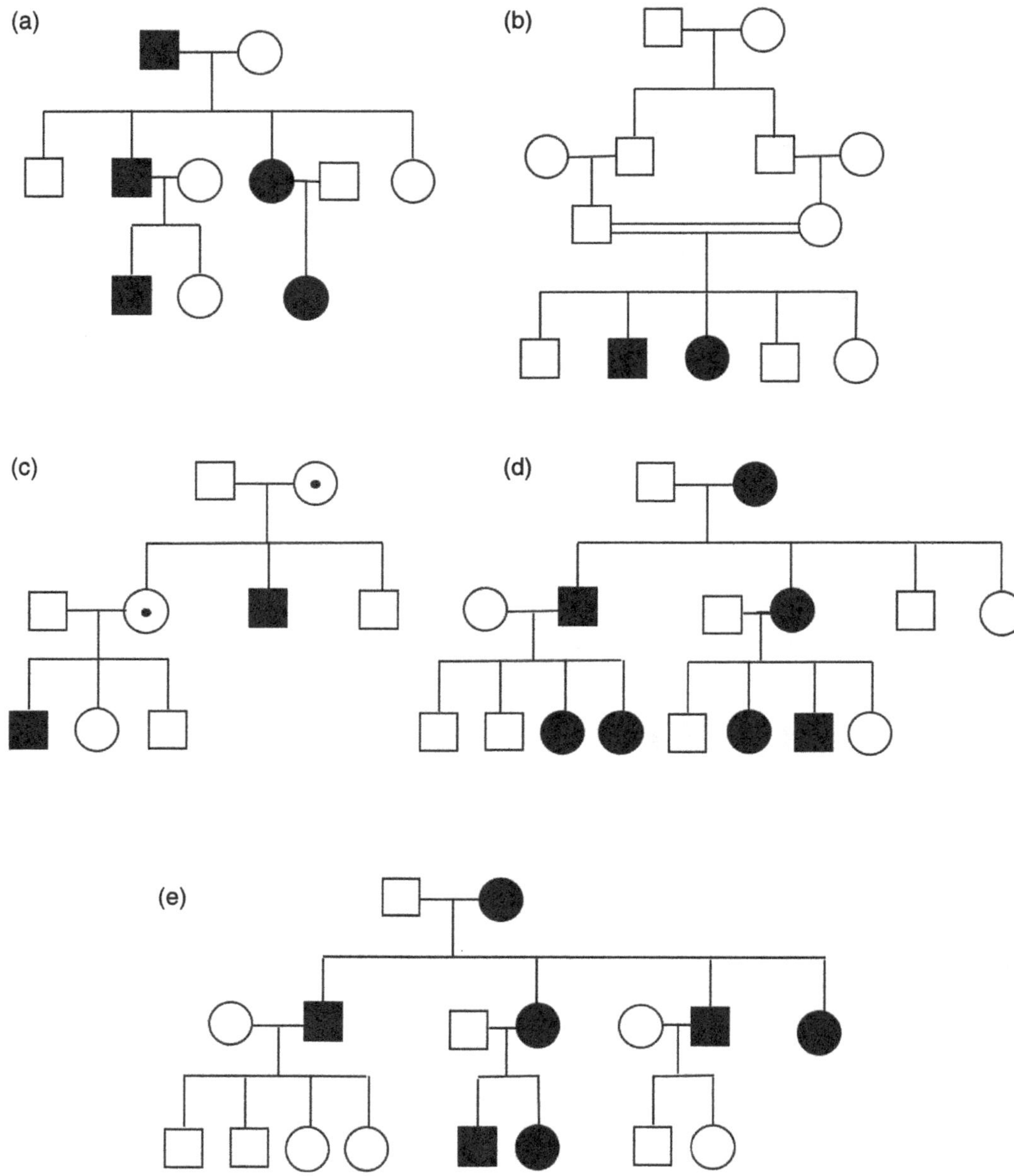

Figure 2.10 Pedigrees consistent with (a) autosomal dominant inheritance, (b) autosomal recessive inheritance, (c) X-linked recessive inheritance, (d) X-linked dominant inheritance, and (e) mitochondrial inheritance. Here and elsewhere squares indicate males, circles indicate females, open symbols indicate unaffected individuals, and solid symbols indicate affected individuals.

X-linked Inheritance

The majority of the genes located on the X chromosome do not have a complementary gene on the Y chromosome. Therefore, males are "hemizygous" at these loci because they have only a single copy of the gene. X-linked conditions were historically described as "X-linked dominant" or "X-linked recessive." In an X-linked recessive condition, a single abnormal allele was sufficient to cause disease in a hemizygous male, while females needed two. On the other hand, in "X-linked dominant" conditions, a single abnormal allele was sufficient to cause a condition in both males and females. Although these patterns of inheritance hold true for many conditions, female carriers may exhibit features of an X-linked recessive condition, due to skewed X-inactivation. For example, in Duchenne muscular dystrophy, approximately 25% of female carriers will exhibit some symptoms of the condition, ranging from muscle weakness to features as severe as seen in males (Hoogerwaard et al. 1999). This has led many researchers to argue that conditions with loci on the X-chromosome should simply be termed, "X-linked."

Mitochondrial Inheritance

Mitochondrial inheritance is a non-classical pattern of single gene inheritance that is observed in conditions in which the causative allele is located in the mitochondrial DNA (mtDNA). In humans, each mitochondrion has approximately 2–10 copies of the mtDNA, which contains 37 genes. A mutation in one of these genes may be present in all copies of the mtDNA in a cell (known as homoplasmy). Alternately, a cell may contain some mtDNA with the mutation and other mtDNA without the mutation, known as heteroplasmy. The degree of heteroplasmy may vary by tissue and can influence the severity of the disease and the risk to future offspring. Mitochondria are inherited from the mother and, therefore, are referred to as having "maternal inheritance." The risk for an affected mother to pass on the genetic defect in the mtDNA can approach 100%.

Y-linked

Alleles located on the Y chromosome are transmitted from affected males to all sons, and in each case, the son's Y-linked phenotype will be identical to that of the father; daughters of males with a Y-linked trait will not inherit the trait, since they receive their father's X chromosome. Very few expressed genes have been localized to the Y chromosome.

Genetic Changes Associated with Disease/Trait Phenotypes

Mutations Versus Polymorphisms

Alterations in the genetic code can be neutral, beneficial, or deleterious. Neutral and beneficial changes contribute to the natural variation among individuals and are not considered to have a negative effect on the organism. Rare changes in the genetic code that lead to an abnormal trait or disease phenotype are typically termed a mutation (or a pathogenic variant). The pathology of a mutation can be the result of either a loss or gain of function of the gene product. Such changes can occur in a number of different ways as highlighted below. A polymorphism, on the other hand, is used to describe a genetic variation in which there are two or more possible alleles at a particular locus. A genetic variation is typically termed a polymorphism if it is found in >1% of the population. A polymorphism may be a change in a single nucleotide, known as a single nucleotide polymorphism, an insertion or deletion (indel), or a duplication or deletion of a large segment of DNA, sometimes referred to as a copy number variant (CNV). Much research is currently underway to evaluate the effects of such polymorphisms on common and complex diseases, such as multiple sclerosis, cancer, and heart disease.

Point Mutations

A point mutation is defined as an alteration in a single bp in a stretch of DNA, thereby changing the 3-bp codon. Since the genetic code is degenerate, many such changes do not necessarily alter the resulting amino acid. However, if the single bp change leads to the substitution of one amino acid for another, the result can significantly affect the final protein product. Point mutations can be classified as transition mutations (purine → purine or pyrimidine → pyrimidine) or as the less common transversion mutations (purine → pyrimidine or pyrimidine → purine). In general, transitions are less likely than transversion mutations to change the resulting amino acid. Five effects of point mutations have been described:

- Synonymous or silent mutations are single bp changes in the DNA that do not affect the resultant amino acid
- Nonsense mutations result in a premature stop codon, leading to a polypeptide of reduced length
- Missense mutations lead to the substitution of one amino acid for another
- Splice site mutations affect the correct processing of the mRNA strand by eliminating a signal for the excision of an intron
- Mutations in regulatory genes alter the amount of material produced

Several examples of point mutations in human diseases are illustrated below.

Sickle Cell Anemia

Sickle cell anemia, an autosomal recessive disorder with a carrier frequency in African Americans of approximately 1/12, is a classic example of a point mutation leading to disease. Sickle cell anemia results from a single nucleotide substitution in the HBB gene of an adenine to a thymine, which changes the resultant amino acid from glutamine to valine at codon 6. The pathogenic variant (mutation) leads to the formation of an abnormal form of hemoglobin, hemoglobin S, which causes the red blood cell to take on an abnormal form that resembles a sickle. This causes red blood cells to break down prematurely, leading to anemia. In addition, the sickled red blood cells may accumulate in blood vessels and lead to episodes of severe pain. Interestingly, the carrier state for sickle cell trait may lead to a selective advantage in certain environments: carriers have a resistance to malaria that is useful in tropical climates, which explains the high carrier rate in certain populations.

Achondroplasia

Achondroplasia, the most common type of short-limbed dwarfism, is an autosomal dominant disorder. About 85% of cases are the result of a new mutation. It has been observed that the rate of new dominant mutations increases with advancing paternal age (Penrose 1955; Stoll et al. 1982). Achondroplasia is now known to result from mutations in the fibroblast growth receptor 3 gene (FGFR3), located on chromosome 4p16.3. Interestingly, over 95% of the mutations are the identical G-to-A transition at nucleotide 1138 on the paternal allele (Rousseau et al. 1994; Shiang et al. 1994; Bellus et al. 1995). This single change causes a gain of function mutation which results in constitutive activation of the receptor, thereby inhibiting the proliferation of cartilage cells, or chondrocytes, and significantly restricting growth in individuals with this condition. Other pathogenic variants in FGFR3 are also responsible for hypochondroplasia and thanatophoric dysplasia, types of dwarfism that are clinically distinct from achondroplasia.

Deletion/Insertion Mutations

Another class of mutations involves the deletion or insertion of DNA into an existing sequence. Deletions or insertions may be as small as 1 bp, or they may involve one or many exons or even the entire gene. Even single bp deletions or insertions can have devastating effects, frequently by altering the reading frame of the DNA strand. A specific type of deletion/insertion mutation is a CNV. A CNV is structural variation in which kilobases to several megabases of DNA have been deleted or added and may encompass numerous genes. It is hypothesized that CNVs may account for 5–13% of the genome (Stankiewicz and Lupski 2010; Zarrei et al. 2015). Most CNVs are expected to be benign, while some are directly tied to a particular disease. Other CNVs may confer an increased risk for a particular condition, and much research is currently taking place about the role of CNVs in complex diseases.

Duchenne and Becker Muscular Dystrophy

Duchenne muscular dystrophy (DMD) is a severe, childhood-onset X-linked muscular dystrophy. Becker muscular dystrophy (BMD) is allelic with DMD but typically has a later age of onset and a milder presentation. Boys with DMD typically have normal development for the first few years of life, after which rapidly progressive muscle deterioration becomes obvious. Affected males usually lose the ability to walk by age 10–12 years. The eventual loss of muscle strength in the cardiac and respiratory muscles leads to death in early adulthood. Duchenne and Becker muscular dystrophy is caused by mutations in the DMD gene, which codes for the protein dystrophin (Koenig et al. 1988). Large deletions in this gene account for approximately 60–70% of cases of DMD and BMD; however, duplications and point mutations have also been reported (Takeshima et al. 2010). Mutations in this gene that alter the reading frame typically cause DMD, while mutations that preserve the reading frame lead to BMD. Researchers are investigating a variety of therapies including exon-skipping and read-through of stop codons.

Cystic Fibrosis

Cystic fibrosis (CF), an autosomal recessive disorder, is the most common hereditary disease among Northern Europeans, with a carrier frequency of between 1/25 and 1/30 (Hamosh et al. 1998). This condition affects the function of the pancreas, lungs, and sweat glands, among other organ systems (Ratjen and Doring 2003). In Northern Europeans, a single mutation in the CFTR gene called ΔF508 accounts for about 70% of the abnormal CF alleles. Three bp (codon 508) are deleted, and the resulting amino acid sequence is missing a phenylalanine. Although the reading frame is preserved in this particular mutation, the deletion results in a block in protein processing. CF is an excellent example of allelic heterogeneity in which different mutations, or alleles, at the same locus can cause a disease. In fact, more than 1900 other deleterious mutations in CFTR have been identified (Cystic Fibrosis Mutation Database 2015).

Charcot-Marie-Tooth Disease

Charcot-Marie-Tooth (CMT) Type 1A and hereditary liability to pressure palsies (HNPP) are caused by an abnormal 1.4 megabase CNV on 17p12. CMT is a peripheral neuropathy most commonly caused by a duplication of PMP22, while a deletion of this same gene will result in the phenotypically distinct condition, HNPP. There are numerous types of CMT, many of which are caused by mutations at many different loci, a scenario described as locus heterogeneity.

Nucleotide Repeat Disorders

Certain loci in the genome have variable numbers of nucleotide repeats, typically of di- or tri-nucleotide length. Most are not associated with expressed genes but can be exploited as genetic markers since they are highly polymorphic. A few loci with tri-nucleotide repeats are located near or within a gene and, by expansion beyond a certain threshold, can disrupt gene expression and cause disease.

More than 20 disorders have been shown to be the result of unstable nucleotide repeats (La Spada & Taylor 2010; Polak et al. 2013). The most common of these conditions are summarized in Table 2.4. Expansions may occur in the exon, intron, or regulatory elements of a gene. The phenotypes of some of these conditions, including myotonic dystrophy, Huntington disease, and Machado–Joseph disease, among others, are associated with anticipation, a clinical phenomenon in which disease severity worsens in each successive generation. Because disease expression can be quite variable and difficult to measure, age of onset is frequently utilized as an analogue of severity and anticipation is then observed as a decreasing mean age of onset with each passing generation. The anticipation seen in these disorders is attributed to the intergenerational instability and resulting expansion of the repeated elements. However, some repeat disorders are stable during mitosis and meiosis and therefore do not exhibit anticipation. Furthermore, certain conditions are more likely to exhibit an expansion only when transmitted from a particular sex. For instance, the CAG repeat in Huntington disease typically only expands in male gametes, while the CTG repeats in Fragile X usually only expands in female gametes.

Susceptibility Versus Causative Genes

As the study of common and genetically complex human diseases identifies the significant contribution of heredity in their development, it is likely that more genes or genetic risk factors will be found to affect susceptibility to disease rather than the more traditionally considered causative genes. Historical successes in the localization of genes have been with diseases whose mode of inheritance is known (as illustrated above). These disorders are often highly or completely penetrant and are due to a defect in a single gene, yet these Mendelian disorders are often relatively rare in the population. However, in recent years, genomic research has uncovered genetic risk factors for many diseases that were suspicious for genetic etiology but were unexplained by traditional Mendelian cause and effect. Some of the most common and deadly diseases of society, such as cardiovascular disease, cancer, and obesity, have significant genetic components that are evident from non-random family clustering. These diseases are termed "complex" because they are likely due to the interaction of multiple factors, both environmental and genetic. Susceptibility genes for such complex disorders are substantially harder to identify than genes responsible for Mendelian disorders.

A well-characterized example of a susceptibility locus is that of the apolipoprotein E (APOE) gene and Alzheimer disease (AD). The APOE gene on chromosome 19 has three different alleles, scored as 2, 3, and 4, which occur with frequencies 6%, 78%, and 16% in most European populations, respectively (e.g. Saunders et al. 1993). These alleles differ in their DNA sequence by only one base at codons 112 and/or 158 (Figure 2.11). The APOE 4 allele increases risk and decreases age of onset in familial and sporadic late-onset AD and early-onset sporadic AD. The 2 allele has been shown to be protective to some extent for risk to develop AD (Corder et al. 1994, 1995a,b;

Table 2.4 Salient features of human repeat expansion diseases.

Condition	Gene symbol	Repeat type	Repeat localization	Repeat number abnormal range	Inheritance pattern	Clinical features
Fragile X syndrome	FRAXA	CGG	5' Untranslated region	200–1000 (premutation range of 52–200)	X-linked	Moderate to severe mental retardation, behavioral abnormalities, macroorchidism, large ears, and prominent jaw
Huntington disease (HD)	HD	CAG	Open reading frame	>36 (premutation range 27–35)	Autosomal dominant; expansion more common in paternal allele	Choreiform movements, dystonia, psychiatric illness, cognitive decline, dementia
Myotonic dystrophy	DM1	CTG	3' Untranslated region	Mild: 50–150 Classic: ~100–1000 Congenital: >2000	Autosomal dominant; expansion more common in maternal allele	Weakness, myotonia, ptosis, cataracts, cardiac arrhythmia; endocrine abnormalities, frontal balding
Friedrich Ataxia	FXN	GAA	Intron	66–1700 uninterrupted repeats (premutation range of 34–65 uninterrupted repeats)	Autosomal recessive	Ataxia, sensory loss, weakness, diabetes mellitus, cardiomyopathy

Exon 4

APOE allele	Codon 112 Sequence	Codon 112 Amino acid	Codon 158 Sequence	Codon 158 Amino acid
2	UGC	Cysteine	UGC	Cysteine
3	UGC	Cysteine	CGC	Arginine
4	CGC	Arginine	CGC	Arginine

Figure 2.11 Single base pair changes in exon 4 of APOE define the 2, 3, and 4 alleles at this locus. (*Source:* Modified from Pericak-Vance and Haines (1995).)

Farrer et al. 1997). It is important to note that for APOE and AD, the 4 allele is not by itself sufficient or necessary for the development of AD but has been shown to be associated with increased susceptibility to AD.

Summary

The study of genes, chromosomes, and patterns of transmission of human traits within families has led to remarkable discoveries that are useful in genetic counseling for recurrence risk, presymptomatic testing, prenatal diagnosis (see Chapter 5), and in the understanding of the pathogenesis of diseases. The genetic basis of Mendelian disease is relatively straightforward and is well understood in many cases. The situation in common, complex diseases is markedly different from the study of Mendelian disease, since more than one gene as well as various non-genetic factors are typically associated with trait phenotype expression. Yet, many of the same principles hold true in complex disease: Mendel's laws regarding the transmission of genes and alleles at loci are as important to the study of resemblance between relatives in genetically complex disease as in Mendelian disease; the same holds true for the extent and result of the differing types of mutation. The potential rewards of localizing disease susceptibility genes for common, complex diseases and their underlying pathways are abundant, especially with respect to the prospects of complex disease prevention and treatment, genetic counseling, and genomic medicine.

References

Bellus, G.A., Hefferon, T.W., Ortiz de Luna, R.I. et al. (1995). Achondroplasia is defined by recurrent G380R mutations of FGFR3. *Am. J. Hum. Genet.* 56: 368–373.

Corder, E.H., Saunders, A.M., Risch, N.J. et al. (1994). Protective effect of apolipoprotein E type 2 allele for late onset Alzheimer disease. *Nat. Genet.* 7: 180–184.

Corder, E.H., Saunders, A.M., Strittmatter, W.J. et al. (1995a). Apolipoprotein E, survival in Alzheimer disease patients and the competing risks of death and Alzheimer disease. *Neurology* 45: 1323–1328.

Corder, E.H., Saunders, A.M., Strittmatter, W.J. et al. (1995b). The apolipoprotein E E4 allele and sex-specific risk of Alzheimer's disease [letter; comment]. *JAMA* 273: 373–374.

Cystic Fibrosis Mutation Database. (2011) www.genet.sickkids.on.ca/cftr/ (accessed 3 March 2015)

Farrer, L.A., Cupples, L.A., Haines, J.L. et al. (1997). Effects of age, sex, and ethnicity on the association between apolipoprotein E genotype and Alzheimer disease. A metaanalysis. APOE and Alzheimer Disease Meta Analysis Consortium. *JAMA* 278: 1349–1356.

Friedman, J.M. (1981). Genetic disease in the offspring of older fathers. *Obstet. Gynecol.* 57 (6): 745–749.

Garrod, A.E. (1902). The incidence of alkaptonuria: a study in chemical individuality. *Lancet* ii: 1616–1620.

Hamosh, A., Fitz-Simmons, S.C., Macek, M. et al. (1998). Comparison of the clinical manifestations of cystic fibrosis in black and white patients. *J. Pediatr.* 132: 255–259.

Hardy, G.H. (1908). Mendelian proportions in a mixed population. *Science* 28: 41–50.

Hoogerwaard, E., van der Wouw, P., Wilde, A. et al. (1999). Cardiac involvement in carriers of Duchenne and Becker muscular dystrophy. *Neuromuscul. Disord.* 9: 347–351.

Jorde, L.B., Carey, J.C., and White, R.L. (eds.) (1995). *Medical Genetics*. St. Louis, MO: C. W. Mosby.

Koenig, M., Monaco, A.P., and Kunkel, L.M. (1988). The complete sequence of dystrophin predicts a rod-shaped cytoskeletal protein. *Cell* 53: 219–228.

La Spada, A. and Taylor, J. (2010). Repeat expansion disease: progress and puzzles in disease pathogenesis. *Nat. Rev. Genet.* 11: 247–258.

Penrose, L.S. (1955). Parental age and mutation. *Lancet* ii: 312–313.

Pericak-Vance, M.A. and Haines, J.L. (1995). Genetic susceptilbility to Alzheimer disease. *Trends Genet.* 11: 504–508.

Polak, U., McIvor, E., Dent, S.Y. et al. (2013). Expanded complexity of unstable repeat diseases. *Biofactors* 39 (2): 164–175.

Ratjen, F. and Doring, G. (2003). Cystic fibrosis. *Lancet* 361 (9358): 681–689.

Rousseau, F., Bonaventure, J., Legeai-Mallet, L. et al. (1994). Mutations in the gene encoding fibroblast growth factor receptor-3 in achondroplasia. *Nature* 371: 252–254.

Saunders, A.M., Strittmatter, W.J., Schmechel, D. et al. (1993). Association of apolipoprotein E allele epsilon 4 with late-onset familial and sporadic Alzheimer's disease. *Neurology* 43: 1467–1472.

Shiang, R., Thompson, L.M., Zhu, Y.-Z. et al. (1994). Mutations in the transmembrane domain of FGFR3 cause the most common genetic form of dwarfism, achondroplasia. *Cell* 78: 335–342.

Stankiewicz, P. and Lupski, J. (2010). Structural variation in the human genome and its role in disease. *Annu. Rev. Med.* 61: 437–455.

Stoll C, Roth M-P, Bigel P (1982): A reexamination of parental age effect on the occurrence of new mutations for achondroplasia. *Prog. Clin. Biol. Res.*, 104:419–426.

Strachan, T. and Read, A.P. (1996). *Human Molecular Genetics*, 1ste. New York: Wiley.

Takeshima, Y., Yagi, M., Okizuka, Y. et al. (2010). Mutation spectrum of the dystrophin gene in 442 Duchenne/Becker muscular dystrophy cases from one Japanese referral center. *J. Hum. Genet.* 55: 379–388.

Thompson, M.W., McInnes, R.R., Willard, H.F., and Thompson, J.S. (eds.) (1991). *Genetics in Medicine*, 5the. Philadelphia, PA: W. B. Saunders Company.

Tjio, J.H. and Levan, A. (1956). The chromosome number of man. *Hereditas* 42: 1–6.

Turner, G., Webb, T., Wake, S., and Robinson, H. (1996). Prevalence of fragile X syndrome. *Am. J. Med. Genet.* 64 (1): 196.

Watson, J.D. (1968). *The Double Helix*. New York: Atheneum.

Watson, J.D. and Crick, F.H.C. (1953). A structure for deoxyribose nucleic acid. *Nature* 171: 737–738.

Weinberg, W. (1908). Uber den nachweis der vererbung beim Menschen. Jahreshefte des vereins fur vaterlandische naturkunde in wurttemberg. *Wurttemberg* 64: 368–382.

Zarrei, M., MacDonald, J.R., Merico, D., and Scherer, S.W. (2015). A copy number variation map of the human genome. *Nat. Rev. Genet.* 16 (3): 172–183.

3

Determining the Genetic Component of a Disease

Allison Ashley Koch[1] *and Evadnie Rampersaud*[2]

[1] *Duke Molecular Physiology Institute, Duke University Medical Center, Durham, NC, USA*
[2] *Center for Applied Bioinformatics, St. Jude Children's Research Hospital, Memphis, TN, USA*

Introduction

Genetics is believed to contribute to the etiology of almost every human trait and condition. Even for infectious diseases that have been traditionally described as environmental, such as tuberculosis and HIV, genetic factors have been implicated either in the susceptibility to infection or in the severity of the condition (Bellamy 1998; Hill 1999; Bellamy et al. 2000; Gonzalez et al. 2001; Shields and Dell 2001). Understanding the role of genetics in disease etiology can allow development of successful therapies that improve the quality of life for affected individuals and their families. However, before embarking on expensive and labor-intensive studies to identify the genetic factors involved in a particular condition, one should evaluate the evidence that genes contribute to that trait or condition.

For Mendelian disorders, such as sickle cell anemia, cystic fibrosis, and Duchenne muscular dystrophy, establishing a role for genetics is straightforward. These conditions have predictable, recognizable inheritance patterns, and their primary etiologies can be attributed to variations in single genes. The phenotypic expression of the condition is highly correlated with the genotype at the disease locus. For example, all individuals who carry two copies of the *Hb S* variant of the β-globin gene are affected with sickle cell anemia (Neel 1949). While there is variation in the severity of the symptoms, all individuals who are homozygous for *Hb S* will exhibit some symptoms of the disorder. In contrast, complex disorders, such as cardiovascular disease, cancers, and psychiatric disorders, usually do not display distinct inheritance patterns. Such disorders have a significant genetic component but are caused by an intricate web of genetic and environmental interactions. As a result, establishing a role for genetics in the etiology of complex disorders can be more difficult. While the genotype at a single locus may contribute to the susceptibility to develop a condition, it is expected that other loci are involved, as well. That is, multiple loci may be interacting with each other or the environment to contribute to disease. Moreover, different loci may be contributing to disease susceptibility in different families. A classic example of a complex disease susceptibility gene is the association between the *APOE* locus and Alzheimer disease. Corder et al. (1993) demonstrated that individuals who carry the four allele of the *APOE* locus have a higher

Genetic Analysis of Complex Diseases, Third Edition. Edited by William K. Scott and Marylyn D. Ritchie.

risk and earlier age of onset for late onset Alzheimer disease when compared with individuals who do not carry the four allele. Furthermore, this association is dose-dependent. Individuals who have two copies of *APOE*-4 are at greater risk for the condition than individuals who carry one copy of *APOE*-4. Some Alzheimer families carry the *APOE*-4 genotype and some do not. Thus, the *APOE* gene alone does not explain the etiology of the disease. In addition to the diminished correlation between genotype and phenotype, due to the high frequency of complex diseases in the general population, often one cannot be certain that two individuals within a family have developed the condition as a result of the same genetic liability or genetic heritability. For example, the family may exhibit bilineality (family history is present on both the maternal and paternal side). Consequently, each side of the family may contribute different genetic susceptibilities to the condition. Even in the absence of bilineality, there are complications. Some individuals in the family may express the condition as a result of exposure to an environmental insult. Therefore, it can be quite challenging to determine which family members carry the same genetic susceptibilities. For all these reasons, it is more difficult to establish a genetic basis for a complex disease than for a Mendelian disease.

However, even in the presence of such complexities, there are methods available to evaluate whether or not genetics plays an important role in the disease etiology. Those methods are the primary topic of this chapter and should be explored prior to embarking on more elaborate analyses such as genome-wide association analyses or linkage analyses. Importantly, before considering any analysis, one must first consider study design.

Study Design

The study design process includes all clinical, molecular, bioinformatics, and analytic components of your study. A study population must be defined, as well as the process for selecting individuals from that population for study participation (ascertainment) and criteria for identifying individuals exhibiting the trait of choice (phenotype definition). One must determine the molecular technology to be used (e.g. genome-wide association study versus rare variant gene burden testing) as well as establish the electronic process for storing and retrieving the clinical and molecular data, and the analysis methods to be used. Finally, one must ensure all ethical, legal, and social issues are addressed. Each component in the study design process should be given careful consideration, as the decisions one makes at each step will impact the conclusions that can be drawn from the results. Furthermore, many of the decisions in the study design process are inter-related. For example, the ascertainment process will impact the analysis methods that can be performed. Since many of the steps in the study design process are covered in detail in other chapters of the book, they will not be discussed further here. Only the process of selecting a study population and an ascertainment scheme will be covered in this chapter and expanded upon in Chapter 4.

Selecting a Study Population

In a well-designed study, the study population is clearly defined and individuals from the study population are systematically selected for participation. The two primary types of study population are population-based and clinic-based. A population-based study is the gold standard for study design, but a clinic-based population is a more realistic design for many genetic studies.

Population-Based

Ideally, one will select a study population that is an unambiguous subset of a larger underlying population. For example, this could include individuals selected from a newborn state screening registry (e.g. determining the frequency of congenital hypothyroidism in individuals from the California Newborn Screening Program (Waller et al. 2000)), children who attend a particular school system (e.g. screening for FRAXA and FRAXE in a special needs population (Meadows et al. 1996)), and individuals in a local cancer registry (e.g. examination of the involvement of genetics to fallopian tube cancer in patients identified from the Ontario Cancer Registry (Aziz et al. 2001)). This epidemiologic approach to selecting a study population is much less susceptible to ascertainment bias than a clinic-based study population, and conclusions made from analysis are usually extendable to the larger underlying population from which the study population was drawn. The disadvantage with this approach is that it may be difficult to achieve the desired patient sample size for rare disorders, and even for common disorders, it can be quite laborious to identify eligible participants. For example, for a condition with a frequency of 1/1 000, one would need to screen 100 000 individuals to identify even 100 cases, a rather small sample size.

Clinic-Based

In reality, most genetic studies rely on clinic-based study populations because they provide the investigators with faster access to patients. There are numerous examples of genetic studies relying on this design. Generally, patients are selected for participation from existing patient populations, such as a specialty clinic (e.g. examining the phenotypic heterogeneity of age-related macular degeneration in patients selected from ophthalmology clinics at two academic centers (De La Paz et al. 1997)). Less effort is required for the initial identification of the patients in this approach compared with the population-based approach. However, the disadvantage of a clinic-based study population is that it can be quite challenging to determine the larger underlying population from which a clinic population was drawn. Identifying appropriate control subjects for clinic-based studies can be difficult for this reason, as discussed later in this chapter. Most large, academic medical centers attract patients not only from the immediate geographic region, but often from all over the world. A further complication of clinic-based samples is that patients ascertained in this manner may exhibit a more severe form of the condition, simply because they have sought medical treatment or have been referred for expert diagnosis and care. For example, some individuals with myotonic dystrophy have cataracts as the only phenotypic manifestation (Meola 2013). Therefore, these individuals would not seek medical attention on a frequent basis. In contrast, other individuals with myotonic dystrophy are severely affected and seek medical treatment regularly. Consequently, ascertaining patients with myotonic muscular dystrophy from a neurology clinic would likely over-sample individuals with the most severe form of the condition and omit ascertainment of the individuals who carry the less severe form of the condition (cataracts only). Thus, while clinic-based populations provide fast access to desired patients, results obtained in such studies may not be applicable to the general population, or even to all individuals with the condition.

Ascertainment

There are three basic designs of case ascertainment for a genetic analysis of a binary (present/absent) trait: collection of a single affected individual (case), relative pairs from a family, and extended families with multiple affected and unaffected individuals. Examples of these ascertainment schemes are shown in Figure 3.1. As shown below, certain sampling schemes limit the types

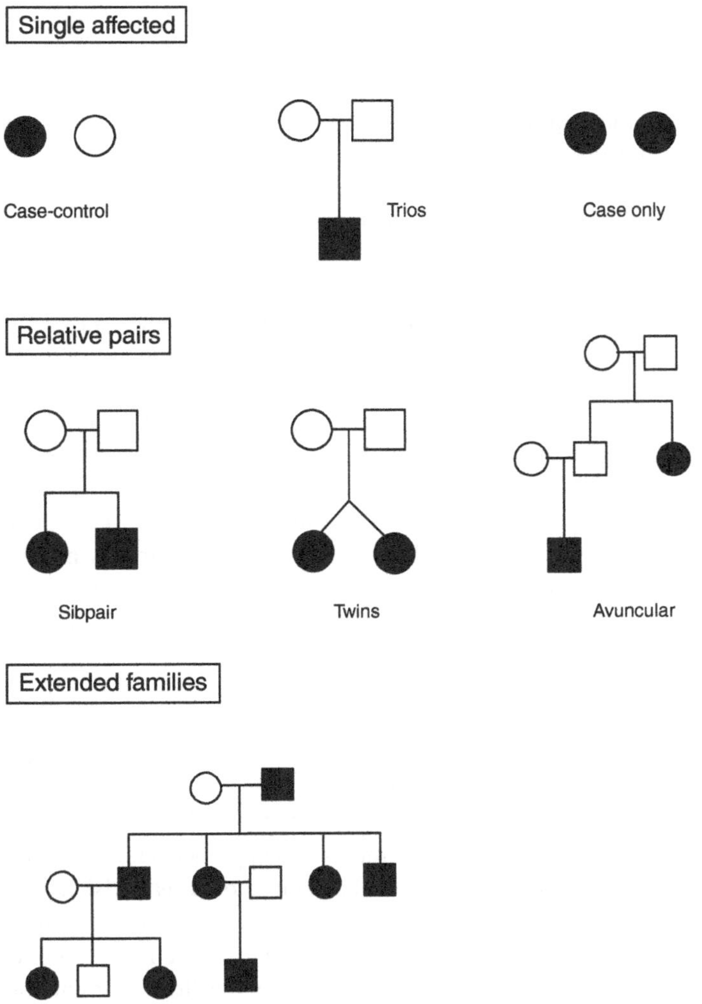

Figure 3.1 Ascertainment schemes for genetic analysis.

of analyses that can be performed. However, one's sampling scheme is often dictated by the natural history of the condition under investigation. For example, with late-onset disorders such as Alzheimer disease, Parkinson disease, and chronic obstructive pulmonary disease, collection of the parents of an affected individual is often not feasible as the parents are often already deceased. In such cases, one may be restricted to collection of affected sibpairs, or a case-control sample.

Single Affected Individual

Collection of a single affected individual can take place in the context of the traditional epidemiologic cross-sectional, case-control, and cohort designs, as well as the case-parent trio design. The case-parent trio design is primarily used in family-based association analysis and includes collection of a case and both their parents. An alternative to the case-control and trio designs is the case-only design (Khoury and Flanders 1996). Similar to the trio design, the case-only approach arose from concerns regarding selection of appropriate controls for the study of genetic factors in the more traditional case-control approach. The case-only approach has been promoted as a particularly useful approach in the examination of gene–environment interactions (Piegorsch et al. 1994). From the ascertainment perspective, collection of single affected individuals is more feasible because in complex disorders, large families with multiple affected individuals are often difficult to identify. A disadvantage of this approach is that it limits the statistical genetic analyses to specific types of association methods (discussed in more detail in Chapter 8). Traditional linkage analysis (described in more detail in Chapter 6) usually cannot be performed on a case-control or trio data set because the necessary family structure is not available for most genetic models.

Relative Pairs

The use of relative pairs has been a common ascertainment design in the genetic analysis of complex disorders. This approach may include the use of sibling pairs that are either concordant for the disorder (affected sibpairs) or discordant (one affected and one unaffected). Monozygotic (identical) and dizygotic (nonidentical) twins are a special case of sibling pairs, and the utility of twins in genetic analysis is described in greater detail later in the chapter. Additionally, there are statistical methods that utilize information from other types of relative pairs, such as parent–child, avuncular pairs (e.g. uncle–niece), cousin pairs, and so on (Weeks and Lange 1988; Davis et al. 1996; Kruglyak et al. 1996). In this approach, relatives other than the analysis pair may also be collected, so that linkage analysis can be performed. Keep in mind that in the case of monozygotic twins, only one of the two individuals may be used in the linkage and association analysis because the twins share 100% of their genetic material.

Extended Families

Extended families refer to large families with many affected individuals in several generations. This study design is optimal for traditional linkage analysis but is often a rare occurrence in complex disorders. If such a family is identified, it is possible that the genetic liability in this particular family is due to a single gene, rather than a more complex etiology. Such a family would provide a unique opportunity to localize a single gene that has a large effect on disease risk in that family but may have a more moderate effect on disease etiology in the general population. Advances in high-throughput sequencing technologies (described more in Chapter 10) have made genome sequencing of small numbers of affected family members feasible, allowing the direct examination of segregation of variants with disease in these pedigrees (described more in Chapter 6). Association methods may also be used with extended families. However, one must ensure that the association method being used considers the within-family dependence (such as the Pedigree Disequilibrium Test (Martin et al. 2000b) or GenABEL (Aulchenko et al. 2007)) or selects only one affected individual from the family to be used in the analysis. A special case can be made for analyzing X-linked variants within families (Choi et al. 2016; Turkmen and Lin 2020).

There are also variations on these three ascertainment schemes. For example, in an analysis of breast cancer in Australia, Hopper and colleagues (1999) employed a "case-control-family" design. In this approach, the cases and controls were selected first and subsequently additional family members were recruited based on the family history. If applied correctly, this approach will have the analytic advantages of a family study, and the results can be placed in the context of an epidemiologic study. Statistical issues associated with this design have been reviewed by other investigators (Liang and Pulver 1996; Seybolt et al. 1997) and will not be discussed here.

Many investigators have explored sampling schemes to determine the optimal ascertainment scheme for genetic analysis of complex disorders. McCarthy and colleagues (1998) considered sampling strategies for affected sibpairs and found that the power to detect a disease gene locus is highly dependent on the larger pedigree structure from which the sibpairs were drawn.

Furthermore, they concluded that imposing a few restrictions on that pedigree structure (such as the presence of at least one unaffected sibling or parent) can provide a modest increase in power, and ascertaining random affected sibpairs (regardless of the larger pedigree structure) tends to be a robust approach under a variety of genetic inheritance models. The advantage of restricting the pedigree structure to one or fewer affected parents is that one can reduce the possibility of bilineality in the pedigree. Terwilliger and Goring (2000) have argued that, even in the case of complex disorders, ascertainment of large pedigrees is a more successful approach for genetic analysis than a case-control approach as the large pedigrees increase the likelihood of genetic homogeneity and additionally, once ascertaining large pedigrees, one has more flexibility with regard to the types of analyses that may be performed. For example, one can analyze the entire pedigree for linkage analysis, and also, by breaking the family structure into smaller units, consider affected sibpair, affected relative pair, or trio approaches as complementary methods for identifying the disease genes. Badner et al. (1998), however, suggest that there is no benefit to collecting large pedigrees under certain genetic models (a qualitative trait with common alleles under single locus, additive and multiplicative inheritance models). In spite of the many elegant theoretical considerations of sampling schemes, there does not appear to be any consensus with regard to an optimal sampling scheme for complex disorders (Baron 1999). The optimal study design for a particular condition is influenced by the underlying genetic model, which is unknown in complex diseases. Consequently, the choice of ascertainment design will be determined primarily by the natural history of the condition under investigation and the available resources (both financial and personnel) rather than theoretical concerns.

Healthy or Unaffected Controls

For some analyses, it is necessary to have control samples to use for comparison with the patient samples. These control samples may include spouses and siblings of affected individuals, classmates, other members of the community, or even untransmitted genetic alleles. Regardless of the relationship of the control sample to the patient sample, one must ensure that the controls are ascertained from the same study population as the patients. Furthermore, the controls can be matched to the patients for confounding factors (any factor that might influence the association between the disease and genotype), such as age, sex, ethnicity, and geographic location. There are two approaches for matching controls to the cases. First, one can select controls such that the overall distribution of cases and controls is comparable with respect to the frequency of the confounders (e.g. for a study of autism spectrum disorders, both cases and controls have a sex ratio of 3 : 1 males to females). This is referred to as frequency or category matching. Alternatively, one or more control individuals may be selected to match each case based on the confounding characteristics (e.g. the case and the control are both African-American females, eight years of age, and reside in Durham County, North Carolina). This approach is called individual matching. An alternative to matching is to consider these potential confounders in statistical analyses, although this may be a less statistically powerful approach. With the increasing availability of publicly accessible data sets, it has become feasible to utilizing existing controls, so long as there is careful consideration of the potential confounding factors. A landmark study by The Wellcome Trust Case Control Consortium (2007) was the first to robustly demonstrate the use of a common set of controls for identifying genetic factors associated with multiple conditions. Subsequently, it has become commonplace to utilize common, publicly available control samples.

It is important to keep in mind that improper selection of controls can lead to incorrect conclusions. For example, if cases and controls are not appropriately matched on ethnicity and the frequency of alleles for the genetic marker differs by ethnicity, an association study can be doomed.

One may falsely conclude an association between a genetic marker and the condition if the "at-risk" marker allele is more prevalent in the predominant ethnicity of the cases versus the controls (e.g. Knowler et al. 1988). Examples of population stratification and approaches for its detection and control of its effects are described more in Chapter 8.

Ascertainment Bias

In genetic studies, research subjects are selected for participation based on the presence or absence of the trait of interest. The family member who comes to the investigator's attention (through admission to a hospital or solicitation of support groups, for example) is called the proband. Most often, the proband is an individual who exhibits the trait of interest. Ascertainment through an affected individual can lead to a bias in the distribution of the numbers of affected and unaffected family members present in the analysis. Because the ascertainment scheme necessitated that the family have at least one affected individual (proband), families that may be carrying the genetic liability of interest but, by chance, do not contain an affected family member will not be ascertained. This phenomenon is referred to as ascertainment bias and is demonstrated in Figure 3.2. Depending on the analysis, ascertainment bias may greatly influence the outcome of the analyses.

In general, ascertainment bias should not affect the ability to accept or reject linkage in linkage analysis (Chapter 6). However, it can affect the estimate of the recombination fraction between the

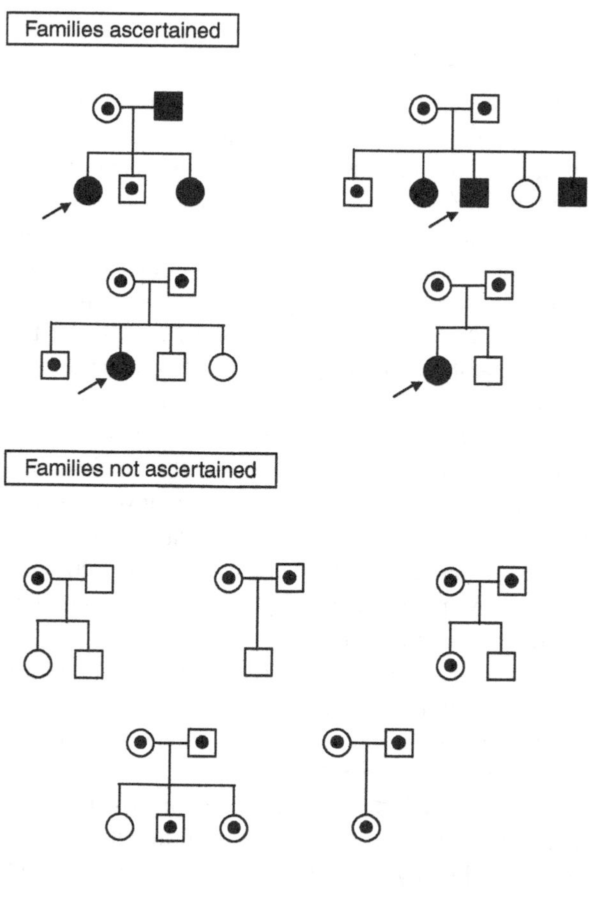

Figure 3.2 Example of ascertainment bias in genetic analysis when ascertaining through an affected individual.

genetic marker and the disease locus (Vieland and Hodge 1996). Ascertainment bias can also influence familial recurrence risk ratios (λ_R) (Guo 1998; Cordell and Olson 2000) and the estimate of the segregation probability of the disease locus in segregation analysis (Ewens and Shute 1986; Greenberg 1986; Stene 1989). Furthermore, it has been argued that in some cases, ascertainment bias may be a reasonable explanation for what appears to be genetic anticipation in some pedigrees (Penrose 1948).

Clearly, ignoring the ascertainment scheme used in an analysis can lead to false conclusions. For example, in the fragile X literature, it had been reported that male offspring of premutation and full-mutation mothers received larger CGG repeat expansions than did female offspring (Rousseau et al. 1994; Loesch et al. 1995). However, once potential sources of ascertainment bias were removed, it was determined that there was no association between sex of the offspring and the size of the mutation inherited from the mother (Ashley-Koch et al. 1998). In this example, there were three types of ascertainment bias that were sequentially removed from the analysis. First, in clinically ascertained families, the transmission to the proband was removed. This is a standard ascertainment correction known as the Weinberg correction scheme (Weinberg 1912). Second, transmissions were excluded where not all individuals within a sibship were tested. That is, in fragile X families, nonsymptomatic females are tested more often than nonsymptomatic males because only the females are at risk to have affected offspring. This ascertainment bias increases the number of transmissions to premutation females compared with premutation males. Finally, the transmissions of the CGG repeat to the mothers of the proband were omitted. Because full-mutation females have reduced fitness (Sherman et al. 1984), mothers of probands are more likely to be premutation carriers merely because they have reproduced. Consequently, including transmissions of the CGG repeat from carrier grandmothers to mothers of probands will also increase the number of transmissions to premutation females. Table 3.1 shows how sequentially correcting for the ascertainment bias in fragile X families changes the conclusion regarding the association between offspring sex and the size of the CGG expansion. Notice that the statistical tests (t-test and logistic regression) become insignificant as the correction schemes are applied. Additionally, it illustrates that in some cases, standard correction schemes, such as the Weinberg method, are not sufficient to abolish the bias in the ascertainment scheme. Thus, all potential sources of bias in the data set must be carefully considered.

Table 3.1 Association between sex of offspring and risk of expansion in fragile X syndrome.

	Proportion of offspring with full mutation (%)		***P*-value**	
Ascertainment scheme (no. of cases)	**Males**	**Females**	**t-test**	**Logistic regression**
Removal of cases associated with ascertainment (434)	0.46	0.38	0.06	0.07
Removal of incompletely ascertained sibships (338)	0.48	0.43	0.38	0.34
Removal of transmissions to proband's mother (298)	0.48	0.50	0.63	0.71

Source: From Ashley-Koch et al. (1998); reprinted with permission.

Approaches to Determining the Genetic Component of a Disease

King et al (1984) described the steps necessary to define the genetic mechanisms involved in a disease or trait many years ago, but those steps are still valid today. First, the evidence for a familial component to the condition must be established. Next, the cause of familial aggregation must be determined. That is, clustering in a disease may result from common environmental factors, rather than genetic factors, and these two hypotheses must be evaluated. Finally, the specific genetic factors must be identified and the manner in which they interact with each other and with environmental factors to contribute to the disease etiology must be defined.

Below, several approaches are presented to evaluate whether or not genes contribute to the etiology of a disease, and to quantitate the contribution of those genes to the disease etiology.

Co-segregation with Chromosomal Abnormalities and Other Genetic Disorders

While complex disorders generally do not exhibit a recognizable inheritance pattern, occasionally in a subset of patients, a complex disorder will segregate with a cytogenetic abnormality or another known genetic disorder. These associations provide valuable information regarding the location of at least one locus involved in the disease etiology. For example, individuals with trisomy 21, or Down syndrome, have an increased risk for developing Alzheimer disease. This increased risk is due to amyloid plaques resulting from an increased dosage of the amyloid precursor protein (APP) (Rumble et al. 1989). Thus, it was not surprising to find that a subset of families with early-onset Alzheimer disease are linked to chromosome 21 and segregate mutations in the *APP* gene (St George-Hyslop et al. 1987; Goate et al. 1991). Another example of co-segregation of a complex disease and cytogenetic abnormality is the association between autistic disorder and 15q11-q13 abnormalities. There are numerous examples of isolated patients with autistic disorder and duplications or inversions involving 15q11-q13 (Wolpert et al. 2000). In many cases, the de novo rearrangements are thought to be maternal in origin (Lindgren et al. 1996). In addition, in the absence of these cytogenetic abnormalities, families with two or more individuals exhibiting autistic disorder display evidence for linkage (Philippe et al. 1999; Bass et al. 2000) and linkage disequilibrium (Cook, Jr. et al. 1998; Martin et al. 2000a; Menold et al. 2000) in this region. Although as yet no gene has been identified, the convergence of cytogenetic, linkage, and association data suggest that a locus involved in susceptibility to autistic disorder is located at 15q11-q13.

Familial Aggregation

One of the characteristics of a genetic disorder is that it aggregates, or clusters, within families. If familial clustering of a disorder is observed, there are several approaches described below to determine if this observation is statistically significant. However, keep in mind that familial clustering may also be due to a common familial environment or simply due to chance. So, while statistical evidence of familial clustering may support the involvement of genetics in the disorder under investigation, it will be necessary to identify the underlying genes to confirm this.

Family History Approach

In this approach, one ascertains the presence or absence of family history in study participants (with and without the condition) and then tests for statistically significant associations between family history and the condition. Khoury et al. (1993) defines three variations on this approach, depending on the level of detail obtained on the relatives. The *Abbreviated Family History Approach* involves simply questioning the study participants on the presence or absence of the condition in other family members. This method is by far the fastest and cheapest, but also the most susceptible to misclassification because no effort is made to confirm the participant responses. That is, the investigator is relying on the accuracy of the informant's memory or perception of the family members, rather than directly observing the condition in the relatives. The *Detailed Family History Approach* requires the investigator to obtain detailed information regarding disease status (including age of onset) and demographic characteristics on each family member. This is the approach routinely utilized in genetic studies. Because the information requested on each family member is more detailed, there is less likelihood of misclassification as with the abbreviated approach. Finally, one can examine and/or interview the relatives directly. This is referred to as the *Family Study Approach*. Often in this approach, medical records are also reviewed in order to confirm the presence or absence of the condition in the relatives. While time-consuming and more expensive, this approach certainly provides the most accurate information regarding the family history of a trait.

Once the necessary information has been collected, there are several methods to test for statistically significant association between family history and the condition. If the study design is case-control, the presence or absence of family history may be treated as a "risk factor" for disease, and the standard epidemiologic 2×2 table may be used:

	Family history	
Disease in study participant	**+**	**–**
+	*a*	*b*
–	*c*	*d*

From here, an odds ratio may be calculated:

$$\text{Odds Ratio} = (a * d)/(b * c) \tag{3.1}$$

The odds ratio derived here is then a measure of the association between family history and the condition. In this case, the odds ratio is the odds of having a positive family history in individuals with the condition compared with the odds of having a positive family history in individuals who are unaffected. This is not to be confused with a risk ratio, which is a prospective measure with respect to an individual's affection status. That is, the risk ratio is the ratio of the incidence of the condition in individuals with a positive family history compared with those who have a negative family history. However, when the incidence of the condition is low, the odds ratio should closely approximate the risk ratio (Rothman and Greenland 1998).

In keeping with the epidemiologic approach, one can also measure the amount of the disease that can be "attributed" to the presence of a positive family history. This in effect provides information regarding what proportion of the disease is due to genetic causes. There are several methods for calculating *attributable fractions*, depending on what information is available. Khoury et al. (1993) reviews three of these formulas (Levin 1953; Kelsey et al. 1986; Miettinen 1974) in their book. The Miettinen formula is given below:

$$\text{Attributable fraction} = \text{fraction of cases with the risk factor} * \frac{(\text{relative risk} - 1)}{\text{relative risk}} \tag{3.2}$$

Example of Calculating Attributable Fraction

In a population-based study examining the contribution of sickle cell disease to the occurrence of developmental disabilities, the Miettinen formula (Equation 3.2) was applied to demonstrate that 34% of the cases of stroke-related developmental disabilities in black children were due to sickle cell disease (Ashley-Koch et al. 2001). In this example, the fraction of cases with stroke-related disabilities with the risk factor (sickle cell disease) was 0.34 (13 children with sickle cell disease among a total of 38 children with stroke-related developmental disabilities), and the relative risk was 130. Thus, the attributable fraction by the Miettinen formula is 34% (0.34 * (129/130)).

For all the ease with the family history approach, there are several pitfalls that may occur. One should be aware that a positive family history is a function of many things, including the frequency of the condition in the general population, the size of the family, the natural history of the condition, the underlying genetic mechanism, and possibly shared environmental exposures. For example, if a condition is relatively common in the population, a positive family history may simply result from the presence of phenocopies in the family. Also, in a condition with late onset, young relatives may be misclassified as "disease-free" (i.e. they may not express the condition simply because they are too young to express it). For example, if one is studying Alzheimer disease and finds that a 55-year-old sibling of an affected proband has no signs of dementia, one should re-examine that sibling at regular intervals to determine if he/she remains asymptomatic over time. It is possible that, at the age of 55 years, the sibling is too young to express symptoms. However, if he/she were re-examined at 60 years of age, he/she would exhibit symptoms. Thus, if the only physical examination for that sibling was performed at 55 years of age, he/she would be classified as "normal." But if another physical examination was performed at the age of 60 years, then the individual would be classified as "affected." Thus, whenever resources allow, follow study participants over time for changes in affection status. In addition to variation in the age of onset of a condition, there may be variable expression of the condition. Therefore, family members who have minor manifestations of the condition or are in the early stages of the disease process may not be recognized as expressing the condition. Furthermore, in complex genetic disorders, individuals in the family may not express the disorder because they possess only a fraction of the necessary genetic and environmental factors to express the condition. For all these reasons, one may want to examine the association of family history with several types of relatives in order to make sure that the same conclusion is reached for all relative types.

Correlation Coefficients

Another simple approach to determining familial aggregation is the use of scatter plots. In the case of a quantitative trait, plotting the trait measurement of one relative against the trait measurement of another relative will provide the correlation of the trait among the pair. The slope of the line that is formed by the data is the square of the correlation coefficient. As shown later in the chapter, the correlation coefficient can also provide information regarding the heritability of a trait. Figure 3.3 examines the correlation of age of onset of Alzheimer disease among affected siblings. These data were randomly selected from a larger study examining the genetic contribution to the age at onset in Alzheimer and Parkinson disease (Li et al. 2002). The proband's age of onset was plotted against the sibling's age of onset. The resulting slope of the line was 0.16, corresponding to a Pearson correlation coefficient of 0.40 ($p < 0.0001$, $n = 400$ sibling pairs), suggesting that there is a significant correlation of age of onset among siblings affected with Alzheimer disease. Khoury and colleagues (1985) demonstrate the use of the loglinear model to determine correlation for qualitative traits.

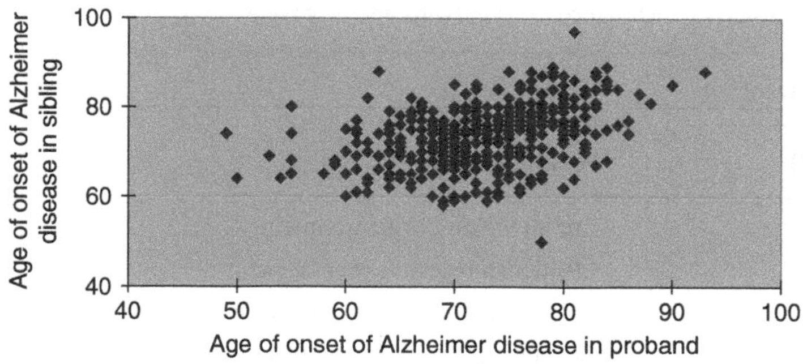

Figure 3.3 Correlation of age of onset among siblings affected with Alzheimer disease.

Twin and Adoption Studies

Twin and adoption studies can be quite useful as they provide an opportunity to tease apart the role of genetics and a common familial environment. The most difficult aspect of these types of studies is obtaining reasonable sample sizes. This is especially true in the United States where twin, adoption, and disease registries are less common than in European countries.

The premise of twin studies is the comparison of the disease concordance in monozygotic (MZ) with dizygotic (DZ) twins. Since MZ twins share 100% of their genetic make-up and DZ twins share on average 50% of their genetic make-up, a greater disease concordance in MZ compared with DZ twins is consistent with the involvement of genetics. An advantage of this approach is that it controls for a common familial environment, but this is generally only applicable for exposures during childhood. It may not entirely control for prenatal exposures because the twins, especially DZ, may not have shared placentas, chorions, and amniotic sacs. However, the intrauterine and extrauterine environments are generally more similar for DZ twins than siblings. It is even less likely that adult exposures are similar among the twins, especially if they reside in different geographic locations. Other possible confounding variables controlled for in the twin study approach include age and sex (provided same-sex DZ twins are utilized). However, keep in mind that for a condition with variable age of onset, it may be necessary to follow the twins for several years in order to conclusively determine that a set of twins is concordant or discordant for the condition. But most importantly, prior to beginning any twin study, one must determine the zygosity of the twins, as misclassification can have a devastating effect on the outcome of the analysis (Ellsworth et al. 1999). Often families will know the zygosity of the twins, but it is prudent to determine this experimentally via genotyping.

Table 3.2 below shows examples of concordance rates that might be observed under various disease etiologies. In practice, the results will not be easily interpreted for complex disorders, much like the results obtained in the last row of this table. In general, however, if the frequency in concordance is greater in MZ compared with DZ twins, it is accepted as evidence for at least a minor role of genetics in the disease etiology.

Adoption studies can also be used to examine the evidence for genetic versus common familial environmental factors. In this approach, cases are ascertained and then the frequency of the condition in the biological parents is compared with the frequency of the condition in the adoptive parents. As shown below in Table 3.3, a higher disease frequency in biological parents argues for the presence of genetic factors, while a higher disease frequency in the adoptive parents argues for the presence of a common familial environmental factor. A variation on adoption studies is the adoptive-twin approach where twins who have been reared apart are

Table 3.2 The association between disease concordance rates in twins and disease etiology.

Frequency of disease concordance in twins		
MZ (%)	**DZ (%)**	**Possible etiology**
85	85	Common familial environment
100	25	Mendelian recessive genetic factor
100	50	Mendelian dominant genetic factor
63	17	Genetic and environmental factors?

Table 3.3 Adoptive studies and disease etiology.

Frequency of disease in parents		
Biological (%)	**Adoptive (%)**	**Possible etiology**
85	4	Genetic factors
4	85	Common familial environment

examined to determine if genetics (high concordance rate) or environment (low concordance rate) plays a role in the disease etiology.

While adoption studies are simple in theory, in reality, they are quite difficult to carry out. Achieving the necessary sample size can be challenging because of the need to identify the biological parents. Moreover, children are often placed in demographic environments similar to that of the biological parents, making it difficult to ensure that one can distinguish between genetics and environment. However, when these issues can be overcome, this approach can provide critical information regarding the etiology of complex traits. For example, this approach has been used to demonstrate that familial aggregation of multiple sclerosis is due to genetics and not a shared environment (Ebers et al. 1995), and to differentiate the role of genes and environment on various psychological measures (Bouchard, Jr. et al. 1990).

Recurrence Risk in Relatives of Affected Individuals

A trio of seminal manuscripts in complex genetic analysis, published by Neil Risch (1990a, b, c), evaluated methods for linkage analysis and determining the mode of inheritance of disease loci using affected relative pairs. The basis for these analyses is familial recurrence risk ratios (λ_R):

$$\lambda_R = \frac{\text{the risk in relatives of type R}}{\text{the risk in the general population}} \tag{3.3}$$

A greater risk for disease in the relatives (of type R) of an affected individual compared with the risk in the general population is consistent with the involvement of genetics in the disease etiology. Because λ_R is a ratio, usually any value over 1.0 is considered to provide evidence for the involvement of a familial component/genetics in a condition. It is generally accepted that the larger the recurrence risk ratio, the stronger the role of genetics in the disease etiology. But this does not always hold true. Notice that the denominator in Equation (3.3) is the risk of the condition in the general population. Thus, for some common diseases, λ_R is not large but that is due to the high frequency of the condition in the general population, not due to the lack of the involvement of genetics. For example, $\lambda_{sibling} = 4 - 5$ for Alzheimer disease (Roses et al. 1995) compared with $\lambda_{sibling} = 100$ for autism spectrum disorder (Bolton et al. 1994; Bailey et al. 1995). In both of these complex disorders, genetics plays a significant role, but the frequency of Alzheimer disease in the general population is much higher than that of autism spectrum disorder, which contributes to the difference in the $\lambda_{sibling}$ values. Recurrence risk ratios are also population-specific as the disease risk in the general population may vary by ethnicity or geographic location. Keep in mind that a high λ_R could result from a common familial environment rather than genetics.

Risch (1990a) also demonstrated that the drop-off in recurrence risk by degree of familial relation is related to the mode of inheritance. For a single-gene model, $\lambda_R - 1$ is expected to decrease by a factor of 2 with each decreasing degree of relation. For a multiplicative model, $\lambda_R - 1$ will drop off more quickly with each degree of relation. Thus, by comparing the λ_R for various types of relatives, one can determine if the disorder is consistent with a particular inheritance pattern. Brown et al. (2000) found that the recurrence risks for ankylosing spondylitis were 8.2% for first-degree relatives, 1.0% for second-degree relatives, and 0.7% for third-degree relatives. Because this drop-off rate is more rapid than a factor of 2, a single-gene model for genetic susceptibility to ankylosing spondylitis is not likely. Using the information on recurrence risks, the authors determined that the most likely mode of inheritance for ankylosing spondylitis was an oligogenic model with a multiplicative interaction between the loci.

One can also use recurrence risks to make a rough estimate of the number of genes involved in a disorder if one assumes a particular inheritance pattern. For example, the $\lambda_{sibling} = 100$ for autism spectrum disorder (Bolton et al. 1994; Bailey et al. 1995). It is clear that autism spectrum disorder does not follow a Mendelian mode of inheritance, and in fact, is more consistent with an epistatic interaction of multiple loci (Jorde et al. 1991). If we assume that there are four genes contributing an equal, additive effect to the genetic susceptibility, the $\lambda_{sibling}$ for each locus would be 25 ($25 + 25 + 25 + 25 = 100$). If there are four genes contributing an equal multiplicative effect, the $\lambda_{sibling}$ for each locus would be approximately 3.2 ($3.2 * 3.2 * 3.2 * 3.2 = 105$). This type of information may be useful in determining the power to detect loci in a data set. That is, with a sample size of 100 sibpairs, the power to detect linkage at a locus with $\lambda_{sibling} = 3.0$ is about 85%, and the power to detect linkage at a locus with $\lambda_{sibling} = 25$ would be approaching 100% (Risch 1990b, figure 1). Thus, in this particular case, under either the additive or multiplicative model, one should have sufficient power to detect loci with 100 sibpairs.

Heritability

Heritability (h^2) is a measurement of the genetic contribution to a trait, or how much of the trait variability can be attributed to genetics rather than other causes. Heritability may be defined in the broad or narrow sense. Heritability in the broad sense is defined as follows:

$$h^2 = \frac{G}{V} = \frac{G}{G+B+E} \tag{3.4}$$

where V is the total phenotypic variance, G is the genetic variance of the phenotype, B is the within family variance of the phenotype, and E is the random environmental variance of the phenotype. The genetic variance can be further broken down into the additive (effects of alleles at a particular locus), dominance (effects of allelic interactions at a particular locus), and epistatic variance (effects of allelic interactions from multiple loci). Heritability in the narrow sense is a measure of the additive genetic variation in a trait. Estimates of heritability will vary between 0 and 1, with $h^2 = 0$ representing traits that are entirely controlled by nongenetic factors and $h^2 = 1$ representing traits that are entirely controlled by genetic factors. Here we will focus simpler methods of calculating heritability in the context of analyses defined in previous sections of the current chapter.

The correlation coefficient that one obtains from plotting a trait value in one relative versus the trait value in another relative may be used to calculate heritability in the narrow sense, provided that the relatives under consideration are not full siblings:

$$h^2 = r^2 / \left(\text{fraction of genes shared identical by descent}\left(\text{IBD}\right)\right) \tag{3.5}$$

For example, if one plots the age of onset for a condition in a proband versus the age of onset in an affected parent, the square of the correlation coefficient (r^2, or the slope of the line) obtained from the plot would be divided by 1/2, the number of alleles shared identical by descent (IBD) for a parent–child pair. This particular approach to calculating heritability does not represent heritability in the narrow sense when considering full siblings because one has to take into account dominance variance, as well. See Hartl and Clark (1997) for a more thorough discussion of using the covariance between two relatives to estimate heritability.

Rice and colleagues (1997) extended this approach to incorporate correlations from multiple relative pairs (r_{sib}, siblings; $r_{\text{p-o}}$, parent–offspring; r_{sp}, spousal):

$$h^2 = \frac{\left(r_{\text{sib}} + r_{\text{p-o}}\right)\left(1 + r_{\text{sp}}\right)}{\left(1 + r_{\text{sp}}\right) + \left(2r_{\text{sp}} * r_{\text{p-o}}\right)} \tag{3.6}$$

Example Using Correlation Coefficients to Calculate Heritability

In a large study that assessed the genetic contribution to normal variation of pulmonary function, Wilk and colleagues (2000) calculated familial correlations for spouses, parent–offspring pairs, and siblings from 455 Caucasian families for forced expiratory volume (FEV1; a measure of airflow), forced vital capacity (FVC; a measure of lung volume), and the ratio of FEV/FVC and using formula 3.6, obtained heritability estimates for the three traits (Table 3.4).

Correlations for a trait among monozygotic and dizygotic twins can also be used to calculate heritability:

$$h^2 = 2\left(r_{\text{MZ}} - r_{\text{DZ}}\right) \tag{3.7}$$

Heritability estimates can be quite useful for data mining of a phenotype prior to initiating a genome scan. For example, if several phenotypic measures for a condition have been collected,

Table 3.4 Familial correlation and heritability estimates of pulmonary function.

Trait	r_{sib}	$r_{\text{p-o}}$	r_{sp}	h^2
FEV1	0.259	0.239	−0.067	0.52
FVC	0.257	0.242	−0.135	0.54
FEV1/FVC ratio	0.27	0.19	0.071	0.45

Source: Wilk et al. (2000, table II).

heritability estimates may provide guidance as to which of those measures will be most useful for genetic analysis. That is, those phenotypes with the highest h^2 values should be prioritized for analysis.

With the advent of Genome-Wide Association Study (GWAS), the concept of "missing heritability" was coined to describe the difference between the estimated heritability and the amount of heritability attributable to genome-wide significant loci detected through GWAS (Manolio et al. 2009). From this observation, methods to estimate heritability using genome-wide SNP data emerged such that one could calculate the "SNP heritability," suggesting that additional SNP loci not meeting genome-wide significance contribute to the missing heritability. The most widely used of these methods include genomic-relatedness-matrix restricted maximum likelihood (GREML) (Yang et al. 2010) which is implemented in GCTA software (Yang et al. 2011), linkage disequilibrium score regression (LDSR) (Bulik-Sullivan et al. 2015), and LD-adjusted kinships (LDAK) (Speed et al. 2017). A good review of genome-wide heritability estimation can be found in Hall and Bush (2016).

Segregation Analysis

Segregation analysis is a modeling tool that is used to examine the patterns of disease in families and determine if the patterns are indicative of traditional genetic inheritance models (such as autosomal dominant, autosomal recessive, and polygenic) or are more consistent with nongenetic (environmental) models (Elston 1981; Morton 1982; Lalouel 1984). The likelihood of the data to fit a particular inheritance model is computed. By comparing the likelihood of several models, one can determine which model provides the "best fit" to the data. Segregation analysis does not prove that a particular inheritance model is correct but will determine if the data are consistent with that inheritance model.

The advantage of segregation analysis is that it can provide an inheritance model and parameters that may be used in parametric linkage analysis. However, generally segregation analysis can only model 1–2 loci, which may not be very useful for most complex diseases. That is, if none of the inheritance models examined in the segregation analysis can adequately accommodate the complexities of the underlying inheritance model of the disorder, even the "best-fitting" model will not provide much information.

Additionally, this approach is extremely sensitive to ascertainment bias. In genetic analysis, families are often collected based on the presence of many affected individuals. Thus, for segregation analysis, there may be a high proportion of families with numerous affected individuals that are used in the analysis, when in reality these types of families only make up a small percentage of the disease population and most cases may be observed in families with only one or two affected individuals. For example, the probability that an affected individual will be ascertained as a proband is π, and when $\pi = 1$, all the individuals in the study population who have the condition have been ascertained. This is referred to as "complete ascertainment." When the probability that an affected individual is a proband is very low (π approaches 0), each sibship is expected to have only a single proband. This is called "single ascertainment." With this latter ascertainment approach, the probability that a family will come to the investigator's attention, and be included in the study, is related to the number of affected individuals in the family. That is, the more affected the individuals in the family, the higher the likelihood that this particular family will be ascertained, thus introducing a biased distribution of family types in the analysis.

If one does not correct for the method in which the families were ascertained, the estimate of the segregation probability of the disease allele can be biased, which can affect the "best-fitting" genetic model. In some cases, the ascertainment bias may be so great that it causes the investigator to incorrectly conclude that the disorder is consistent with a single-gene model (Greenberg 1986). For a discussion on approaches for correcting ascertainment bias, see Khoury et al. (1993).

There are several analytic approaches to segregation analysis, each with its strengths and weaknesses. The most commonly used methods include the mixed model (Morton and MacLean 1974; Lalouel and Morton 1981; MacLean et al. 1984), the transmission probability model (Elston and Stewart 1971), the unified model (Lalouel et al. 1983) which draws on the strengths of both the mixed and transmission probability models, and the regressive model (Bonney 1984). All of these approaches are computationally intensive but are available for use in several software packages such as POINTER (Lalouel and Morton 1981), SAGE (http://darwin.cwru.edu/sage/) and PAP (Hasstedt 1993).

Segregation analysis is not widely used in the evaluation of complex diseases because it is susceptible to the presence of genetic heterogeneity, phenocopies, gene–gene and gene–environment interactions, which are quite difficult to model. Consequently, segregation analysis results that support the involvement of a major gene in a condition are much easier to interpret than results that do not support such an effect. This is especially true if there are other data to support the involvement of genetics, such as twin data or familial clustering. In spite of these difficulties, there are examples of the successful application of segregation analysis to complex disorders, even to psychiatric disorders that tend to have the additional complexity of defining a precise phenotype for investigation (evidence for a Mendelian genetic factor contributing to obsessive compulsive disease and (Nestadt et al. 2000; Notarnicola et al. 2000).

Summary

The analyses presented in this chapter will aid in determining whether genes play an important role in a condition or trait. It is important to establish a genetic basis before embarking on more elaborate analyses such as whole genome scans using linkage and/or association analysis because of the ethical, financial, and labor challenges associated with them. However, the first step in the analytic process should be a thorough examination of the literature. Often, because complex disorders are common and of public health importance, many of the analyses discussed in this chapter will have been previously completed by other investigators. Therefore, there may already exist sufficient evidence to support the involvement of genetics in the etiology of the disorder under investigation but evaluate the literature critically, keeping in mind how the study population (including age and ethnicity) and phenotype was defined in each of the published analyses. Significant deviations from the definition of one's own study population and phenotype may suggest that prior results do not generalize to the present study and may necessitate repeating of the analyses described in this chapter.

If one determines that the types of analyses presented in this chapter (or any of the other chapters) must be performed, give careful consideration to study design prior to collecting and analyzing the data. Study design will influence the types of biases introduced into the data and can limit the types of analyses that may be performed.

Pay careful attention to the types of bias in the data. Ignoring ascertainment biases and confounding factors (age, ethnicity, etc.) may lead one to the wrong conclusions. If control samples are utilized, one must ensure that they are properly matched to the cases for confounding factors. When historical data are obtained on family members, follow up with examination of medical records or physical examination of those individuals to corroborate the information whenever resources allow. If a late-onset disorder is being studied, re-examine "unaffected" individuals after a period of time to confirm that they are truly unaffected and not misclassified because they were too young to express the disorder.

Keep in mind that for complex genetic disorders, there may be a high percentage of the patient population for whom the disorder is sporadic (no other affected family members), and the genetic mechanisms that lead to complex disorders may or may not be the same in familial and sporadic cases. For example, the *BRCA1* gene that has been implicated in familial cases of breast cancer is not frequently mutated in cases of sporadic breast cancer (Easton et al. 1993; Futreal et al. 1994). This caveat is also important when comparing genetic mechanisms for early-onset versus late-onset familial versions of a disorder.

Finally, recognize that statistical associations for familial clustering may not represent a genetic etiology for a disorder but may be due to a common familial environment, or simply chance. The definitive evidence that a disorder is genetic is the identification of the specific genetic variations that lead to the condition.

References

Ashley-Koch, A.E., Robinson, H., Glicksman, A.E. et al. (1998). Examination of factors associated with instability of the FMR1 CGG repeat. *Am. J. Hum. Genet.* 63: 776–785.

Ashley-Koch, A., Murphy, C.C., Khoury, M.J., and Boyle, C.A. (2001). Contribution of sickle cell disease to the occurrence of developmental disabilities: a population-based study. *Genet. Med.* 3: 181–186.

Aulchenko, Y.S., Ripke, S., Isaacs, A., and van Duijn, C.M. (2007). GenABEL: an R library for genome-wide association analysis. *Bioinformatics* 23: 1294–1296.

Aziz, S., Kuperstein, G., Rosen, B. et al. (2001). A genetic epidemiological study of carcinoma of the fallopian tube. *Gynecol. Oncol.* 80: 341–345.

Badner, J.A., Gershon, E.S., and Goldin, L.R. (1998). Optimal ascertainment strategies to detect linkage to common disease alleles. *Am. J. Hum. Genet.* 63: 880–888.

Bailey, A., Le Couteur, A., Gottesman, I. et al. (1995). Autism as a strongly genetic disorder: evidence from a British twin study. *Psychol. Med.* 25: 63–77.

Baron, M. (1999). Optimal ascertainment strategies to detect linkage to common disease alleles. *Am. J. Hum. Genet.* 64: 1243–1248.

Bass, M.P., Menold, M.M., Wolpert, C.M. et al. (2000). Genetic studies in autistic disorder and chromosome 15. *Neurogenetics* 2: 219–226.

Bellamy, R. (1998). Genetic susceptibility to tuberculosis in human populations. *Thorax* 53: 588–593.

Bellamy, R., Beyers, N., McAdam, K.P.W.J. et al. (2000). Genetic susceptibility to tuberculosis in Africans: a genome-wide scan. *Proc. Natl Acad. Sci. USA* 97: 8005–8009.

Bolton, P., Macdonald, H., Pickles, A. et al. (1994). A case-control family history study of autism. *J. Child Psychol. Psychiatry Allied Discip.* 35: 877–900.

Bonney, G.E. (1984). On the statistical determination of major gene mechanisms in continuous human traits: regressive models. *Am. J. Med. Genet.* 18: 731–749.

Bouchard, T.J. Jr., Lykken, D.T., McGue, M. et al. (1990). Sources of human psychological differences: The Minnesota study of twins reared apart. *Science* 250: 223–228.

Brown, M.A., Laval, S.H., Brophy, S., and Calin, A. (2000). Recurrence risk modelling of the genetic susceptibility to ankylosing spondylitis. *Ann. Rheum. Dis.* 59: 883–886.

Bulik-Sullivan, B.K., Loh, P.R., Finucane, H.K. et al. (2015). LD score regression distinguishes confounding from polygenicity in genome-wide association studies. *Nat. Genet.* 47: 291–295.

Choi, S., Lee, S., Qiao, S. et al. (2016). FARVATX: Family-Based Rare Variant Association Test for X-linked genes. *Genet. Epidemiol.* 40 (6): 475–485.

Cook, E.H. Jr., Courchesne, R.Y., Cox, N.J. et al. (1998). Linkage-disequilibrium mapping of autistic disorder, with 15q11-13 markers. *Am. J. Hum. Genet.* 62: 1077–1083.

Cordell, H.J. and Olson, J.M. (2000). Correcting for ascertainment bias of relative-risk estimates obtained using affected-sib-pair linkage data. *Genet. Epidemiol.* 18: 307–321.

Corder, E.H., Saunders, A.M., Strittmatter, W.J. et al. (1993). Gene dose of apolipoprotein E type 4 allele and the risk of Alzheimer's disease in late onset families. *Science* 261: 921–923.

Davis, S., Schroeder, M., Goldin, L.R., and Weeks, D.E. (1996). Nonparametric simulation-based statistics for detecting linkage in general pedigrees. *Am. J. Hum. Genet.* 58: 867–880.

De La Paz, M.A., Pericak-Vance, M.A., Haines, J.L., and Seddon, J.M. (1997). Phenotypic heterogeneity in families with age-related macular degeneration. *Am. J. Ophthalmol.* 124: 331–343.

Easton, D., Ford, D., and Peto, J. (1993). Inherited susceptibility to breast cancer. *Cancer Surv.* 18: 95–113.

Ebers, G.C., Sadovnick, A.D., and Risch, N.J. (1995). A genetic basis for familial aggregation in multiple sclerosis. *Nature* 377: 150–151.

Ellsworth, R.E., Ionasescu, V., Searby, C. et al. (1999). The CMT2D locus: refined genetic position and construction of a bacterial clone-based physical map. *Genome Res.* 9: 568–574.

Elston, R.C. (1981). Segregation analysis. *Adv. Hum. Genet.* 11: 63–120.

Elston, R.C. and Stewart, J. (1971). A general model for the genetic analysis of pedigree data. *Hum. Hered.* 21: 523–542.

Ewens, W.J. and Shute, N.C. (1986). A resolution of the ascertainment sampling problem I. Theory. *Theor. Popul. Biol.* 30: 388–412.

Futreal, P.A., Liu, Q., Shattuck-Eidens, D. et al. (1994). BRCA1 mutations in primary breast and ovarian carcinomas. *Science* 266: 120–126.

Goate, A., Chartier-Harlin, M.C., Mullan, M. et al. (1991). Segregation of a missense mutation in the amyloid precursor protein gene with familial Alzheimer's disease. *Nature* 349: 704–706.

Gonzalez, E., Dhanda, R., Bamshad, M. et al. (2001). Global survey of genetic variation in CCR5, RANTES, and MIP-1α: impact on the epidemiology of the HIV-1 pandemic. *Proc. Natl. Acad. Sci. U. S. A.* 98: 5199–5204.

Greenberg, D.A. (1986). The effect of proband designation on segregation analysis. *Am. J. Hum. Genet.* 39: 329–339.

Guo, S.W. (1998). Inflation of sibling recurrence-risk ratio, due to ascertainment bias and/or overreporting. *Am. J. Hum. Genet.* 63: 252–258.

Hall, J.B. and Bush, W.S. (2016). Analysis of heritability using genome-wide data. *Curr. Protoc. Hum. Genet.* 91: 1.30.1–1.30.10.

Hartl, D.L. and Clark, A.G. (1997). *Principles of Population Genetics*. Sunderland, MA: Sinauer Associates, Inc.

Hasstedt, S.J. (1993). Variance components/major locus likelihood approximation for quantitative, polychotomous, and multivariate data. *Genet. Epidemiol.* 10: 145–158.

Hill, A.V. (1999). Defence by diversity. *Nature* 398: 668–669.

Hopper, J.L., Chenevix-Trench, G., Jolley, D.J. et al. (1999). Design and analysis issues in a population-based, case-control-family study of the genetic epidemiology of breast cancer and the Co-operative Family Registry for Breast Cancer Studies (CFRBCS). *J. Natl. Cancer Inst. Monogr.* 26: 95–100.

Jorde, L.B., Hasstedt, S.J., Ritvo, E.R. et al. (1991). Complex segregation analysis of autism. *Am. J. Hum. Genet.* 49: 932–938.

Kelsey, J.L., Thompson, W.D., and Evans, A.S. (1986). *Methods in Observational Epidemiology*. New York: Oxford University Press.

Khoury, M.J. and Flanders, W.D. (1996). Nontraditional epidemiologic approaches in the analysis of gene environment interaction: case-control studies with no controls! *Am. J. Epidemiol.* 144: 207–213.

Khoury, M.J., Beaty, T.H., Tockman, M.S. et al. (1985). Familial aggregation in chronic obstructive pulmonary disease: use of the loglinear model to analyze intermediate environmental and genetic risk factors. *Genet. Epidemiol.* 2: 155–166.

Khoury, M.J., Beaty, T.H., and Cohen, B.H. (1993). *Fundamentals of Genetic Epidemiology*. New York: Oxford University Press.

King, M.C., Lee, G.M., Spinner, N.B. et al. (1984). Genetic epidemiology. *Annu. Rev. Public Health* 5: 1–52.

Knowler, W.C., Williams, R.C., Pettitt, D.J., and Steinberg, A.G. (1988). Gm3; 5, 13, 14 and type 2 diabetes mellitus: an association in American Indians with genetic admixture. *Am. J. Hum. Genet.* 43: 520–526.

Kruglyak, L., Daly, M.J., Reeve-Daly, M.P., and Lander, E.S. (1996). Parametric and nonparametric linkage analysis: a unified multipoint approach. *Am. J. Hum. Genet.* 58: 1347–1363.

Lalouel, J.M. (1984). Segregation analysis: a gene or not a gene. In: *Genetic Epidemiology of Coronary Heart Disease: Past, Present, and Future* (eds. D.C. Rao, R.C. Elston, L.H. Kuiler, et al.), 217–243. New York: Alan R. Liss, Inc.

Lalouel, J.M. and Morton, N.E. (1981). Complex segregation analysis with pointers. *Hum. Hered.* 31: 312–321.

Lalouel, J.M., Rao, D.C., Morton, N.E., and Elston, R.C. (1983). A unified model for complex segregation analysis. *Am. J. Hum. Genet.* 35: 816–826.

Levin, M.L. (1953). The occurrence of lung cancer in man. *Acta Unio Int. Contra Cancrum* 9: 531–541.

Li, Y.J., Scott, W.K., Hedges, D.J. et al. (2002). Age at onset in two common neurodegenerative diseases is genetically controlled. *Am. J. Hum. Genet.* 70: 985–993.

Liang, K.Y. and Pulver, A.E. (1996). Analysis of case-control/family sampling design. *Genet. Epidemiol.* 13: 253–270.

Lindgren, V., Cook, E.H. Jr., Leventhal, B.L. et al. (1996). Maternal origin of proximal 15q duplication in autism. *Am. J. Hum. Genet.* 39 (supp): 688.

Loesch, D.Z., Huggins, R., Petrovic, V., and Slater, H. (1995). Expansion of the CGG repeat in fragile X in the FMR1 gene depends on the sex of the offspring. *Am. J. Hum. Genet.* 57: 1408–1413.

MacLean, C.J., Morton, N.E., and Yee, S. (1984). Combined analysis of genetic segregation and linkage under an oligogenic model. *Comput. Biomed. Res.* 17: 471–480.

Manolio, T.A. 1, Collins, F.S., Cox, N.J. et al. (2009). Finding the missing heritability of complex diseases. *Nature* 461 (7265): 747–753. https://doi.org/10.1038/nature08494.

Martin, E.R., Menold, M.M., Wolpert, C.M. et al. (2000a). Analysis of linkage disequilibrium in γ-aminobutyric acid receptor subunit genes in autistic disorder. *Am. J. Med. Genet.* 96: 43–48.

Martin, E.R., Monks, S.A., Warren, L.L., and Kaplan, N.L. (2000b). A test for linkage and association in general pedigrees: the pedigree disequilibrium test. *Am. J. Hum. Genet.* 67: 146–154.

McCarthy, M.I., Kruglyak, L., and Lander, E.S. (1998). Sib-pair collection strategies for complex diseases. *Genet. Epidemiol.* 15: 317–340.

Meadows, K.L., Pettay, D., Newman, J. et al. (1996). Survey of the fragile X syndrome and the fragile XE syndrome in a special education needs population. *Am. J. Med. Genet.* 64: 428–433.

Menold, M.M., Bass, M.P., Gilbert, J.R. et al. (2000). SNP analysis of $GABA_A$ receptor subunit in autistic disorder. *Am. J. Hum. Genet.* 67 (339): 1886A.

Meola, G. (2013). Clinical aspects, molecular pathomechanisms and management of myotonic dystrophies. *Acta Myol.* 32: 154–165.

Miettinen, O.S. (1974). Proportion of disease caused or prevented by a given exposure, trait or intervention. *Am. J. Epidemiol.* 99: 325–332.

Morton, N.E. (1982). Segregation and linkage analysis. In: *Human Genetics, Part B: Medical Aspects* (eds. B. Bonne-Tamir, T. Cohen and R.M. Goodman), 3–14. Alan R. Liss.

Morton, N.E. and MacLean, C.J. (1974). Analysis of family resemblance. 3. Complex segregation of quantitative traits. *Am. J. Hum. Genet.* 26: 489–503.

Neel, J.V. (1949). The inheritance of sickle cell anemia. *Science* 110: 64–66.

Nestadt, G., Lan, T., Samuels, J. et al. (2000). Complex segregation analysis provides compelling evidence for a major gene underlying obsessive-compulsive disorder and for heterogeneity by sex. *Am. J. Hum. Genet.* 67: 1611–1616.

Notarnicola, M., Cavallini, A., Cardone, R. et al. (2000). K-ras and p53 mutations in DNA extracted from colonic epithelial cells exfoliated in faeces of patients with colorectal cancer. *Dig. Liver Dis.* 32: 131–136.

Penrose, L.S. (1948). The problem of anticipation in pedigrees of dystrophia myotonica. *Ann. Eugenics* 14: 125–132.

Philippe, A., Martinez, M., Guilloud-Bataille, M. et al. (1999). Genome-wide scan for autism susceptibility genes. *Hum. Mol. Genet.* 8: 805–812.

Piegorsch, W.W., Weinberg, C.R., and Taylor, J.A. (1994). Non-hierarchical logistic models and case-only designs for assessing susceptibility in population-based case-control studies. *Stat. Med.* 13: 153–162.

Rice, T., Despres, J.P., Daw, E.W. et al. (1997). Familial resemblance for abdominal visceral fat: the HERITAGE family study. *Int. J. Obes. Relat. Metab. Disord.* 21: 1024–1031.

Risch, N. (1990a). Linkage strategies for genetically complex traits. I. Multilocus models. *Am. J. Hum. Genet.* 46: 222–228.

Risch, N. (1990b). Linkage strategies for genetically complex traits. II. The power of affected relative pairs. *Am. J. Hum. Genet.* 46: 229–241.

Risch, N. (1990c). Linkage strategies for genetically complex traits. III. The effect of marker polymorphism on analysis of affected relative pairs. *Am. J. Hum. Genet.* 46: 242–253.

Roses, A.D., Devlin, B., Conneally, P.M. et al. (1995). Measuring the genetic contribution of APOE in late-onset Alzheimer disease (AD). *Am. J. Hum. Genet.* 57: A202.

Rothman, K.J. and Greenland, S. (1998). *Modern Epidemiology*. Philadelphia, PA: Lippincott-Raven.

Rousseau, F., Heitz, D., Tarleton, J. et al. (1994). Higher rate of transition from fragile X permutations into full mutation in males than in females suggest post-conceptional expansion of the CGG repeats. *Am. J. Hum. Genet.* 55: A240.

Rumble, B., Retallack, R., Hilbich, C. et al. (1989). Amyloid A4 protein and its precursor in down's syndrome and Alzheimer's disease. *N. Engl. J. Med.* 320: 1446–1452.

Seybolt, L.M., Vachon, C., Potter, K. et al. (1997). Evaluation of potential sources of bias in a genetic epidemiologic study of breast cancer. *Genet. Epidemiol.* 14: 85–95.

Sherman, S.L., Morton, N.E., Jacobs, P.A., and Turner, G. (1984). The marker (X) syndrome: a cytogenetic and genetic analysis. *Ann. Hum. Genet.* 48 (Pt 1): 21–37.

Shields, R. and Dell, H. (2001). Genes for HIV susceptibility. *Trends Genet.* 17: 19.

Speed, D., Cai, N., the UCLEB Consortium et al. (2017). Reevaluation of SNP heritability in complex human traits. *Nat. Genet.* 49: 986–992.

St George-Hyslop, P.H., Tanzi, R.E., Polinsky, R.J. et al. (1987). The genetic defect causing familial Alzheimer's disease maps on chromosome 21. *Science* 235: 885–890.

Stene, J. (1989). The incomplete, multiple ascertainment model: assumptions, applications, and alternative models. *Genet. Epidemiol.* 6: 247–251.

Terwilliger, J.D. and Goring, H.H. (2000). Gene mapping in the 20th and 21st centuries: statistical methods, data analysis, and experimental design. *Hum. Biol.* 72: 63–132.

Turkmen, A.S. and Lin, S. (2020). Detecting X-linked common and rare variant effects in family-based sequencing studies. *Genet. Epidemiol.* 45 (1): 36–45. https://doi.org/10.1002/gepi.22352. Epub 2020 Aug 30.

Vieland, V.J. and Hodge, S.E. (1996). The problem of ascertainment for linkage analysis. *Am. J. Hum. Genet.* 58: 1072–1084.

Waller, D.K., Anderson, J.L., Lorey, F., and Cunningham, G.C. (2000). Risk factors for congenital hypothyroidism: an investigation of infant's birth weight, ethnicity, and gender in California, 1990–1998. *Teratology* 62: 36–41.

Weeks, D.E. and Lange, K. (1988). The affected-pedigree-member method of linkage analysis. *Am. J. Hum. Genet.* 42: 315–326.

Weinberg, W. (1912). Zur Verebung der Anlage der Bluterkrankheit mit methodol. Erganzungen meiner Geschwistermethode. *Arch. Rass. Ges. Biol.* 6: 694–709.

Wellcome Trust Case Control Consortium (2007). Genome-wide association study of 14,000 cases of seven common diseases and 3,000 shared controls. *Nature* 447: 661–678.

Wilk, J.B., Djousse, L., Arnett, D.K. et al. (2000). Evidence for major genes influencing pulmonary function in the NHLBI family heart study. *Genet. Epidemiol.* 19: 81–94.

Wolpert, C.M., Menold, M.M., Bass, M.P. et al. (2000). Three probands with autistic disorder and isodicentric chromosome 15. *Am. J. Med. Genet.* 96: 365–372.

Yang, J., Benyamin, B., McEvoy, B.P. et al. (2010). Common SNPs explain a large proportion of the heritability for human height. *Nat. Genet.* 42 (7): 565–569.

Yang, J., Lee, S.H., Goddard, M.E., and Visscher, P.M. (2011). GCTA: a tool for genome-wide complex trait analysis. *Am. J. Hum. Genet.* 88 (1): 76–82.

4

Study Design for Genetic Studies

Dana C. Crawford[1] *and Logan Dumitrescu*[2]

[1] *Department of Population and Quantitative Health Sciences, Case Western Reserve University, Cleveland, OH, USA*
[2] *Department of Neurology, Vanderbilt University, Nashville, TN, USA*

Introduction

The study design process includes all clinical, molecular, bioinformatics, and analytic components of your study. A study population must be defined, as well as the process for selecting individuals from that population for participation (this chapter) and criteria for identifying individuals exhibiting the trait of choice. One must determine the molecular technology to be used (Chapter 10) as well as establish the electronic process for storing and retrieving the clinical and molecular data (Chapter 7) and the analysis methods to be used (e.g., linkage versus association, Chapters 6, and 8–10). Finally, one must ensure that all ethical, legal, and social issues are addressed (Chapter 5). Each component in the study design process should be given careful consideration, as the decisions one makes at each step will impact the conclusions that can be drawn from the results. Furthermore, many of the decisions in the study design process are interrelated. For example, the ascertainment process will impact the analysis methods that can be performed. Since many of the steps in the study design process are covered in detail in other chapters of the book, they will not be discussed further here. Selecting a study population and collecting biological samples will be covered in this chapter.

Selecting a Study Population

The application of epidemiological methods to the study of genetics in complex diseases has become widespread. Population- and family-based study designs can be used to test for relationships between genotypes and phenotypes. However, determination of the type(s) of individuals to collect (e.g. using cohorts of unrelated individuals or family-based approaches, see below) is based on knowledge of phenotype and any evidence for the genetic model (as discussed in Chapter 3). For example, if a single gene is suspected, large extended families are likely to prove most efficient. For a more complex genetic model, large numbers of unrelated individuals may prove most efficient. This section discusses the basic study designs that are used in genetic epidemiology with present-day examples and an emphasis on advantages and disadvantages of each study design.

Genetic Analysis of Complex Diseases, Third Edition. Edited by William K. Scott and Marylyn D. Ritchie.

Family-Based Studies (Linkage)

For human diseases with evidence of Mendelian inheritance patterns (dominant, recessive, etc.) and single gene mutations with relatively high penetrance, the linkage study is a powerful approach to disease gene mapping. Linkage is based on the observation that an assayed marker cosegregates with disease within a family (Chapter 6). The marker is said to cosegregate with the disease if no recombination occurs between the two during the meiotic event. Recombination is the physical exchange of DNA that occurs during the meiotic process and can be measured as the recombination fraction (θ) where 0 is no recombination (i.e., perfect linkage between markers) and 0.50 is recombination (i.e., markers are independent). Once recombination between the two occurs, linkage between the assayed marker and the disease gene is lost. In practice, each meiotic event (offspring; N) is counted in a linkage study as either a recombinant (R) event or a nonrecombinant event (NR) between the assayed marker and disease gene (whose proxy is disease status), and the likelihood of linkage can be estimated in a likelihood ratio (LR) setting by comparing the empirical data assuming linkage (with an assumed recombination fraction) to the assumption that the pedigree occurred by chance:

$$\mathrm{LR} = \left(\theta^{\mathrm{R}}\left(1-\theta\right)^{\mathrm{NR}}\right) / \left(\theta = 0.5\right)^{N}$$

Linkage between the assayed marker and presumed disease gene is traditionally declared when the odds of the likelihood is 1000:1 (or log of the odds [LOD] score = 3) (Morton 1955).

In a linkage study, families are typically recruited through an affected individual in a clinic setting. Families can also be recruited through an unaffected individual such as an individual referred to the clinic for carrier testing. Within the family unit, the individual first recruited or seeking medical attention as part of the study is known as the "proband". A single large, extended family can be the basis for a linkage study. Classic examples of large families used successfully in linkage analysis include large American and Venezuelan families with Huntington's disease (Gusella et al. 1983). With greater than 1800 individuals in a single pedigree, the Venezuelan family with Huntington's disease from the Lake Maracaibo region is an extreme example of use of a large family for linkage studies. In the absence of a single large family, multiple smaller families can be included in linkage studies given that LOD scores are additive. This latter strategy was used extensively and successfully in studies that linked the putative cystic fibrosis gene to specific markers on chromosome 7 (Knowlton et al. 1985; Tsui et al. 1985; Wainwright et al. 1985) and early onset breast cancer to a region of 17q21 (Hall et al. 1990). While access to a large family can offer reduced genetic and phenotype heterogeneity compared with the collection and analysis of smaller families, the ability to ascertain such an ideal family is dependent on the disease and the population in which it is found.

An ideal family structure for linkage includes three generations of a family (grandparents, parents, and proband with siblings) with many offspring. The third generation (grandparents) is used to phase parental genotypes to determine the combination of alleles along the chromosomes (termed "haplotypes"). If phase is known (i.e., both grandparents are available), the task of determining and counting recombinants and nonrecombinants is relatively straightforward for a highly penetrant mutation. If phase is unknown, recombinants from among the offspring are scored with an assumed phase and an alternative phase. For every phase-known meiosis, the maximum possible LOD score is 0.30 whereas it is less for phase-unknown meiosis. Therefore, more meioses are required for pedigrees with unknown phase to detect linkage with the disease gene compared with pedigrees with known phase.

Even when the ideal pedigree or set of pedigrees is available for a linkage analysis, there are several factors that impact the ability to detect linkage between an assayed marker and the disease gene. Commonly known challenges impeding a successful linkage study include misspecification of parameters (allele frequencies, genetic models, penetrance, and/or mutation frequency), scoring errors (incorrect phenotyping, genotyping, and/or family relationships), and locus heterogeneity (particularly when using multiple families) (see more detail in Chapter 6).

Several nonparametric linkage study designs and methods were developed to directly address many of the limitations commonly faced in implementing traditional parametric (model-based) linkage analysis. In these studies, cotransmission or coinheritance of a marker and disease gene are not directly observed in a large multigenerational family. Instead, family members with disease are evaluated for increased allele sharing compared with expected sharing based on degree of relatedness. For example, for a pair of siblings, at each marker the two can share 0, 1, or 2 alleles, and the expected distribution of this sharing is expected to be 0.25, 0.50, and 0.25, respectively, based on the pair's familial relationship. Significant deviations from the expected distribution signal evidence for the disease gene's location.

Common study designs for these allele-sharing methods include affected relative pairs (Risch 1990) or sib-pairs (Haseman and Elston 1972) such as extremely concordant and discordant sib-pairs (Risch and Zhang 1995). These study designs are particularly suited for late-onset disorders or situations in which parents and grandparents cannot be ascertained for traditional linkage study designs. These study designs were also touted as more powerful to detect linkage for complex diseases and traits compared with parametric linkage analysis approaches. However, despite promising power calculations early in the development of methods applied to these study designs (Risch and Zhang 1996), the number of sib-pairs required for complex traits with small to moderate genetic effect sizes (such as type 2 diabetes and hypertension) proved unrealistic compared with other study designs (Risch and Merikangas 1996).

Family-Based Studies (Association)

The most common study design for family-based association studies is the trio: mother, father, and affected offspring. With this study design, alleles at each marker are examined in parents and offspring to determine which alleles were "transmitted" or "untransmitted" to the affected child. A commonly applied test statistic to the family-based association study is the transmission disequilibrium test (TDT) (Spielman et al. 1993), which has been modified over the years to accommodate various family-based study designs such as extended pedigrees (pedigree disequilibrium test or PDT) (Martin et al. 2000). The statistical tests for family-based association study designs effectively test for both linkage (transmission to offspring) and association (specific alleles of each marker) (Chapter 8).

A major advantage of the family-based association study design over population-based study designs is they are robust to population stratification, which when left unaddressed in population-based study designs can lead to false-positive associations (Cardon and Bell 2001). Another advantage compared with linkage studies under specific scenarios is that family-based association study designs are more powerful to identify loci or genomic regions for common, complex disease (Risch and Merikangas 1996) and only the offspring requires phenotyping. Major disadvantages of family-based association study designs are often related to the feasibility of ascertaining the required family members. And, although both linkage and family-based association study designs offer genotyping quality control opportunities such as identifying markers violating Mendelian patterns of inheritance not possible with population-based designs, these study designs require genotyping more individuals (resulting in an overall increase in budget) compared with studies of unrelated individuals. Furthermore, statistical tests such as the TDT only use data from parents that are heterozygous at a given marker. Given the lower polymorphism information content of the popular diallelic single nucleotide polymorphisms (SNPs) compared with less popular multiallelic short tandem repeat markers, more markers must be genotyped in all family members.

While the required increase in markers was initially prohibitively expensive, the relatively recent development of cost-effective fixed-content genotyping arrays targeting 750 000 to >1 million SNPs across the genome makes study designs such as family-based association study economically feasible again.

Studies of Unrelated Individuals (Association)

The population-based studies described here are defined as studies ascertaining unrelated individuals as opposed to related individuals with specific family structures. As defined here, the population-based study may not necessarily be representative of a larger population from which it was sampled, although there are many examples of representative population-based studies that will be mentioned below. Indeed, many of the population-based study designs described here in the context of genetic studies borrow heavily from epidemiology, a field devoted to identification of risk factors in a representative sample drawn from a larger population.

The population-based study is necessarily a genetic association study as opposed to a linkage study given that transmission of genetic markers from parent to offspring cannot be directly observed. Instead, the genetic association study compares the distribution of allele or genotype frequencies between cases and controls. Because specific alleles are tested for association with disease (case status), the genetic association study design is less powerful in the presence of both allelic and locus heterogeneity. However, as will be illustrated below, a major advantage of the population-based association study design is the sample size given the relative ease of ascertainment compared with family-based study designs.

Cohort Studies

The "gold standard" of epidemiological study designs is the prospective cohort study. Typically, study subjects are selected on the basis of their exposure status, are disease free at the start of the follow-up period, and are followed forward in time to onset of disease. Thus, the cohorts accessed for genetic epidemiology studies are recruited based on membership in a population known to be at risk for common complex diseases, not genetic susceptibility.

There are several advantages of a prospective study design. Perhaps one of the most valuable advantages of the cohort study is the ability to establish a temporal relationship between exposure and disease. In the context of genetic epidemiology studies, the cohort study with its temporal relationships is the best study design for gene–environment interaction studies. Cohort studies are also suitable for the study of rare exposures and provide the opportunity to study multiple diseases related to a single exposure. Additionally, prospective data may be used to provide direct estimates of the relative risk (RR) and to calculate measures of impact, such as attributable risk.

It is important to emphasize that while the typical image of a cohort study is the collection of data prospectively, a true cohort study can be designed with either retrospective or prospective data collection. When distinguishing studies as cohorts or not, the emphasis is placed on the direction of inference, not the timing of the data collection with respect to study subject ascertainment.

Despite the attractive statistical properties and inferences possible, prospective cohort studies do have considerable disadvantages compared with other study designs. Cohort studies are powerful for common diseases but are not efficient or powerful for the investigation of relatively rare disease outcomes such as birth defects or rare cancers. The biggest deterrent to the use of cohorts in studies is that cohorts can be costly and time consuming as they often require a long follow-up period and a large number of participants. Indeed, most genetic epidemiology studies access existing cohort studies that have collected DNA samples as opposed to establishing new cohorts for genetic studies. This strategy instantly gratifies the need for individual-level genetic data associated with health outcomes for a cost-effective study (Willett et al. 2007). However, it must be recognized that this strategy does not optimize study design for the specific scientific questions or hypotheses to be tested. Also, the existing cohorts, such as those in the United States, often do not represent the full range of ages or racial/ethnic groups found in the general population (Collins and Manolio 2007). Furthermore, even established cohorts have difficulty in sustaining sufficient funding for continued data collection (Straight talk with. . .Daniel Levy 2013). The shortcomings and challenges of accessing existing cohorts for genetic association studies have contributed to the popularity of newer models of obtaining biospecimens and health-related data, such as biorepositories linked to electronic medical records, are further discussed below.

There are numerous examples of cohort studies that can be accessed for genetic association studies. The Women's Health Initiative (WHI) Observational Study is an example of a prospective cohort study (The Women's Health Intitiative Study Group 1998; Hays et al. 2003). WHI includes 93 676 postmenopausal women between the ages of 50 and 79 years who were recruited at 40 clinical centers throughout the United States between September 1994 and December 1998. The health and lifestyle of these participants were tracked over an average of eight years. DNA samples are available on the vast majority of WHI participants, and individual-level genotype data linked to phenotype data for 12 008 African American and Hispanic participants enrolled in one or more of the WHI program components (including the Observational Study) can been accessed by qualified investigators for secondary analyses through the Database of Genotypes and Phenotypes (dbGaP) (Mailman et al. 2007; Tryka et al. 2014).

WHI by design only ascertained postmenopausal women. Other cohorts in the United States with DNA samples that represent both men and women include the Coronary Artery Risk in Young Adults (CARDIA), the Atherosclerosis Risk in Communities (ARIC), and the Cardiovascular Health Study (CHS). CARDIA is a multicenter longitudinal study for cardiovascular disease risk factors and outcomes among 5155 young adults (18–30 years at baseline) ascertained between 1985 and 1986 (Friedman et al. 1988). Both men and women as well as European Americans and African Americans were ascertained from four cities in the United States (Birmingham, AL; Chicago, IL; Minneapolis, MN; and Oakland, CA). Participants were reexamined at years 2, 5, 7, 10, 15, and 20 after baseline. Like CARDIA, ARIC is mainly a study of cardiovascular risk factors and outcomes in both males and females and European Americans and African Americans. With 15 792 participants from four communities (Forsyth County, NC; Jackson, MS; suburban areas of Minneapolis, MN, and Washington County, MD) aged 45–64 years at baseline between 1987 and 1989 (The ARIC Investigators 1989), ARIC is larger and older than CARDIA but has fewer follow-ups (1990–1992; 1993–1995; and 1996–1998). Older yet is CHS, which ascertained mostly European Americas ≥65 years of age from four centers (Forsyth County, NC; Sacramento County, CA; Washington County, MD; and Pittsburgh, PA) between 1989 and 1990 with an additional recruitment event for African Americans between 1992 and 1993 for a total of 5888 participants (Fried et al. 1991).

There are several cohorts with available DNA samples that emphasize diverse populations or extremes of the age distribution. For diversity, DNA samples linked to health outcomes and lifestyle data collected as part of a prospective cohort are available for the Multiethnic Cohort (MEC) and the Multi-ethnic Study of Atherosclerosis (MESA). The MEC is a population-based prospective cohort of 215 251 men and women aged 45–75 years at the time of ascertainment between 1993 and 1996 (Kolonel et al. 2004). The MEC collected health and lifestyle data related to cancer outcomes and included several self-reported racial/ethnic groups including European Americans, African Americans, Japanese Americans, Latinos, and Native Hawaiians. Biospecimens were collected on approximately 67 000 MEC participants between 1995 and 2004. Similarly, MESA is a cohort study of several self-reported racial/ethnic groups including European Americans, African Americans, Hispanics, and Asians of Chinese-descent ascertained from six US communities. Approximately 6000 participants were ascertained for MESA between 2000 and 2002 at ages between 45 and 84 years at baseline to study subclinical cardiovascular disease (Bild et al. 2002).

An example of epidemiologic cohorts with biospecimens that have sampled from the extremes of the age distribution is the Health Aging and Body Composition (Health ABC) Study. Health ABC ascertained approximately 3000 participants aged 70–79 between 1997 and 1998 as was designed to assess the relationship between body composition, long-term conditions, and incident mobility limitations. Participants included both men and women as well as members of various self-reported racial/ethnic groups. At baseline, eligible participants were English speaking Medicare recipients residing near Pittsburgh, Pennsylvania or Memphis, Tennessee, and participants were required to be free of mobility and activities of daily living disability. The main health measures collected annually for study are aging-related such as bone mineral density content and muscle mass (Goodpaster et al. 2001).

Other cohorts emphasize or sample based on demographics other than sex, age, or race/ethnicity. One example of such a cohort is the Southern Community Cohort Study (SCCS). Like the other cohorts described here, the SCCS ascertained both men and women and included both European Americans and African Americans. Approximately 86 000 participants between the ages of 45 and 79 were ascertained between 2002 and 2009 in 12 southern US states for studies of racial disparity and common diseases such as various cancers and type 2 diabetes (Signorello et al. 2010). The SCCS sets itself apart from other large cohort studies in that ascertainment targeted community health centers in the South, which resulted in a cohort of participants with low income and low education, demographics which are often underrepresented in other cohorts. Additional exposures of interest include occupation, tobacco and alcohol use, diet and nutrition, physical activity, anthropometrics, prevalent medical conditions, medication use, family history, psychosocial factors, and health service utilization. SCCS participant enrollment data were submitted to the Social Security Administration (SSA) for vital status, and this is updated annually with submissions to SSA and the National Death Index. SCCS participant enrollment data are also linked to cancer registries in the 12 southern states where ascertainment occurred, and incident cancer cases are being collected on a continuous basis (with the acknowledgement that reporting within a state can lag one to two years). Additional follow-up data were collected approximately five years after enrollment using a questionnaire, and subsequent follow-ups are planned every five years.

As mentioned previously, population-based genetic association study designs assume participants are unrelated. Cryptic relatedness can lead to spurious genetic associations if left uncorrected or unaccounted (Voight and Pritchard 2005). The presence and extent of cryptic relatedness should be determined in the data set being accessed ideally using genome-wide genetic data as part of quality control practices prior to analysis. For some cohorts, the presence of cryptic relatedness is not in question because families and households were intentionally surveyed. However, the extent of relatedness is often not known for these cohorts because pedigree data were not explicitly collected. Examples of these cohorts include the Jackson Heart Study (JHS) and the Study of Latinos. The JHS, a subset of the larger ARIC cohort, is a longitudinal study of 5301 African Americans ascertained from the Jackson, Mississippi area for cardiovascular disease outcomes and risk factors (Taylor 2005; Shurin 2012). Nested within the larger JHS is the JHS Family Study where relatives of the index JHS participant were recruited for the family-based cohort resulting in 264 genetically estimated pedigrees (Wilson et al. 2005). Similarly, the Hispanic Community Health Study (HCHS)/Study of Latinos (SOL) is a multicenter community-based cohort study in the United States that ascertained greater than 16 000 Hispanic adults aged 18–74 years between 2006 and 2009 (Sorlie et al. 2010). Recruitment targeted households near four recruitment centers (San Diego State University, Northwestern University in Chicago, Albert Einstein College of Medicine in the Bronx area of New York, and the University of Miami) likely to represent different groups of origins for Hispanics (Central Americans, Cubans, Dominicans, Mexicans, Puerto Ricans, and South Americans). In addition to confounding by cryptic relatedness, the analysis of HCHS/SOL in the context of a population-based genetic association study must also account for the extensive population stratification expected based on the ascertainment of multiple, differing genetic ancestries.

There are several cohorts outside of the United States active in epidemiologic and genomics studies. One such cohort is the Rotterdam Elderly Study (also known as the Rotterdam Study), a prospective cohort of 7983 men and women ≥55 years of age established in 1990 in Ommoord, a suburb of Rotterdam (Hofman et al. 1991). The Rotterdam Study is a single-site cohort that was designed to study the prevalence and incidence of risk factors for chronic diseases in older adults such as cardiovascular diseases, neurologic diseases, locomotor diseases, and ophthalmologic diseases. The Rotterdam Study emphasizes biomarkers and imaging data in the study of risk factors impacting various diseases. Participants from the original cohort continue to be examined every three to four years. Since the establishment of the original cohort, two other enrollments were offered: (i) participants aged ≥55 years ascertained between 2000 and 2001 and (ii) participants aged ≥45 years ascertained in 2006 (Hofman et al. 2007). The additional enrollment added almost 7000 participants to the Rotterdam Study for a total close to 15 000 available for epidemiologic and genetic studies.

Another well-known non-US cohort is the **Ko**operative Gesundheitsforschung in der **R**egion **A**ugsburg (Cooperative Health Research in the Augsburg Region) or KORA surveys. Established in 1984, KORA surveys ascertained a random sample of the citizens of the Augsburg region in Germany between the ages of 25 and 74. The original focus of the surveys was cardiovascular disease (as a registry of acute myocardial infarction known as MONItoring trends and determinants in Cardiovascular or MONICA) (Lowel et al. 2005), but other chronic diseases have been included in subsequent survey years (Holle and Happich 2005). Participants in KORA have been reexamined four times in the past 20 years augmenting the baseline data which included a detailed medical exam and a survey that collected sociodemographic variables, risk factors (such as smoking, alcohol consumption, physical activity, to name a few), medical history and family history of chronic diseases, and medication use. DNA samples are available for approximately 18 000 MONICA/KORA participants.

Probably the most well-known prospective cohort study turned family study is the Framingham Study, also known as the Framingham Heart Study (FHS). Named after the town in Massachusetts in which participants were residents, Framingham was established in 1948 to identify factors that contribute to cardiovascular disease (CVD). Approximately 5200 men and women aged 30–62 years were ascertained at baseline between 1948 and 1953 and were reexamined every two years for health and lifestyle measures (Dawber et al. 1951; Mahmood et al. 2014).

Between 1971 and 1975, the FHS enrolled approximately 5000 adult children and spouses of the original FHS cohort for similar CVD-related studies (Feinleib et al. 1975). And, between 2002 and 2005, approximately 4000 participants representing the third generation of the original FHS were ascertained (Splansky et al. 2007). The first cohort of its time, the FHS helped establish risk factors associated with cardiovascular events such as high blood pressure and coronary heart disease (Dawber et al. 1957) as well as helped develop risk prediction tools for disease prevention. The most well-known FHS risk prediction tool is the Framingham Risk Score, which uses an individual's age, gender, total cholesterol levels, high-density lipoprotein cholesterol, smoking status, and systolic blood pressure to assess the individual's 10-year risk of a having a heart attack (Wilson et al. 1998). The FHS has more recently contributed to numerous population-based genetic association studies for a variety of quantitative traits and disease outcomes (Govindaraju et al. 2008; Fox 2010).

Despite the history and continued productivity and utility of prospective cohorts such as the FHS, the justification of their continued existence has been difficult given their expense (Fox 2013). Cohorts in general are expensive and require sustained funding such that when funding agencies experience fluctuations in the ability to support these studies, the cohorts might not be sustainable. An example of this sustainability problem is the National Children's Study established by the United States Congress as part of the Children's Health Act of 2000. The original study design aimed to ascertain 100 000 children and to follow them from the womb to adulthood (21 years) (Committee on National Statistics, Division on Behavioral and Social Sciences and Education, Board on Children, Youth and Families, National Research Council, and Institute of Medicine 2013). This study promised to be a unique resource for the study of multiple health measures among infants and children and the impact of carefully measured environmental exposures (Guttmacher et al. 2013). However, full-scale ascertainment of the study participants was halted after the pilot study centers reported that ascertainment was more difficult and expensive than previously planned (Baker et al. 2014). The pilot study centers' experiences, coupled with a 2014 National Academy Study assessment of the status of the National Children's Study (Panel on the Design of the National Children's Study and Implications for the Generalizability of Results, Committee on National Statistics, Division of Behavioral and Social Sciences and Education, Board on Children, Youth and Families, National Research Council, and Institute of Medicine 2014), resulted in NIH halting the study until further review. Both the National Children's Study and the FHS are recent examples of the difficulties in establishing or maintaining cohort studies for future epidemiologic and genetic epidemiologic research.

Despite the challenges associated with establishing a new, large cohort, the NIH recently launched a new national cohort as part of the Precision Medicine Initiative (PMI) announced in 2015 (Collins & Varmus, 2015). The cohort, now known as All of US, officially began ascertainment in 2018 in the United States. *All of Us* is a key element of the NIH's PMI, which was funded in 2016 with a $130 million investment to build a national cohort reflecting the rich diversity of the US, including participants as partners, and be a catalyst for positive change in research. The goal is to build a cohort of one million or more people and to use this cohort to accelerate health research and medical breakthroughs in precision medicine. This study utilizes a clinic-based recruitment strategy through health systems with longitudinal examination data captured in electronic health records and is thus, not a random population sample (All of Us Research Program Investigators, 2019). However, the study will attempt to enroll a sample as representative of the national population as is possible.

Similar to NIH, the United States Department of Veterans Affairs (VA) has also made substantial investments in establishing a new observational cohort and biobank. The Million Veteran Program (MVP), as its name implies, aims to ascertain one million United States veterans to create a longitudinal VA-based mega-biobank linked to veterans' electronic health records; self-reported health, lifestyle, and exposures; and a variety of 'omic data (including genome-wide genotyping data and now whole genome sequencing data) for genomic and nongenomic studies (Gaziano et al., 2016). The MVP is one of the largest of today's biobanks linked to electronic health records available for research purposes, many of which are further described below. In addition to its size, the MVP has several other features not available in other cohorts such as racial/ethnic diversity and phenotypes and exposures specific to or overrepresented in veterans. Also, veterans receive most of their healthcare with the VA system, making the data available in the electronic health records more complete compared with data from open systems that non-veterans access. The MVP by design draws from an older and mostly male population. Regardless of their ability to generalize to all groups, both the MVP and All of Us demonstrate a resurgence in interest in establishing large observational longitudinal cohorts given this is the best study design to identify genetic and non-genetic risk factors relevant to human health.

Cross-Sectional Studies

Studies in which disease and exposure status are measured simultaneously in a given population are known as cross-sectional studies. Participants in cross-sectional studies are not followed over time as in cohort studies. Rather, repeated cross-sectional studies are conducted independently of one another. Cross-sectional studies allow for a "snapshot" of the prevalence and characteristics of a disease or phenotype in a particular population at a particular point in time. When conducted repeatedly, cross-sectional studies are most often used to describe trends. Cross-sectional studies can also be resources for case–control genetic association studies further described below.

One of the most well-known cross-sectional studies in the United States is the population-based National Health and Nutrition Examination Surveys (NHANES) conducted by the Centers for Disease Control and Prevention (CDC). NHANES was established in the early 1960s as a series of surveys conducted over several years to collect demographic, health, and lifestyle data on noninstitutionalized residents in the United States regardless of health status. In 1999, NHANES became a yearly survey. More than 50-year lifespan of NHANES makes it possible to track health trends over time such as the increase in body mass index and obesity (Ng et al. 2014; Ogden et al. 2014) as well as the trends in lipid levels before and after lipid-lowering medication became widely available (Carroll et al. 2012).

The major advantage of a cross-sectional study is the potentially large sample size and breadth of health data spanning several years or decades (if the survey is continuous like NHANES). A major disadvantage of the cross-sectional study design is that the temporal relationship between the exposure and disease status is unknown. Also, case–control studies drawn from cross-sectional surveys may suffer from survival bias given that prevalent as opposed to incident cases are available for study. Therefore, in the context of genetic association studies, cross-sectional study designs are tools for hypothesis generation of possible genetic factors and interactions with other genes and the environment that should be further examined in longitudinal data sets.

Case–Control Studies

Case–control study designs are inarguably the most frequent study design for genetic association studies. As the name implies, the design of the case–control study involves ascertaining participants that have a particular disease of interest (cases) and participants free of that disease (controls). The major advantage of case–control studies is the ability to amass very large sample sizes with the outcome of interest for study. Also, the case–control study is often less expensive and more feasible to conduct compared with a cohort study. As described above, while the temporal relationship between exposures and disease are documented in the cohort study design, such studies are generally ambitious undertakings. Notably, however, genotypes do not change over time, thereby allowing for easy retrospective assessment of this exposure variable even in a noncohort study design.

Cases of the disease of interest can be ascertained from a variety of sources. A common source of cases is direct recruitment from a clinic specializing in that disease. Cases can also be participants from a cohort study. Indeed, a case–control study can be conducted nested within a larger cohort study if there are sufficient numbers of participants who have developed the disease of interest over the course of the study (termed "incident" cases). Cases can also be participants in a cross-sectional study (termed "prevalent" cases). Most large case–control genetic association studies consist of both prevalent and incident cases. Prevalent cases, however, come with biases such as survival bias and therefore may not be appropriate for all genetic association studies.

While much attention must be put into defining case status, ascertaining the appropriate control population is just as important. Indeed, difficulties in selecting controls make case–control studies more susceptible to spurious associations and lack of replication compared to prospective studies (Cardon and Bell 2001). Ideally, controls are ascertained from the same study population as the cases and are similar to cases in every way possible except for the presence of disease. In many epidemiologic studies, control samples are often matched to case samples for confounding factors (i.e., any factor that might influence the association between the disease and exposure), such as age, sex, race/ethnicity, and geographic location. There are two approaches for matching. First, controls can be selected such that the overall distribution of cases and controls is comparable with respect to the frequency of the confounders (termed "frequency" or "category" matching). Alternatively, one or more control individuals may be selected to match an individual case based on the confounding characteristics (termed "individual" matching). In the absence of matching, confounding may be controlled by including potential confounding factors as covariates for adjustment in statistical models.

For genetic association studies, the one confounder that must be matched or otherwise corrected is genetic ancestry. As already discussed, confounding by genetic ancestry or population stratification can lead to spurious associations (Cardon and Palmer 2003). Self-reported race/ethnicity is often used to match cases and controls by genetic ancestry. Self-reported race/ethnicity is highly correlated with genetic ancestry for some race/ethnic groups, but not all. And, even within seemingly homogenous populations, subtle population stratification can be present (Campbell et al. 2005; Novembre et al. 2008). Therefore, it is highly advised that all DNA samples be genotyped either with ancestry informative markers (AIMs) or genome-wide arrays. These data can then be used to check for outliers within self-reported racial/ethnic groups compared with reference DNA samples such as the International HapMap samples, 1000 Genomes Project samples, or other reference samples of similar genetic ancestry (Tian et al. 2008). If outliers are detected, these samples should be removed from the case–control analysis. Alternatively, data from AIMs or genome-wide arrays can be used to generate variables describing genetic ancestry (using principal components or other clustering methods) that can then be used as covariates for adjustments in statistical models (Price et al. 2006).

In the preceding paragraphs, the general recommendation is that controls should be ascertained from the same population or geographic area as the cases. While this is good, standard epidemiologic practice, many case–control genetic association studies use convenience controls to save both time and costs. Commonly used convenience controls include birth cohorts, cross-sectional studies, and controls from other studies. In genetic association studies, investigators also use data from publically available reference samples such as the International HapMap Project (The International HapMap Project 2003; The International HapMap Consortium 2005; A second generation human haplotype map of over 3.1 million SNPs 2007; Nature 2010) and the 1000 Genomes Project (a map of human genome variation from population-scale sequencing 2010) where the participants are presumed disease free. While seemingly cost effective, the inclusion of convenience controls can lead to both false-positive associations from population stratification (due to genetic background, technical differences in genotyping, sample type and processing, and time/year of sampling) as well as false-negative associations from misclassification of disease status.

An example of a case–control genetic association study that also used convenience control samples is the Wellcome Trust Case Control Consortium (WTCCC) (2007). The WTCCC performed a genetic association study of 2000 cases for each of seven diseases (bipolar disorder, coronary artery disease, Crohn's disease, hypertension, rheumatoid arthritis, type 1 diabetes, and type 2 diabetes) and 3000 shared controls. The controls were individuals from the 1958 British Birth Cohort and selected blood donors recruited for the WTCCC studies. Detailed phenotyping was not available on the controls. Nevertheless, the WTCCC discovered significant genetic associations even for common diseases such as type 2 diabetes. The WTCCC investigators speculated that the case definitions used in their study encouraged the ascertainment of extreme cases (by requiring extreme phenotypes and/or strong family history) thereby increasing the statistical power of the study and minimizing the misclassification of controls despite not phenotyping them (WTCCC consortium 2007).

As already noted, sample size, relative ease of ascertainment/implementation, and overall lower costs are the major advantages of the case–control study design compared with cohorts (Manolio et al. 2006). Another advantage of the case–control study design is that it is the only study design appropriate for rare outcomes such as birth defects. The CDC's National Birth Defects Prevention Study is an example of a multistate case–control of birth defects in the United States (Yoon et al. 2001). Genetic risk estimates calculated as odds ratios approximate risk ratios when the outcomes are rare in the general population. However, for more prevalent diseases, the genetic effect estimates from case–control studies are odds ratios that might not reflect the true risk estimate as can be determined from cohort studies. That is, for common outcomes of interest (>10% incidence), the odds ratio overestimates the risk ratio when it is greater than 1.0 and underestimates the risk ratio when it is less than one, and this over/underestimation increases as the incidence of disease increases in the study population.

Other Study Designs

Beyond cohorts, cross-sectional studies, and case–control studies, there are other study designs being used for genetic studies albeit with lower frequency. One such design is the disease outbreak or the public health emergency. Epidemiologic research is common for major outbreaks and public health emergencies such as the study of first responders and exposure to the World Trade Center debris in 2001 (Prezant et al. 2002; Lurie et al. 2013) and the Gulf Long-term Follow-up Study (GuLF) STUDY, a health study for individuals who participated in the oil spill cleanup or cleanup training after the 2010 Deepwater Horizon Gulf of Mexico disaster that affected coastal Louisiana and Alabama (Reardon 2011). Less common is the collection of DNA samples in these emergency response protocols, although steps have been taken to include provisions for collection of biospecimens for DNA extraction (Rubin et al. 2007; Steinberg et al. 2007).

Disease outbreaks and other public health emergencies can offer unique data for environmental exposures not routinely collected by other epidemiologic study designs. However, these study designs have not been used often in genetic association studies as they are often very small in sample size and therefore inconclusive in a genetic association study setting. The emergence of severe acute respiratory syndrome coronavirus 2 (SARS-CoV-2) and associated coronavirus disease 2019 (COVID-19) is proving to be an exception, with powerful genome-wide association and sequencing studies enabled by rapid, world-wide sample sharing and near real-time genomic data dissemination (COVID-19 Host Genetics Initiative, 2020).

Another study design that can be used in genetic studies is the randomized clinical trial (RCT) study design. The RCT is relatively free of bias, and this study design is considered hypothesis-testing rather than hypothesis generating and would be ideal for the assessment of gene–environment interactions and pharmacogenomics. In recognition of the potential utility of RTCs in genetic association studies, the National Human Genome Research Institute (NHGRI) at the National Institutes of Health (NIH) funded the Genomics and Randomized Trials Network (GARNET), a cooperative agreement of three RCTs with DNA samples (GARNET Study, n.d.). The sample sizes of the three RCTs were moderate (2164–5420) for genome-wide association studies. Apart from genome-wide studies, RCTs with DNA samples are currently being conducted for specific pharmacogenomics variants considered in genotype-guided drug prescriptions in comparison with standard medical practices (Kimmel et al. 2013; Pirmohamed et al. 2013).

Biobanks

The final study design discussed here is the biobank or biorepository. Biobanks are not a study design per se but rather a resource of health data and biological materials that can be used in a genetic study. Once the biorepositories have been established and data have been collected, genetic studies can be conducted using the already described family-based and unrelated sample designs. Much interest of late has been paid to the establishment of biobanks given the already realized expense of establishing and maintaining cohorts such as the Framingham Heart Study. Several models of biobanking have emerged, each with strengths and weaknesses impacting downstream genetic studies.

In the United States, there are several large biorepositories linked to electronic medical records that are being used for genetic research. Many of these biorepositories participate in an NHGRI-funded collaborative cooperative agreement known at the electronic MEdical Records and GEnomics (eMERGE) Network (McCarty et al. 2011; Gottesman et al. 2013; Crawford et al. 2014). Among these eMERGE biorepositories is a case–control study of Alzheimer's disease in Seattle, Washington, linked to the patients' electronic medical records at Kaiser Permanente Washington(formerly Group Health Cooperative). The advantage of this model of establishing a biobank is that genetic data, along with structured and harmonized medical and exposure data have been collected on cases and controls for a specific disease outcome, and these patients are linked to their electronic health records as potential sources of new, supplementary data. A major drawback to this approach is that any subsequent studies that repurpose the Alzheimer's disease case and control data to study another disease outcome will suffer ascertainment bias in downstream analyses.

Another model of establishing biobanks utilized within the eMERGE network is Vanderbilt University Medical Center's BioVU. This biorepository extracts DNA from blood samples discarded after routine clinical care, and this DNA sample is then linked to a deidentified version of the patient's medical record (Roden et al. 2008). In its initial inception, BioVU did not seek informed consent from participants; rather, patients had to explicitly opt-out of the program if he/she wished to exclude the DNA sample from being extracted and used for research purposes (Pulley et al. 2010). Subsequently, Vanderbilt implemented the BioVU consent in the consent-to-treat process and patients treated at Vanderbilt sign the BioVU consent form upon check-in at an outpatient clinic. This form gives the patient a choice about participating in BioVU. BioVU began DNA collection in 2007 and, as of winter 2019, had over 285 000 DNA samples linked to the deidentified version of the electronic health records available for research.

Despite the sample size and potential statistical power offered by this collection approach, there are several disadvantages associated with the opt-out model. First, as opposed to the existing case–control or cohort model, the phenotype data available are not collected for research purposes and require extensive resources to clean and create the actual data sets used in analysis (Kohane 2011). The effort expended on developing and validating phenotyping algorithms, in fact, is a major activity of the eMERGE network (Newton et al. 2013; Pathak et al. 2013) and the aforementioned MVP (Knight et al., 2020). Another disadvantage is that recontact is not possible in deidentified data sets like BioVU. BioVU is clinic-based as is representative of the Vanderbilt University Medical Center patient population (Crawford et al. 2015). Also, it is not possible to easily add variables not routinely collected by clinics such as environmental exposure data. Finally, BioVU like other medical centers in the United States serve patients that have a variety of health care options and come and go within any given medical system over time. Because the health care system is not necessarily a closed system, there will be missing data and loss of follow-up that directly impact downstream analyses. These reasons are likely a major part of the decision for BioVU to convert from an opt-out consent to an opt-in (consented) model.

Other biobanks in the United States use a consented model to ascertain participants. One such example is from the Marshfield Clinic Personalized Medicine Research Project (PMRP). The Marshfield Clinic PMRP is a population-based cohort of approximately 20 000 adult participants with DNA, serum, and plasma samples linked to each participant's electronic health record (McCarty et al. 2005). Given this is a consented biobank, participants can be re-contacted for further data collection such as the recent efforts to collect data on dietary intake (Strobush et al. 2011). Another example of a biobank using a consented model is Kaiser Permanente's Research Program on Genes, Environment, and Health (RPGEH) in California (Hoffmann et al. 2011). Kaiser Permanent's RPGEH Health through Understanding of Genetics and Environment has collected DNA samples and questionnaire data for exposures on 200 000 of the targeted 500 000 patients (Scott et al. 2012). Because Kaiser Permanente is a health maintenance organization, it is much more likely that their patients receive all or almost all of their health care within the system, which makes the electronic health record data more complete and less prone to loss of follow-up compared with biobanks established at other centers such as BioVU. Also, compared with other US biobanks such as BioVU, Kaiser Permanente located in northern and southern California provides health care to a diverse patient population often underrepresented in genetic studies.

There are many more biobanks in the United States not described here, and there are also many larger biobanks outside the United States (Kohane 2011). Of international note is the UK BioBank, which has consented and ascertained 500 000 participants. The UK Biobank is an ambitious prospective study of adults in the United Kingdom promising to contribute toward the understanding of genes, the environment, and lifestyle in disease risk (Ollier et al. 2005). The scale of this prospective biobank required innovations in data collection (Manolio et al. 2012), and perhaps as innovative is the biobank's concept of open access for the study data (Allen et al. 2014).

Other Biobanks

The current excitement surrounding biobanks is due, in part, to the inclusion of electronic health record data as described above. In addition to these biobanks, there are other biorepository collections useful and available for research settings. Coriell Cell Repositories is maintained by the Coriell Institute for Medical Research, a nonprofit biomedical research institution founded in 1953 and based in New Jersey. Coriell is the host biobank for several NIH agencies' biospecimens including more than 11 300 cell lines and 5700 DNA samples funded by the National Institute of Genomic Medical Science (Tang et al. 2013). The NHGRI's biorepository within Coriell includes the International HapMap (The International HapMap Consortium 2005) DNA samples and corresponding cell lines. Coriell also serves as a biobank for non-NIH agencies such as the CDC and the Wistar Institute. The phenotype or health data associated with the samples, however, are not as dense as other biobanks with electronic health records or epidemiologic data, potentially limiting their use in research studies.

One of the largest biobanks in the United States is 23andMe, a for-profit direct-to-consumer genetic testing company. 23andMe has offered genetic testing since 2007 and has amassed approximately 900 000 DNA samples. In addition to genome-wide-genotyping, 23andMe genotypes targeted rare mutations and collects self-reported health and trait data. Initially, 23andMe provided customers with reports on ancestry, relatedness, and potential genetic risk factors related to health. For a period of time, 23andMe no longer reported any genetic risk factors to customers as they waited for approval from the Food and Drug Administration (Annas and Elias 2014). As of 2018, 23andMe provides a set of reports including genetic risk, wellness reports, carrier status, and an ancestry report.

With its massive sample size and self-reported health data, 23andMe has been active in research. 23andMe has demonstrated that self-reported web-based data coupled with genome-wide-genotyping data can replicate known genotype–phenotype associations as well as discover novel associations (Eriksson et al. 2010; Li et al. 2012; Hinds et al. 2013). The self-reported health data collected by 23andMe is broad, but the company also has specific medical research initiatives focusing on Parkinson's disease (Do et al. 2011), inflammatory bowel disease, and myeloproliferative neoplasms. Despite 23andMe's history of collaborating with other research groups for genetics association studies, it is unclear if these data will be accessible for analyses either in aggregate or as de-identified individual-level data. 23andMe had announced the development of a Research Portal for outside instigators to conduct real-time genome-wide association studies as part of a larger Research Platform service that included access to research survey tools and clinical research recruitment. The Research Platform service is not yet available, and it is unclear when the infrastructure will be built to enable access to these data.

Similar to 23andMe, Ancestry is a for-profit company collecting DNA on consenting customers. Also like 23andMe, Ancestry reports ancestry and relatedness to customers, a natural extension of the access to genealogy data provided by the company. Interestingly, Ancestry has begun collecting health data and family health histories from consenting customers under the Ancestry Human Diversity Project as part of AncestryDNA. Unlike 23andMe, Ancestry has not established a history of medical research in the literature, so it remains to be seen who accesses these data and for what purposes.

Not all for-profit companies follow the models of the likes of 23andMe and Ancestry. Case in point is Illumina, a sequencing and array-based company known for technology development and services that result in the production of genetic data for analysis. Less well known are Illumina's services to provide individual level control data for genetic association studies. Illumina's iControlDB is an online database of genotype and phenotype data that can be accessed by Illumina customers for use in genetic association studies. Similar control data sets are available as propriety data sets as well as through dbGaP (such as the Wellcome Trust 1958 birth cohort (WTCCC consortium 2007)).

Biospecimens for Biobanks

Biobanks established for genetic studies typically collect biospecimens for DNA extraction and storage. The most common biospecimens collected for DNA studies are blood, saliva, and cheek swabs. Blood samples can be used to extract abundant amounts of DNA sufficient for many genotyping and sequencing experiments, a major advantage over other biospecimens that yield less DNA. Blood samples can also be used to establish immortalized cell lines (Gilbert 2001), which, in turn, offer a perpetual supply of DNA for the individual who donated the blood sample (Gilbert and Vance 2001).

In some studies, however, the collection of saliva or buccal (cheek) swabs is preferred. While phlebotomy is considered routine, the procedure does require trained personnel and can cause discomfort to the participant donating the sample. For studies of children or participants who cannot give blood easily, noninvasive collection of biospecimens for DNA extraction is preferred. Also, saliva and cheek swabs offer a cost-effective method to collect and extract DNA, particularly if participants do not live near the biobank DNA extraction facility. Shipment of blood samples requires special packaging and handling especially if these samples are to be used to establish cell lines. Saliva and cheek swabs, in contrast, require less packaging and handling and are the preferred method of collection for direct-to-consumer biobanks such as 23andMe (Eriksson et al. 2010).

Beyond DNA extraction, the collection of other biospecimens is desirable to enable broader genomic studies. Epidemiologic collections such as NHANES have routinely collected and stored blood plasma and serum samples for use in current biomarker protocols and for anticipated use in new protocols for novel biomarkers or biomarker development. Blood plasma is the liquid part of the blood and requires the centrifugation of the blood sample for collection. Serum, in contrast, is isolated by allowing the blood sample to clot followed by extraction of the remaining liquid (sans the clotting factors) and further centrifugation. Collection of blood plasma and serum samples requires the use of the appropriate vacutainer at the time of the blood draw to ensure that anticoagulants such as EDTA are not introduced to blood samples destined for serum extraction. Given the more complex collection protocols, few biobanks drawing from routine patient care include blood plasma and serum samples. BioVU has begun a pilot study of blood plasma collection from discarded blood samples drawn for routine clinical care. BioVU plans to collect 3600 blood plasma samples for biomarker studies of congestive heart failure, medication adherence in patients with resistant hypertension, and mechanisms underlying breast cancer immunotherapy.

Even less common than the collection of blood plasma and serum in biobanking is the extraction of RNA from blood. RNA extraction is valuable for gene expression studies. Unlike DNA, however, RNA is known to be unstable. Large-scale automated collection of RNA from whole blood drawn clinically would require an RNA stabilization and extraction system (Duale et al. 2012). Also challenging is the fact that different tissues express different genes at different levels, and in some cases, in response to different environments. To capture data for these different scenarios would require the extraction of RNA from different tissues under different conditions, a feat that is not easily automated or compatible with a discarded tissue model of collection. In recognition of the challenges associated with collecting gene expression data, the Genotype-Tissue Expression (GTEx) Project was initiated by the NIH Common Fund Project to collect human tissues within

24 hours of a donor's death for DNA and RNA extraction, banking, and phenotyping (Lonsdale et al. 2013). Although linked to medical history data, GTEx was designed for the study of gene expression among healthy donors (Keen and Moore 2015). The most recent release of GTEx (version 8) consists of 838 donors for which both genotype data and RNA sequencing data of up to 52 tissues are available (PMID 32913098).

While gene expression studies are popular experiments associated with human tissue collection, tissue samples are also critical resources for the establishment of experimental cells lines and induced pluripotent stem cells (iPS cells or iPSCs). As already mentioned, blood samples can be used to create immortalized cells lines for a perpetual supply of DNA for a study participant. These cell lines termed "lymphoblastoid cell lines" can also be used for gene expression studies along with other genomic studies such as metabolomics and proteomics (Sie et al. 2009). Cell lines can be established using different sources such as fibroblasts from biologics such as skin. Human fibroblasts have also been used to create iPSCs (Takahashi et al. 2007; Yu et al. 2007), a process that induces fully differentiated adult cells to become pluripotent stem cells. Cells such as iPSCs hold great promise for regenerative medicine and underscore the need for tissue collection in large-scale biobanking efforts.

In addition to blood, there are other biological samples that could be useful in biobank research. Bodily fluids such as cerebrospinal fluid (CSF) and urine could be collected for the development of novel biomarkers such as CSF tau for Alzheimer's disease (Fagan et al. 2014) or the measurement of environmental exposures in urine such as heavy metals measured in NHANES (Adams and Newcomb 2014). Fecal or stool samples could be collected (Kumar et al. 2014) from clinic patients for downstream gut microbiome studies as proposed for prospective cohort studies (Mai and Morris 2013). Discarded biologic materials from biopsies and surgeries are routinely collected for biobanking and linkage to cancer registries; however, the quality of collection and preservation of the material can vary dramatically hampering research efforts accessing the sample (Biorepositories 2012). Targeted collection of specific tissues of interest is less common, but such collections are quite valuable for studies of gene and protein expression in clinically relevant tissue. One example of such a collection is the NIH-supported NeuroBioBank network (http://neurobiobank.nih.gov), which collects central nervous system tissue (brain, spinal column) for studies of neurodevelopmental, neuropsychological, and neurodegenerative diseases.

Given the complexity in collection, extraction, and storage, no one biobank or epidemiologic study can be expected to collect and measure all known biologic materials and environmental exposures. Often the type of sample collected and the specific conditions of its collection will depend on the research question of the study or the individual biobank's goals. Also of consideration are the participant and the burden associated with donating multiple samples (further discussed in Chapter 5). Participant burden must be balanced with the potential data gleaned from samples stored for future use as well as the cost of extensive collection and storage.

Summary

In this chapter, we describe common study designs in genetics, many of which borrow heavily from epidemiology. We also provide specific examples of epidemiologic studies with DNA samples as well as emerging biobanks. Finally, we give an overview of the biospecimens typically collected for biobanking with notes on their collection and storage. Overall, both biospecimen collection and study design choice will depend on the research question and available resources, the latter of which continues to evolve and expand with changing technologies and the emergence of rich data repositories ripe for genetic studies.

References

A map of human genome variation from population-scale sequencing (2010). *Nature* 467: 1061–1073.

A second generation human haplotype map of over 3.1 million SNPs (2007). *Nature* 449: 851–861.

Adams, S.V. and Newcomb, P.A. (2014). Cadmium blood and urine concentrations as measures of exposure: NHANES 1999-2010. *J Expo Sci Environ Epidemiol* 24: 163–170.

All of Us Research Program Investigators. (2019). The "All of Us" Research Program. *N Engl J Med* 381: 668–676.

Allen, N.E., Sudlow, C., Peakman, T. et al. (2014). UK biobank data: come and get it. *Sci Transl Med* 6: 224ed4.

Annas, G.J. and Elias, S. (2014). 23andMe and the FDA. *N Engl J Med* 370: 985–988.

Baker, D., Park, C., Sweeney, C. et al. (2014). Recruitment of women in the National Children's study initial vanguard study. *Am J Epidemiol* 179: 1366–1374.

Bild, D.E., Bluemke, D.A., Burke, G.L. et al. (2002). Multi-ethnic study of atherosclerosis: objectives and design. *Am J Epidemiol* 156: 871–881.

Biorepositories, B.M. (2012). Building better biobanks. *Nature* 486: 141–146.

Campbell, C.D., Ogburn, E.L., Lunetta, K.L. et al. (2005). Demonstrating stratification in a European American population. *Nat Genet* 37: 868–872.

Cardon, L.R. and Bell, J.I. (2001). Association study designs for complex diseases. *Nat Rev Genet* 2: 91.

Cardon, L.R. and Palmer, L.J. (2003). Population stratification and spurious allelic association. *Lancet* 361: 598–604.

Carroll, M.D., Kit, B.K., Lacher, D.A. et al. (2012). Trends in lipids and lipoproteins in us adults, 1988-2010. *JAMA* 308: 1545–1554.

Collins, F.S. and Manolio, T.A. (2007). Merging and emerging cohorts: necessary but not sufficient. *Nature* 445: 259.

Collins FS, Varmus H., (2015). A new initiative on precision medicine. *N Engl J Med.* 372(9): 793–5.

Committee on National Statistics, Division on Behavioral and Social Sciences and Education, Board on Children, Youth and Families, National Research Council, and Institute of Medicine (2013). *Design of the National Children's Study: A Workshop Summary*. Washington, DC: National Academies Press (US).

COVID-19 Host Genetics Initiative. (2019). The COVID-19 host genetics initiative, a global initiative to elucidate the role of host genetic factors in susceptibility and severity of the SARS-CoV-2 virus pandemic. *Eur J Hum Genet* 28(6): 715–718.

Crawford, D.C., Crosslin, D.R., Tromp, G. et al. (2014). eMERGEing progress in genomics---the first seven years. *Front Genet*: 5.

Crawford DC, Goodloe R, Farber-Eger E, Boston J, Pendergrass SA et al. (2015) Leveraging epidemiologic and clinical collections for genomic studies of complex traits. *Hum Hered*. In press.

Dawber, T.R., Meadors, G.F., and Moore, F.E. (1951). Epidemiological approaches to heart disease: the Framingham study. *Am J Public Health Nations Health* 41: 279–281.

Dawber, T.R., Moore, F.E., and Mann, G.V. (1957). Coronary heart disease in the Framingham study. *Am J Public Health Nations Health* 47: 4–24.

Do, C.B., Tung, J.Y., Dorfman, E. et al. (2011). Web-based genome-wide association study identifies two novel loci and a substantial genetic component for Parkinson's disease. *PLoS Genet* 7: e1002141.

Duale, N., Brunborg, G., Ronningen, K. et al. (2012). Human blood RNA stabilization in samples collected and transported for a large biobank. *BMC Res Notes* 5: 510.

Eriksson, N., Macpherson, J.M., Tung, J.Y. et al. (2010). Web-based, participant-driven studies yield novel genetic associations for common traits. *PLoS Genet* 6: e1000993.

Fagan, A.M., Xiong, C., Jasielec, M.S. et al. (2014). Longitudinal change in CSF biomarkers in autosomal-dominant Alzheimer's disease. *Sci Transl Med* 6: 226ra30.

Feinleib, M., Kannel, W.B., Garrison, R.J. et al. (1975). The Framingham offspring study. Design and preliminary data. *Prev Med* 4: 518–525.

Fox, C.S. (2010). Cardiovascular disease risk factors, type 2 diabetes mellitus, and the Framingham heart study. *Trends Cardiovasc Med* 20: 90–95.

Fox, J.L. (2013). Sequestration to slash research grants and delay biosimilars. *Nat Biotechnol* 31: 271.

Fried, L.P., Borhani, N.O., Enright, P. et al. (1991). The cardiovascular health study: design and rationale. *Ann Epidemiol* 1: 263–276.

Friedman, G.D., Cutter, G.R., Donahue, R.P. et al. (1988). CARDIA: study design, recruitment, and some characteristics of the examined subjects. *J Clin Epidemiol* 4111: 1105–1116.

GARNET Study. (n.d.) Genomics and randomized trials network (GARNET). www.garnetstudy.org (accessed on 4 Decemeber 2021).

Gaziano, J.M., Concato, J., Brophy, M., et al. (2016). Million Veteran Program: A mega-biobank to study genetic influences on health and disease. *J Clin Epidemiol* 70: 214–23.

Gilbert, J. (2001). Establishment of permanent cell lines by Epstein-Barr virus transformation. *Curr Protoc Hum Genet* 2: A.3H.1–A.3H.5.

Gilbert, J.R. and Vance, J.M. (2001). Isolation of gneomic DNA from mammalian cells. *Curr Protoc Hum Genet* 19: A.3B.1–A.3B.6.

Goodpaster, B.H., Carlson, C.L., Visser, M. et al. (2001). Attenuation of skeletal muscle and strength in the elderly: the health ABC study. *J Appl Physiol* 90: 2157–2165.

Gottesman, O., Kuivaniemi, H., Tromp, G. et al. (2013). The electronic medical records and genomics (eMERGE) network: past, present, and future. *Genet Med* 15: 761–771.

Govindaraju, D.R., Adrienne Cupples, L., Kannel, W.B. et al. (2008). Chapter 2 genetics of the Framingham heart study population. In: *Advances in Genetics* (ed. C.H. Jeffrey), 33–65. Academic Press.

Gusella, J.F., Wexler, N.S., Conneally, P.M. et al. (1983). A polymorphic DNA marker genetically linked to Huntington's disease. *Nature* 306: 234–238.

Guttmacher, A.E., Hirschfeld, S., and Collins, F.S. (2013). The National Children's study–a proposed plan. *N Engl J Med* 369: 1873–1875.

Hall, J.M., Lee, M.K., Newman, B. et al. (1990). Linkage of early-onset familial breast cancer to chromosome 17q21. *Science* 250: 1684–1690.

Haseman, J.K. and Elston, R.C. (1972). The investigation of linkage between a quantitative trait and a marker locus. *Behav Genet* 2: 3–19.

Hays, J., Hunt, J.R., Hubbell, F.A. et al. (2003). The women's health initiative recruitment methods and results. *Ann Epidemiol* 13: S18–S77.

Hinds, D.A., McMahon, G., Kiefer, A.K. et al. (2013). A genome-wide association meta-analysis of self-reported allergy identifies shared and allergy-specific susceptibility loci. *Nat Genet* 45: 907–911.

Hoffmann, T.J., Zhan, Y., Kvale, M.N. et al. (2011). Design and coverage of high throughput genotyping arrays optimized for individuals of East Asian, African American, and Latino race/ethnicity using imputation and a novel hybrid SNP selection algorithm. *Genomics* 98: 422–430.

Hofman, A., Grobbee, D.E., de Jong, P.T., and van den Ouweland, F.A. (1991). Determinants of disease and disability in the elderly: the Rotterdam elderly study. *Eur J Epidemiol* 7: 403–422.

Hofman, A., Breteler, M.M.B., van Duijn, C.M. et al. (2007). The Rotterdam study: objectives and design update. *Eur J Epidemiol* 22: 819–829.

Holle, R., Happich, M., Löwel, H. et al. (2005). KORA - a research platform for population based health research. *Gesundheitswesen* 67: 19–25.

Keen, J.C. and Moore, H.M. (2015). The genotype-tissue expression (GTEx) project: linking clinical data with molecular analysis to advance personlized medicine. *J Pers Med* 5: 22–29.

Kimmel, S.E., French, B., Kasner, S.E. et al. (2013). A pharmacogenetic versus a clinical algorithm for warfarin dosing. *N Engl J Med* 369: 2283–2293.

Knight, K.E., Honerlaw, J., Danciu, I., et al. (2020). Standardized architecture for a mega-biobank phenomic library: the Million Veteran Program (MVP). *AMIA Jt Summits Transl Sci Proc* 30: 326–334.

Knowlton, R.G., Cohen-Haguenauer, O., Van Cong, N. et al. (1985). A polymorphic DNA marker linked to cystic fibrosis is located on chromosome 7. *Nature* 318: 380–382.

Kohane, I.S. (2011). Using electronic health records to drive discovery in disease genomics. *Nat Rev Genet* 12: 417–428.

Kolonel, L.N., Altshuler, D., and Henderson, B.E. (2004). The multiethnic cohort study: exploring genes, lifestyle and cancer risk. *Nat Rev Cancer* 4: 519–527.

Kumar, R., Eipers, P., Little, R.B. et al. (2014). Getting started with microbiome analysis: sample acquisition to bioinformatics. *Curr Protoc Hum Genet* 82: 18.8.1–18.8.29.

Li, R., Brockschmidt, F.F., Kiefer, A.K. et al. (2012). Six novel susceptibility loci for early-onset androgenetic alopecia and their unexpected association with common diseases. *PLoS Genet* 8: e1002746.

Lonsdale, J., Thomas, J., Salvatore, M. et al. (2013). The genotype-tissue expression (GTEx) project. *Nat Genet* 45: 580–585.

Lowel, H., Doring, A., Schneider, A. et al. (2005). The MONICA Augsburg surveys – basis for prospective cohort studies. *Gesundheitswesen* 67: 13–18.

Lurie, N., Manolio, T., Patterson, A.P. et al. (2013). Research as a part of public health emergency response. *N Engl J Med* 368: 1251–1255.

Mahmood, S.S., Levy, D., Vasan, R.S., and Wang, T.J. (2014). The Framingham heart study and the epidemiology of cardiovascular disease: a historical perspective. *Lancet* 383: 999–1008.

Mai, V. and Morris, J.G. (2013). Need for prospective cohort studies to establish human gut microbiome contributions to disease risk. *J Natl Cancer Inst* 105: 1850–1851.

Mailman, M.D., Feolo, M., Jin, Y. et al. (2007). The NCBI dbGaP database of genotypes and phenotypes. *Nat Genet* 39: 1181–1186.

Manolio, T.A., Bailey-Wilson, J.E., and Collins, F.S. (2006). Genes, environment and the value of prospective cohort studies. *Nat Rev Genet* 7: 812–820.

Manolio, T.A., Weis, B.K., Cowie, C.C. et al. (2012). New models for large prospective studies: is there a better way? *Am J Epidemiol* 175: 859–866.

Martin, E.R., Monks, S.A., Warren, L.L., and Kaplan, N.L. (2000). A test for linkage and Association in General Pedigrees: the pedigree disequilibrium test. *Am J Human Genet* 67: 146–154.

McCarty, C.A., Wilke, R.A., Giampietro, P.F. et al. (2005). Marshfield clinic personalized medicine research project (PMRP): design, methods and recruitment for a large population-based biobank. *Pers Med* 2: 49–79.

McCarty, C., Chisholm, R., Chute, C. et al. (2011). The eMERGE network: a consortium of biorepositories linked to electronic medical records data for conducting genomic studies. *BMC Med Genomics* 4: 13.

Morton, N.E. (1955). Sequential tests for the detection of linkage. *Am J Hum Genet* 7: 277–318.

The International HapMap 3 Consortium (2010). Integrating common and rare genetic variation in diverse human populations. *Nature* 467: 52–58.

Newton, K.M., Peissig, P.L., Kho, A.N. et al. (2013). Validation of electronic medical record-based phenotyping algorithms: results and lessons learned from the eMERGE network. *J Am Med Inform Assoc* 20: e147–e154.

Ng, M., Fleming, T., Robinson, M. et al. (2014). Global, regional, and national prevalence of overweight and obesity in children and adults during 1980–2013: a systematic analysis for the Global Burden of Disease Study 2013. *Lancet*.

Novembre, J., Johnson, T., Bryc, K. et al. (2008). Genes mirror geography within Europe. *Nature* 456: 98–101.

Ogden, C.L., Carroll, M.D., Kit, B.K., and Flegal, K.M. (2014). Prevalence of childhood and adult obesity in the United States, 2011–2012. *JAMA* 311: 806–814.

Ollier, W., Sprosen, T., and Peakman, T. (2005). UK biobank: from concept to reality. *Pharmacogneomics* 6: 639–646.

Panel on the Design of the National Children's Study and Implications for the Generalizability of Results, Committee on National Statistics, Division of Behavioral and Social Sciences and Education, Board on Children, Youth and Families, National Research Council, and Institute of Medicine (2014). *The National Children's Study 2014*. Washington, DC: National Academies Press (US).

Pathak, J., Kho, A.N., and Denny, J.C. (2013). Electronic health records-driven phenotyping: challenges, recent advances, and perspectives. *J Am Med Inform Assoc* 20: e206–e211.

Pirmohamed, M., Burnside, G., Eriksson, N. et al. (2013). A randomized trial of genotype-guided dosing of warfarin. *N Engl J Med* 369: 2294–2303.

Prezant, D.J., Weiden, M., Banauch, G.I. et al. (2002). Cough and bronchial responsiveness in firefighters at the world trade center site. *N Engl J Med* 347: 806–815.

Price, A.L., Patterson, N.J., Plenge, R.M. et al. (2006). Principal components analysis corrects for stratification in genome-wide association studies. *Nat Genet* 38: 904–909.

Pulley, J., Clayton, E., Bernard, G.R. et al. (2010). Principles of human subjects protections applied in an opt-out, De-identified biobank. *Clin Transl Sci* 3: 42–48.

Reardon, S. (2011). Ten months after Deepwater horizon, picking up the remnants of health data. *Science* 331: 1252.

Risch, N. (1990). Linkage strategies for genetically complex traits. II. The power of affected relative pairs. *Am J Hum Genet* 46: 229–241.

Risch, N. and Merikangas, K. (1996). The future of genetic studies of complex human diseases. *Science* 273: 1516–1517.

Risch, N. and Zhang, H. (1995). Extreme discordant sib pairs for mapping quantitative trait loci in humans. *Science* 268: 1584–1589.

Risch, N.J. and Zhang, H. (1996). Mapping quantitative trait loci with extreme discordant sib pairs: sampling considerations. *Am J Hum Genet* 58: 836–843.

Roden, D.M., Pulley, J.M., Basford, M.A. et al. (2008). Development of a large-scale De-identified DNA biobank to enable personalized medicine. *Clin Pharmacol Ther* 84: 362–369.

Rubin, C.S., Holmes, A.K., Belson, M.G. et al. (2007). Investigating childhood leukemia in Churchill County. *Nevada Environ Health Perspect* 115: 151–157.

Scott, C.T., Caulfield, T., Borgelt, E., and Illes, J. (2012). Personal medicine—the new banking crisis. *Nat Biotechnol* 30: 141–147.

Shurin, S.B. (2012). Foreword: celebrating the first ten years of research excellence from the Jackson heart study. *Ethn Dis* 22: 1.

Sie, L., Loong, S., and Tan, E.K. (2009). Utility of lymphoblastoid cell lines. *J Neurosci Res* 87: 1953–1959.

Signorello, L.B., Hargreaves, M.K., and Blot, W.J. (2010). The southern community cohort study: investigating health disparities. *J Health Care Poor Underserved* 21: 26–37.

Sorlie, P.D., Aviles-Santa, L.M., Wassertheil-Smoller, S. et al. (2010). Design and implementation of the Hispanic community health study/study of latinos. *Ann Epidemiol* 20: 629–641.

Spielman, R.S., McGinnis, R.E., and Ewens, W.J. (1993). Transmission test for linkage disequilibrium: the insulin gene region and insulin-dependent diabetes mellitus (IDDM). *Am J Hum Genet* 52: 506–516.

Splansky, G.L., Corey, D., Yang, Q. et al. (2007). The third generation cohort of the National Heart, Lung, and Blood Institute's Framingham heart study: design, recruitment, and initial examination. *Am J Epidemiol* 165: 1328–1335.

Steinberg, K.K., Relling, M.V., Gallagher, M.L. et al. (2007). Genetics studies of a cluster of acute lymphoblastic leukemia cases in Churchill County. *Nevada Environ Health Perspect* 115: 158–164.

Levy, D. (2013). Straight talk with. . .Daniel Levy. *Nat Med* 19: 1197.

Strobush, L., Berg, R., Cross, D. et al. (2011). Dietary intake in the personalized medicine research project: a resource for studies of gene-diet interaction. *Nutr J* 10: 13.

Takahashi, K., Tanabe, K., Ohnuki, M. et al. (2007). Induction of pluripotent stem cells from adult human fibroblasts by defined factors. *Cell* 131: 861–872.

Tang, Z., Berlin, D.S., Toji, L. et al. (2013). A dynamic database of microarray-characterized cell lines with various cytogenetic and genomic backgrounds. G3: genes|genomes|. *Genetics* 3: 1143–1149.

Taylor, H.A. (2005). The Jackson heart study: an overview. *Ethn Dis* 15: 1–3.

The ARIC Investigators (1989). The atherosclerosis risk in communities (ARIC) study: design and objectives. *Am J Epidemiol* 129: 687–702.

The International HapMap Consortium (2005). A haplotype map of the human genome. *Nature* 437: 1299–1320.

The International HapMap Consortium (2003). The International HapMap Project. *Nature* 426: 789–796.

The Women's Health Intitiative Study Group (1998). Design of the Women's health initiative clinical trail and observational study. *Control Clin Trials* 19: 61–109.

Tian, C., Gregersen, P.K., and Seldin, M.F. (2008). Accounting for ancestry: population substructure and genome-wide association studies. *Hum Mol Genet* 17: R143–R150.

Tryka, K.A., Hao, L., Sturcke, A. et al. (2014). NCBI's database of genotypes and phenotypes: dbGaP. *Nucleic Acids Res* 42: D975–D979.

Tsui, L.-C., Buchwald, M., Barker, D. et al. (1985). Cystic fibrosis locus defined by a genetically linked polymorphic DNA marker. *Science* 230: 1054–1057.

Voight, B.F. and Pritchard, J.K. (2005). Confounding from cryptic relatedness in case-control association studies. *PLoS Genet* 1: e32.

Wainwright, B.J., Scambler, P.J., Schmidtke, J. et al. (1985). Localization of cystic fibrosis locus to human chromosome 7cen-q22. *Nature* 318: 384–385.

Willett, W.C., Blot, W.J., Colditz, G.A. et al. (2007). Merging and emerging cohorts: not worth the wait. *Nature* 445: 257–258.

Wilson, P.W.F., DΓÇÖAgostino, R.B., Levy, D. et al. (1998). Prediction of coronary heart disease using risk factor categories. *Circulation* 97: 1837–1847.

Wilson, J.G., Rotimi, C.N., Ekunwe, L. et al. (2005). Study design for genetic analysis in the Jackson heart study. *Ethn Dis* 15: 30–37.

WTCCC Consortium (2007). Genome-wide association study of 14,000 cases of seven common diseases and 3,000 shared controls. *Nature* 447: 661–678.

Yoon, P.W., Rasmussen, S.A., Lynberg, M.C. et al. (2001). The national birth defects prevention study. *Public Health Rep* 116: 32–40.

Yu, J., Vodyanik, M.A., Smuga-Otto, K. et al. (2007). Induced pluripotent stem cell lines derived from human somatic cells. *Science* 318: 1917–1920.

5

Responsible Conduct of Research in Genetic Studies

Susan Estabrooks Hahn[1], *Adam Buchanan*[2], *Chantelle Wolpert*[3], *and Susan H. Blanton*[4]

[1] *Genomic Services, Quest Diagnostics, North Andover, MA, USA*
[2] *Genomic Medicine Institute, Geisinger, Danville, PA, USA*
[3] *Physician Assistant Program, Thomas Jefferson University, Philadelphia, PA, USA*
[4] *Dr. John T. Macdonald Foundation Department of Human Genetics, University of Miami Miller School of Medicine, Miami, FL, USA*

Introduction

The genetic analysis methods described elsewhere in this book are extraordinary tools for identifying genes associated with human traits and disorders. Although studies offer investigators opportunities to identify and study disease genes, they also present unique challenges and ethical quandaries not encountered in traditional medical research. It is necessary for investigators conducting such research to be aware of these when designing and conducting their studies.

The issues discussed in this chapter pertain to gene discovery research (hereafter referred to as "genetic research") discussed in this book, as opposed to clinical trial research. A clinical trial is typically designed to determine the effectiveness and safety of a medical intervention such as a drug, surgical procedure, or medical device. Most clinical trials do not involve the use of gene products and therapies. The distinction is an important one because developments of research regulations, procedures, and codes of conduct have historically favored clinical trial research. In addition, the study design and associated risks and benefits between the two types differ substantially.

For these reasons, it is often difficult for investigators to apply existing research regulations and policies to the new and emerging issues associated with the unique attributes of gene discovery research. The codes of conduct for performing genetic research are still evolving. As a result, there is considerable variability in study design, particularly with regard to quantifying and managing risk, and handling research results and incidental findings (IFs).

Clinical trials and genetic research pose different risks for participants. While clinical research may be associated with a risk of individual physical harm, the risks of participating in a genetic research study are mainly psychological and social and may also affect other family members.

This revised chapter contains material contributed by Amy Crunk, in the previous edition of this book.

Genetic Analysis of Complex Diseases, Third Edition. Edited by William K. Scott and Marylyn D. Ritchie.

For instance, individuals with or at risk for a genetic disorder may have feelings of low self-esteem, guilt, and anxieties about the future health status of themselves, their children, or other family members that may be brought to the forefront as a result of participating. In addition, there is a theoretical risk of discrimination due to the predictive nature of genetic information and to the conceivable use of personal genetic information as a basis for stigmatization and discrimination.

Furthermore, clinical trials typically offer participants more immediate benefits than genetic research. In turn, individuals are usually motivated to participate in clinical trials because the treatment or therapy may personally benefit them. Genetic research, on the other hand, has rarely correlated with clinical improvement. Families and individuals have historically participated in genetic research for altruistic reasons, hoping to help future generations through the knowledge gained from their participation in the genetic study. However, advances in genomic technology are increasing the speed and likelihood of identifying genetic variants associated with disease. In turn, increasingly, participants are expecting genetic research results, both those individual research results (IRRs) that contribute to the disease under investigation, as well as IFs, which are not related to the disease of study, but are uncovered in the research process.

In this chapter, we address topics important to the responsible conduct of genetic research to help the investigator provide high-quality genetic research services while minimizing potential problems. This includes ethical, legal, and social issues (ELSI) pertinent to the conventional practices of genetic research. In addition, practical methods are provided for maximizing participation and increasing the efficiency of the research. Note that this chapter primarily uses research regulations in the United States as examples and points of discussion. Genetics researchers in other regions should consult their local regulations. Web resources that provide more content in many of the topics covered in the following sections are provided in Table 5.1.

Research Regulations and Genetics Research

In the United States, regulatory oversight of human subjects research is conducted by the Office of Human Research Protection (OHRP) under the U.S. Department of Health and Human Services (HHS). Human subjects regulations are part of the Code of Federal Regulations (CFR) established by Congress at 45 CFR part 46 and contains five subparts. Subpart A is the basic set of protections for all human subjects and is most commonly referred to as the Common Rule. Details in this chapter pertain to 45 CFR part 46. However, it should be noted that on 22 July 2011, the federal government issued an Advanced Notice of Proposed Rule Making (ANPRM) to the common rule. The purpose is to, "request comment on how current regulations for protecting human subjects who participate in research might be modernized and revised to be more effective. This ANPRM sought comments on how to better protect human subjects who are involved in research, while facilitating valuable research and reducing burden, delay, and ambiguity for investigators." As a result, changes were made to the Common Rule, including the addition of new categories of exempt research, alterations to patient consent requirements, a new provision for broad consent for future research, and multiple changes to Institutional Review Board ("IRB") requirements. These changes were to be implemented on 19 January 2018. However, this was delayed until January 2019 to allow IRBs and investigators time to consider how to implement the changes.

In accordance with federal regulations, research studies involving human subjects funded by federal agencies (e.g., National Institutes of Health (NIH), Department of Energy [DOE],

Table 5.1 Resources regarding human subjects genetics research and regulations.

Topic	URL[a]
Human subjects regulations 45 CFR 46	http://www.hhs.gov/ohrp/humansubjects/index.html https://www.hhs.gov/ohrp/regulations-and-policy/regulations/finalized-revisions-common-rule/index.html
Office of Human Research Protections Decision Tools	http://www.hhs.gov/ohrp/policy/checklists/decisioncharts.html
Guidance on Engagement of Institutions in Human Subjects Research	http://www.dhhs.gov/ohrp/policy/engage08.html
Guidance on Withdrawal of Subjects from: Data Retention and Other Related Issues	http://www.hhs.gov/ohrp/policy/subjectwithdrawal.html http://www.fda.gov/downloads/regulatoryinformation/guidances/ucm126489.pdf
Guidance on Research Involving Coded Private Information or Biological Specimens	https://www.hhs.gov/ohrp/regulations-and-policy/guidance/research-involving-coded-private-information/index.html
Advanced Notice of Proposed Rule Making for Revision to Common Rule	http://www.gpo.gov/fdsys/pkg/FR-2011-07-26/pdf/2011-18792.pdf
Health Insurance Portability and Accountability Act of 1996	http://www.hhs.gov/ocr/privacy/index.html http://privacyruleandresearch.nih.gov http://privacyruleandresearch.nih.gov/research_repositories.asp
Certificate of Confidentiality	https://grants.nih.gov/policy/humansubjects/coc.htm
Informed Consent Policy Guidance	http://www.hhs.gov/ohrp/policy/index.html#informed
NIH Data Sharing Policy	https://osp.od.nih.gov/wp-content/uploads/NIH_GDS_Policy.pdf https://osp.od.nih.gov/wp-content/uploads/Supplemental_Info_GDS_Policy.pdf
Risk Management Guide for Information Technology Systems National Institute for Standards and Technology	https://www.nist.gov/publications (search for "risk management guide for information technology systems" to get most updated version)
Privacy and Progress in Whole Genome Sequencing	Report archived at: https://bioethicsarchive.georgetown.edu/pcsbi/sites/default/files/PrivacyProgress508_1.pdf

[a] All URLs accessed and active September 2021.

Food and Drug Administration) or covered by an applicable OHRP approved assurance created under 45 CFR 46.103 must be reviewed by an IRB (U.S. Department of Health and Human Services 2018). Most often, investigators have an IRB at their facility or medical center where they will be conducting the research. Research facilities without an existing IRB must either establish an IRB or submit their project proposals for review by an external IRB that satisfies federal requirements.

Although the majority of privately funded companies and institutions are connected to federal agencies via funding or facilities, those that are not connected are not subject to the Common Rule or IRB review (U.S. Department of Health and Human Services 2009). However, other Federal, State, or local laws or regulations may apply. Independent investigators typically choose to contract with an external IRB for protocol review to ensure that participants' interests are protected. In addition, evidence of IRB approval is also increasingly required for publication in peer-reviewed journals.

Composed of a multidisciplinary group of professionals, an IRB protects the rights and welfare of human research subjects who participate in research activities. Each IRB is regulated by, and adheres to, research regulations set forth by the OHRP. They also have the ability to waive or approve an alteration of the Privacy Rule's Authorization requirement of the Health Insurance Portability and Accountability Act of 1996 (HIPAA). HIPAA applies to the access, transmission, and storage of all medical information (http://www.hhs.gov/ocr/privacy/index.html), including that used in research. In addition to the informed consent required by the Common Rule, investigators also need to obtain HIPAA Authorization (Privacy Rule Authorization) from participants, or IRB approval for a partial or complete waiver of such authorization. This authorization can be combined with the research informed consent document to streamline the process for participants, but some institutions prefer to keep them separate despite the additional paperwork for participants (http://privacyruleandresearch.nih.gov/pr_02.asp). Investigators should work with their respective IRBs to ensure that genetic research projects are being conducted according to HIPAA regulations.

IRBs critically review protocols for scientific and ethical validity before a study begins, and the process is designed to be, as much as possible, beneficial to both the investigator and the study participant. In accordance with protecting the rights and welfare of participating families, IRB members offer additional insight into the study design and strengthen the research by assuring that genetic research studies are conducted in accordance with federal and state regulations. It should be noted, however, that given the complexity and rarity (in some institutions) of genetic research, IRB members may lack awareness of and/or experience with related techniques and issues. As such, it may be necessary for the investigator to educate the IRB. A common mistake of investigators is not entering into a dialog with IRB members, especially when they suspect IRB requests may be the result of misinformation and/or are not in the best interest of the participant. While IRB members may ultimately disagree, it is important that investigator and IRB work together as a team to protect participants.

There are three levels of IRB review: exempt, expedited, and full board review. Typically, genetic research studies do not qualify for exempt review. This category is reserved for studies that pose the lowest risk of harm to subjects. They include things such as educational practices, some surveys and qualitative or observational studies, and existing data and specimens that are either publicly available or are anonymous, among others. Most genetic research studies meet the criteria for expedited or full board review, depending on the risk level and other factors. Even if a study qualifies for an expedited review per criteria, often IRBs will require a full board review due to the special nature of genetic information. While full board review involves all or most of the IRB committee members, expedited review is usually performed by a designated member (e.g., the committee chair) or a subset of members. The OHRP provides clear and easy-to-use decision charts to not only help investigators determine the level of IRB review but also if research qualifies as research involving human subjects, and whether certain elements of informed consent meet criteria for a waiver (http://www.hhs.gov/ohrp/policy/checklists/decisioncharts.html).

Typically, the review process involves providing the IRB with a detailed description of all study procedures from recruitment to data analysis and storage, details about risks and benefits, documentation of funding, informed consent documents, and data collection materials. In addition, the investigator should present a protocol for maintaining identifying information, if any, on study participants and the methods, such as data coding, used to ensure participants' anonymity.

It should be noted that these regulations apply to research conducted in the United States or by US investigators outside the United States. Other countries have their own sets of regulations. As with the United States, their policies were often based initially on the Nuremberg Code (Shuster et al. 1997). This was a code for conducting human subjects research developed in 1947 at the end of the Doctor's portion of the Nuremberg Trials. When conducting research outside the United States, investigators need to familiarize themselves with the laws and policies of the relevant region, as they may differ greatly from those discussed here.

Addressing Pertinent ELSI in Genetic Research

Genetic Discrimination

Fear of genetic discrimination has been a consistent and significant concern of the lay public and the medical community (Hudson 2007; Lowstuter et al. 2008; Plichta et al. 2016; Wauters and Van Hoyweghen 2016; Hann et al. 2017). Genetic discrimination is the unfair treatment of a person based on his/her real or perceived genetic status. The most commonly cited forms of genetic discrimination pertain to insurance and employment. In some instances, fear of genetic discrimination has kept individuals from undergoing clinically useful testing (Hudson 2007; Wauters and Van Hoyweghen 2016; Hann et al. 2017). However, it was not until May 2009 that the Genetic Information Non-discrimination Act, also known as GINA, was passed in the United States (GINA 2008). While a tremendous step forward, GINA's protections only apply to health insurance and employment discrimination (in companies with 15 or more employees). They are also limited to those at risk, but not symptomatic. Table 5.2 outlines more details about what protections GINA does and does not afford (Hudson 2007). In addition, it does not protect those with U.S. military health coverage or those who receive services from Indian Health Services, The federal health program for American Indians and Alaska Natives.

Other countries and regions have enacted their own laws and/or guidelines. For example, Canada enacted the Genetic Non-Discrimination Act in 2017 (http://laws-lois.justice.gc.ca/eng/acts/G-2.5/index.html). This law prevents the use of genetic information to deny health or life insurance, employment, and housing or to influence adoption or child custody decisions. In 2010, England enacted the Equality Act 2010 which restricts employers to asking prospective employees only questions related to their ability to perform the duties associated with the job. In addition, a voluntary moratorium on the use of predictive genetic testing in life insurance decisions is in place (https://www.abi.org.uk/data-and-resources/tools-and-resources/genetics/code-on-genetic-testing-and-insurance/). Insurers can only request genetic test results for individuals who have had Huntington's disease predictive tests and are applying for life insurance policies worth in excess of £500 000. Other countries with genetic discrimination regulations include Germany, Poland, Austria, Norway, and Switzerland (Soini 2012). These generally restrict the use of genetic testing for employment purposes and regulate when insurers may request/require genetic testing.

Table 5.2 Information on the Genetic Information Nondiscrimination Act (GINA).

What GINA does:	What GINA does not do:
Prohibits use of an individual's genetic information in setting eligibility or premium or contribution amounts by group and individual health insurers.	Does not prohibit medical underwriting based on current health status. Does not mandate coverage for any particular medical tests or treatments.
Prohibits health insurers from requesting or requiring an individual to take a genetic test.	Does not interfere with the ability of a treating health care professional to request that an individual or family member undergo a genetic test. Nor does it limit the authority of a health care professional who is employed by or affiliated with a health plan or issuer from notifying an individual about genetic tests or providing information about a genetic test as part of a wellness program.
Prohibits use of an individual's genetic information by employers in employment decisions such as hiring, firing, job assignments, and promotions.	Does not subject employers to remedies and procedures that are any different from those in other civil rights laws such as Title VII and the Americans with Disabilities Act.
Prohibits employers from requesting, requiring, or purchasing genetic information about an individual employee or family member.	Does not prohibit workplace collection of genetic information for toxic monitoring programs, employer-sponsored wellness program, administration of federal and state Family and Medical Leave laws, and in certain cases of inadvertent acquisition of information. However, the employer may not use or disclose the information.

GINA table: Hudson, K.L. (2007). Prohibiting genetic discrimination. N Engl J Med. 356(20): 2021–2023.

Privacy and Confidentiality

GINA's limitations combined with the potentially stigmatizing nature of genetic information necessitate that all information collected and generated as part of the genetic study be kept confidential. Several steps can be taken to protect a research participant's identity. One measure is to refrain from documenting genetic study participation in a participant's medical record. Unlike clinical trials that affect the treatment of the participant, genetic research studies generally do not affect the individual's medical care. Therefore, there is no medical benefit to including such information in the record. Instead, documenting this information in the record increases the likelihood that a third party will learn about their participation when they should not. Since it is the investigator's responsibility to protect privacy, maintain confidentiality, and minimize harm whenever possible (per human subject protection regulations), it is difficult to justify including such information in the medical record (Wolf and Beskow 2019). Of note, there are no regulations stipulating whether or when research informed consent documents should be placed in the medical record. This is at the discretion of the IRB and/or institution where the research occurs, but it is a common practice when research impacts or may impact clinical care. Educating the IRB about the risks associated with including information from non-clinical gene identification studies in the medical record may be necessary. Additionally, all data obtained for or generated by the research study, paper or electronic, must be housed securely via HIPAA and regulatory standards to maintain individual confidentiality (http://www.hhs.gov/ocr/privacy/index.html, 2012).

Certificate of Confidentiality

Additional privacy protection may be afforded through a Certificate of Confidentiality (CoC). This protection, authorized under Section 301(d) of the Public Health Service Act (42 U.S.C. 241), permits research "investigators and others who have access to research records to refuse to disclose identifying information in any civil, criminal, administrative, legislative, or other proceeding, whether at the federal, state, or local level. Identifying information is broadly defined as any item or combination of items in the research data that could lead directly or indirectly to the identification of a research subject." However, the certificate does not protect information that is voluntarily provided by research participants to third parties or information that is requested by the Department of Health and Human Services (DHHS) for use in an audit or program evaluation or if it is required by the Federal Food, Drug and Cosmetics Act. In addition, it does not prevent the investigator from reporting child abuse, communicable diseases, or threatened violence.

CoCs are only issued for research projects which have IRB approval, and which collect identifiable, sensitive information and the research is collected or used in the United States. As of 1 October 2017, a CoC is automatically granted to all NIH-funded research. Investigators with non-NIH DHHS funding may request one from their respective funding agency. For example, the CDC and the FDA both issue CoCs. If the research is not federally funded nor involves any federal agency, investigators may still request a CoC from the NIH institute or Center that supports similar research. There is an online application (https://grants.nih.gov/policy/humansubjects/coc/how-to-apply.htm) which greatly simplifies the application process. More information can be obtained from the NIH website https://grants.nih.gov/policy/humansubjects/coc.htm.

Coding Data and Samples

The majority of genetic research studies necessitate the use of identifying information in some capacity. In addition to using it in the course of the research, some investigators maintain identifying information so that participants can be re-contacted for a variety of reasons as discussed later (see below, Maintaining Contact with Participants). Per the OHRP regulations (Section 46.102 (f)), a living individual becomes a human research subject when an investigator conducting research obtains data through intervention or interaction with the individual or obtains identifying private information about that individual. This includes any information that can be used to directly or indirectly link an individual with his or her sample (e.g., numerical code, name, address, pedigree information, and medical records).

In an effort to protect the privacy of research subjects, investigators should minimize the amount of identifying information they collect to the greatest extent possible without compromising the research (Stewart et al. 2019). Because it is rare for an investigator to use anonymous data without severely compromising the quality of the research, coding techniques are used to protect privacy while retaining identifying information. By labeling families and individuals with a numbering system, they can be referred to at meetings and in publications without revealing the participant's identity. It is also ideal to label and store biological samples and subsequent data with a different code that only the investigator and other key members of the research team can break. This prevents ascertainment staff from having routine access to genetic research data, and laboratory personnel and analysts from having access to identifying information. Most importantly, it also makes it more difficult for a third party to access the information from the outside, which poses the greatest risk to participants.

A more robust form of coding is de-identification. The term de-identified emerged with the introduction of HIPAA and a desire to create a central government repository of samples and data collected using taxpayer dollars. Data sent to the government repository must be "de-identified," which means it lacks all of the 18 HIPAA identifiers – either because the data were collected that way or the identifiers were removed prior to use or sending. Without these identifiers, the identity of the individual is not readily identifiable to a user. When a key to the code still exists at the originating institution, the data cannot be truly anonymous, but, from the perspective of the receiving institution, it is as close as it can be. In this way, only those at the originating institution should be able to break the code offering the greatest privacy protection possible under the circumstances.

When sharing data and samples with collaborators outside the originating department or institution, it has become common to de-identify them. Because of the additional protection it affords, doing otherwise must typically be justified to and approved by the originating IRB. Of note, if the potential exists to share data and samples, this should be addressed in the informed consent document. Interestingly, while shipment of de-identified samples to an outside institution is under the purview of the IRB of the originating institution, receipt of such information by another institution is not technically subject to human subjects' regulations. This is because the investigator at the recipient institution does not obtain data through intervention or interaction with the individual, nor does he/she obtain identifying private information about that individual; the definition of a human subject (Title 45 Code of Federal Regulations, Part 46 2018; U.S. Department of Health and Human Services 2018; U.S. Department of Health and Human Services OHRP 2008b; U.S. Department of Health and Human Services GCPP 2008a). Another similar situation is possible within the same institution. The term "Human Non-Subjects Research" has been coined when referring to research with de-identified samples (Wolf and Kohane 2017). As described in detail in Chapter 4, there is a repository at Vanderbilt University called BioVU where data and samples collected in the course of routine clinical care are de-identified by software for use in research without the informed consent of the source (Roden et al. 2008; Wolf and Kohane 2017). The repository is approved by the Vanderbilt IRB. In the early implementation, opt in informed consent was not necessary because samples and data were not obtained by the investigator through intervention or interaction with the individual, and the investigator did not obtain identifying private information about that individual. They did permit individuals to opt out, but this is not required per regulations. Some question whether this is an ethical practice based on a flaw in the human subjects protection policy (Rothstein 2010). In the more recent implementation, BioVU has changed to an informed opt-in consent process (more detail in Chapter 4).

Secondary Subjects

Research regulations may also apply to "secondary subjects" who meet the definition of a human subject. Secondary subjects include non-consented family members on whom private health information is collected. In genetic research, this most often occurs during a family history interview. With regard to regulations regarding secondary research subjects, the IRB must determine whether informed consent is necessary from family members who will not otherwise be formally enrolled. This determination is often based on the extent of identifying information collected and the potential risk of harm if breached (Clayton et al. 2018). The IRB's discussion, particularly the rationale for deciding whether or not informed consent is necessary, must be clearly documented (NHRPAC 2002; Clayton et al. 2018). For further guidance regarding what constitutes minimal risk and when a decision should be made to obtain informed consent on secondary subjects, please see the OHRP website (Table 5.1).

It should be emphasized that despite efforts to uphold confidentiality such as securing data and files, using CoCs, and data coding, a possibility still exists that information will be inadvertently disclosed to, or accessed inappropriately by third parties (e.g., indirectly, through a notation in a medical record alluding to the study). Such information could put the research participant at risk for genetic discrimination. Therefore, it is recommended that investigators include information regarding the theoretical risk to future discrimination, particularly future insurability, in the informed consent process and form (Presidential Commission for the Study of Bioethical Issues 2012). Furthermore, investigators should be aware that disclosure of personal information to third parties without a participant's consent might be subject to legal action (Palley et al. 2017).

Future Use of Samples/Data Sharing

Many researchers are increasingly seeking the ability to use samples and data collected as part of their research for other unrelated studies in the future. This is referred to as future or secondary use of samples. This frequently involves sharing the data, as well. Consent for future use is problematic since it is technically impossible to protect the participant's autonomy (i.e. there is no way to make an informed and voluntary decision to participate in research when no details are available). However, many participants are willing to "blindly" consent to such uses. Furthermore, some genomic techniques and approaches require very large cohorts that could be facilitated by combining several smaller cohorts (Greely 2007; Wijmenga and Zhernakova 2018; Wright et al. 2019). Given the benefits of sharing data and future uses to society by furthering science at a more rapid pace, inserting language regarding such uses into consent forms is now fairly standard. In 2003, the NIH implemented a policy on sharing of research data. As of the receipt date of 1 October 2003, "investigators submitting an NIH application seeking $500,000 or more in direct costs in any single year are expected to include a plan for data sharing or state why data sharing is not possible." (Final NIH Statement on Sharing Research Data, February 26, 2003 – http://grants.nih.gov/grants/guide/notice-files/NOT-OD-03-032.html). This policy was expanded in 2014 with the National Institutes of Health Genomic Data Sharing Policy (https://osp.od.nih.gov/wp-content/uploads/NIH_GDS_Policy.pdf). Under this policy, sharing of large-scale genomic data, including genome-wide association studies (GWAS), single nucleotide polymorphisms arrays, and genome sequence, transcriptomic, metagenomic, epigenomic, and gene expression data, is required for all studies, irrespective of funding mechanism or funding level. Guidance on what constitutes a large-scale study is provided in *Supplemental Information to the NIH Genomic Data Sharing Policy* (https://osp.od.nih.gov/wp-content/uploads/Supplemental_Info_GDS_Policy.pdf). Examples of studies which require sharing are sequence data from more than one gene or region of comparable size in the genomes of more than 1000 human research participants or sequence data from more than 100 genes or region of comparable size in the genomes of more than 100 human research participants. Data sharing requirements of other countries or entities need to be considered when conducting research outside the United States.

It should be noted that any future use of a sample is subject to human research regulations under the future research study using it. Despite the consent to future use, an IRB may mandate re-consent of the individual under this different protocol. This is more likely to occur in studies of greater than minimal risk, and/or that are ethically controversial where the participant may have a moral objection to use of his/her material in this way. In addition, in October 2012, the Presidential Commission for the Study of Bioethical Issues released a report titled, *Privacy and Progress in Whole Genome Sequencing* (Presidential Commission for the Study of Bioethical Issues 2012). In it, they recommended that policies be established that prohibit "unauthorized whole genome sequencing without the consent of the individual from whom the sample came." This implies that re-consent is necessary when using research samples that did not specify this analysis method. At present, such policies have not been established, but could be a significant barrier to future use using retrospective samples that pre-date genome sequencing. Prospectively, mentioning the potential for genome sequencing in the consent form is prudent for those studies that anticipate future uses.

Within the informed consent, future use may be posed as a condition of the primary research, or as a separate option, but the latter is recommended and often required by the IRB. More than one option reflecting different conditions may also be posed. For example, there may be an option to consent to future use only on the disease specified in the originating study, or for research on any disease. It is important for investigators to decide in advance whether they would like the option to use samples in the future for other purposes and, if so, work with their IRB to determine a permissible method for achieving this goal.

Handling of Research Results

> *If investigators are aware of identities of the individuals whose samples they have received, do they have a moral or legal obligation to notify the participants of information that may save, or at least change, [the participants'] lives?* —Researcher of Huntington disease, quoted in Callier et al. (2016)

Investigators need to consider how they will handle three different types of results related to their research: clinical results, individual research results (IRRs), and incidental findings (IFs). Investigators disclosing results must define a policy detailing which results will be disclosed and the method of disclosure, and this should be transparent to the participant during the informed consent process. Clinical results are those that are derived as part of the phenotyping process in the study. Examples include cholesterol testing done in a study of cardiovascular disease, results from a mental status exam in a study of dementia, or results from an Autism Diagnostic Inventory performed in an autism study. When results are available from tests or procedures that are standard in clinical care, the investigator needs to decide if and how such results will be communicated. This decision may depend on method used to generate the results (i.e. testing in a Clinical Laboratories Improvement Amendments [CLIA]-certified laboratory), and/or financial, logistic, legal, and/or ethical factors, among others. If it is decided to give such results back, some investigators provide an opportunity for the participant to choose during the informed consent process whether to receive the results, which provides the participant more autonomy. This is discussed more in the section addressing confirmation of a diagnosis.

IRRs are those that pertain to the aim of the research (e.g. contribute to the disease under investigation). For example, determining a genetic variant believed to cause a person's spina bifida in a study aimed at finding genes that contribute to spina bifida. Again, the investigator needs to decide if such results will be communicated. Factors which often complicate this decision are whether the research results are generated in a CLIA-certified laboratory or whether they can be replicated in one (see more discussion below), the clinical validity and utility of the finding, funding, and logistics, among others. If available, these results must be offered in a manner sensitive to participants' desire, or lack thereof, to receive it. While participants may volunteer for family studies understanding the potential for identification of causative genes, they may not have fully anticipated the consequences and concerns that arise from the reality of the availability of testing, especially presymptomatic or predictive testing.

The last and most complicated result is the IF. IFs are research results that are unrelated to the aim of the study (which typically pertains to a specific disease) and therefore, unexpected. (Wynn et al. 2018; Wong et al. 2018). Considering the spina bifida research example above, discovery of a Lynch syndrome mutation (that highly predisposes an individual to colon and other cancers), or carrier status for an autosomal recessive condition that puts the individual at reproductive risk would be considered IFs. While several studies suggest participants would like to receive IFs, decisions regarding whether to and how to return IFs are fraught with ethical and logistical issues that are relatively new (Kaufman et al. 2008; Murphy et al. 2008; Shalowitz and Miller 2008; Angrist 2011). With the rapid inclusion of genome sequencing, the likelihood of finding IFs is relatively high and increasing and research participants should be aware of this (Presidential Commission for the Study of Biomedical Issues 2012; Saelaert et al. 2018; Ibuki et al. 2020). What criteria would be used to decide what results should ethically be returned? Many researchers support the right of participants to receive results of clinical significance (Bookman et al. 2006; Wynn et al. 2018; Wong et al. 2018). Are participants aware that they may be contacted about IFs? Can the study financially support the responsible release of IFs? Is the laboratory CLIA certified (see more on this below)? What would the legal liability be for returning and not returning the results (Roberts et al. 2017; Wong et al. 2018)? Recommendations were recently published that address the release of IFs (Wynn et al. 2018; Wong et al. 2018). While the details are too numerous to discuss here, an overarching theme is that investigators need to make decisions up front and these decisions should be reflected clearly in the informed consent document and the IRB protocol.

In 2013, the American College of Medical Genetics and Genomics issued guidelines for the return of results from **clinical** exome and genome sequencing and updated it in 2017 (Green et al. 2013; Kalia et al. 2017). The current guidelines provide for the return of actionable variants in 59 genes in a clinical setting. Again, investigators need to decide whether or not to follow the same recommendations for research data as for clinical data prior to initiating the study. In 2018, the National Academy of Science, Engineering, and Medicine issued a report where they state that "investigators and their institutions should routinely consider whether and how to return IRRs on a study-specific basis through an informed decision-making process". More details on this report can be found here: http://nationalacademies.org/hmd/reports/2018/returning-individual-research-results-to-participants.aspx.

IFs, such as false paternity or adoption status, are another type of IF unrelated to the purpose of the original research. Such information is both potentially harmful when revealed to unsuspecting family members (Hindorff et al. 2018), but potentially helpful when used in conjunction with a clinical protocol. For example, if false paternity was uncovered in a family in which the father had Huntington disease; such information could at once be devastating to and welcomed by the children, who were assumed to carry a 50% risk for developing this neurodegenerative disease. In the end, cases that are not clearly defined under the study protocol or involve serious or immediate medical implications should be referred to an ethics committee or an IRB on an individual basis to determine appropriate action.

CLIA Regulations: Separation of Research and Clinical Laboratories

The CLIA of 1988 is U.S. federal legislation enacted to regulate the quality assurance of testing and reporting of results by clinical laboratories. Under CLIA, laboratories that disclose test results to patients for the purposes of the diagnosis, prevention, or treatment of a disease are subject to regulation. If the research laboratory itself is not CLIA certified, then a mechanism needs to be in place to offer testing in a CLIA-certified laboratory. It is essential that research laboratories define the extent of information that will be disclosed as part of the informed consent process to determine whether the CLIA regulations are applicable and to develop a policy about releasing results to review with research participants. Many research laboratories follow a protocol of complete nondisclosure of information to families (i.e., they do not provide research results to individuals). However, some research laboratories may desire to be authorized to release results under CLIA regulations. Research laboratories that disclose results or other information to participants must consider additional issues as discussed above.

Releasing Children's Genetic Research Results

Genetic testing of an apparently unaffected child in the absence of any direct medical benefit is a controversial practice (Wertz, et al. 1994) and should be done only within the context of genetic counseling and according to specific guidelines (ASHG/ACMG Report 1995; Council on Ethical and Judicial Affairs 1995). Concern arises over the potential risk of stigmatization of a child known by family members, future employers, and insurers to carry a genetic mutation (OHRP). In addition, questions about the child's autonomy and his or her right not to know personal genetic information may conflict with parental autonomy. Parents wishing to have their children tested for genetic disorders should be referred to a genetic counselor to discuss the potential risks and benefits of testing.

DNA Ownership

In a landmark decision in 1980, the U.S. Supreme Court ruled that biological materials could be patented, opening the door for the patenting of genes and gene products (Diamond v. Chakrabarty, cited in Winkler & Wiemann 2016). Ten years later, the California Supreme Court ruled that a person has no ownership rights to his or her cells once they have been removed from his or her body (Moore v. Regents of University of California 1990). Therefore, although the legal status of DNA ownership has yet to be completely resolved, one could argue that the DNA derived from discarded tissue or from research participants is the property of the investigator.

The right of a research participant to control the fate of his or her sample remains a sensitive and ambiguous issue. Several reports suggest that in addition to DNA storage and accessibility issues, participants should be offered a choice in determining the fate of their samples in terms of research and commercial uses (Clayton et al. 2018). Guidelines for DNA banking as well as legislation establishing an individual's ownership rights to his or her DNA have been proposed (Goodman 2016; Paul et al. 2018; Majumder et al. 2020). Although the dispute over DNA ownership is an important one, it appears that unless a sample proves to have a commercial value (Moore v. Regents of University of California), research participants are generally more concerned that the information derived from genetic research be secure than about the fate of their individual sample material. Still, it may be helpful to include a clause in the informed consent form specifying that the individual's DNA becomes the property of the Institution at the time of the blood draw.

DNA Banking

Genetic research laboratories are inherently DNA banking facilities. Therefore, research laboratories should develop policies that address banking issues, including length of DNA storage and accessibility to banked samples. These policies should be incorporated into the informed consent process (Hall et al. 1991; Stoeklé et al. 2016). Some practical guidelines for informing the participants about DNA banking options include the following:

1) Incorporation of the research center's policy on DNA banking into the informed consent process. This document should state clearly whether DNA samples collected for research purposes will be available to the research participants or their surviving family members.

2) Advising research participants about the availability of commercial DNA banking facilities. Commercial DNA banking enables an individual, or his or her family members as designated by the individual, to access a DNA sample to pursue diagnostic genetic testing when such testing becomes available. This advice may be particularly relevant for affected elderly family members or when a sole individual in a family has an unknown or terminal disorder. It is advantageous for both the family and the research team to have interested family members deposit DNA samples with a commercial DNA bank.

Family Coercion

Although all studies involving human subjects are subject to potential coercion, there is heightened concern that family studies are more prone to coercion by family members, especially when families are ascertained as a group. Therefore, it is especially important to remind research participants prior to and during the visit that their participation is voluntary. To minimize the possibility of family coercion to participate in the study (Parker and Lidz 1994; NHRPAC Working Group on Genetics 2002), it is a good idea whenever possible to obtain informed consent from individual family members before visiting the family. It is vital that the informed consent process include an explanation of what should and should not be expected during the visit. For instance, while a directed physical examination may be performed to confirm the diagnosis, clinical services, like medical treatment, are usually not provided by investigators, especially in the field. In addition, individual decisions about whether to participate in the study should not be discussed with other family members unless permission has been given to do so. A written summary of these discussions needs to be sent to all the members of the family who will participate.

Practical Methods for Efficient High-Quality Genetic Research Services

The numerous and diverse responsibilities associated with conducting genetic research include the administration of the study, recruitment of proband, and in many instances, family members, obtaining informed consent from research participants, coordination of sample collection and field trip visits to families, and working with the IRB for study approval. Tasks may also include genetic counseling or other clinical activities. One professional should be designated as the genetic research coordinator to handle these diverse, but vital responsibilities. Because of the diverse skills required, the study coordinator should ideally be a genetic counselor, social worker, psychologist, or other health professional with graduate training in counseling and medical genetics. When such professionals are not available, individuals with an undergraduate degree can often be trained to perform these duties. However, these individuals will require supervision by a qualified professional.

The study coordinator plays an important role in facilitating research while addressing the research participants' needs (McConkie-Rosell and Markel 1995). Serving as the participant's contact person with the research team, the study coordinator works to establish rapport. In particular, the study coordinator helps foster realistic expectations about the study. In addition, participants may have more questions related to the research study or to the genetic aspects of their condition, or they may need to be re-sampled or re-examined at a later date. By being available to answer questions and offer support, the study coordinator helps meet the needs of participants, including the need for information, the need to be informed of the study progress, and the need to be understood. The study coordinator is essential in providing referrals to participants that would benefit from clinical services.

In addition to working with participants, the study coordinator serves as the laboratory's representative to referring professionals and facilities. By being available to answer clinicians' questions about the research study, the study coordinator builds credibility among clinical professionals who may serve as a referral source for participants. Also, the coordinator provides current and accurate information on the status of genetic research for a particular disease. For example, the coordinator may provide referring professionals with regular bulletins or notifications of research discoveries with clinical implications. Thus, by maintaining a rapport with outside institutions, the genetic studies coordinator enhances referrals and participation.

The Investigator as the Genetic Study Coordinator

Those who do research in genetics are often among the few experts who can provide patients and professionals with accurate and comprehensive information about the disorder they are studying. While providing such information to research participants and referring professionals can be a time-consuming process, it is a vital part of conducting genetic research. Many participants with genetic disorders, especially rare disorders, may have little, incorrect, or incomplete information about the condition. Additionally, newly diagnosed individuals may need more support and information about the disorder, its mode of inheritance, and genetics in general. Individuals who require formal counseling and genetic testing should be referred for these services when they are not provided as part of the research protocol.

A single investigator with limited funding may not be able to hire a professional for participant ascertainment. One option in such instances is to consider collaborating with a research group that has a study coordinator who can assist with participant ascertainment. If, alternatively, the investigator chooses to serve as the study coordinator, he or she should allot time for communicating with participants, recognize the necessity of maintaining ongoing communication with participants, and be aware of the participants' needs and their reasons for participating in the study. Since the needs of a participant may go beyond the expertise of the investigator, it is helpful for all investigators to have a resource list of genetic counselors and other health care professionals as well as disease support groups for family referral.

Time Spent

Ascertaining participants and performing sample collection for genetic research can be extremely time-consuming. A coordinator spends a large amount of time talking with potential research participants on the telephone or in-person; interviewing potential participants, family members, and physician's offices; visiting outpatient clinics to recruit and enroll research participants; traveling on field studies to collect samples and information; and reviewing medical records to either identify potential subjects, or confirm a diagnosis. There is also time involved in writing letters, entering data, constructing pedigrees, documenting research activities, and organizing field trips.

How a coordinator's time is spent will vary depending on the criteria of the research study and the data being collected. Studies requiring data that are best or often collected off-site may require a larger percentage of coordinators' time to be spent performing field studies. Other studies may require that the data be collected on-site, resulting in larger amounts of the coordinators' time being spent in outpatient clinics. For some study designs, such as a disease focused one, a study coordinator can ascertain and perform sample collection on as few as 14 participants a month (Feliciano-Astacio et al. 2019). Other study designs, such as enrolling individuals in a "biobank", require much less time from the study coordinator. Ultimately, rate of participant enrollment will vary depending on the ascertainment criteria and amount of information needed for the particular study.

Recruitment

Participants can be ascertained using several methods, each with its own advantages and disadvantages. The difficulty in finding individuals interested in research studies will vary depending on the incidence of the disorder being investigated, the methods used to publicize the study, the severity of the disorder, and the belief in genetic influence. The main sources of referrals are disease support groups and similar organizations, health care providers, public databases and websites, medical clinics, and the social media such as Facebook. Whatever strategies will be employed for recruitment, they should be outlined in the recruitment section of the IRB protocol.

Support Groups and Organizations

Support groups for specific genetic disorders can serve as an excellent referral source. Research studies endorsed by their advocacy/support group may be viewed as more credible (Melvin et al. 1999). Not only does the support group lend integrity to the research but can also provide general information about research studies to its membership and refer interested individuals. In recent years, support groups have begun to take a more active role in research involvement. Some lay advocacy groups have already taken a stance to become equal partners in research (Payne et al. 2018; Riazi and Kraeva 2020). Many of the support organizations will promote research projects on their website. They may also aid by highlighting the importance of participating in general. Terry et al. (2000) presented case examples of how lay advocacy groups have worked with investigators as partners. She concludes that some of the most obvious benefits to the investigator of this method are the education of the research participants, the media, funding agencies, and policy makers about the importance of the research. In addition, these groups are optimally positioned to directly collect epidemiological and phenotypic data through recruitment.

Likewise, investigators can refer participants to support groups as a valuable resource for individuals seeking further information and services. Among the many genetic support groups in (and out of) the United States, there are three umbrella organizations that can facilitate finding a specific disorder support group: the Genetic Alliance (http://www.geneticalliance.org), the National Organization of Rare Disorders (NORD) (https://rarediseases.org), and the Self-Help Clearinghouse (a federally funded program that creates an information office in each US state).

Referrals from Health Care Providers

Advertising in medical journals, announcements at medical conferences, or word-of-mouth referrals among professional colleagues are additional methods for reaching families. Focus groups conducted with research participants by Melvin et al. (1999) revealed that one of the primary ways that individuals would like to be informed about research studies is from their personal physician. According to the OHRP, an institution that informs or refers its clients to a research project is not considered "engaged" in research (http://www.dhhs.gov/ohrp/policy/engage08.html, 16 October 2008). Professionals have indicated a desire for information regarding the disclosure of results, the study protocol, cost, informed consent, and any travel that would be required by the participant before making a referral (Melvin et al. 2000). When a health care professional makes the referral, the family should be given the investigator's name and contact information or, alternatively, the referring professional should obtain the family's permission to be contacted by the investigator directly. Again, it is very important that individuals never be contacted directly unless permission to do so has been obtained through a referring professional or support group.

Research Databases and the Internet

Investigators can register and list their study descriptions and contact information directly with several databases. Health care professionals seeking out investigators to whom they can refer their patients usually access these databases. As an integral part of communication, the Internet is a valuable mechanism to promote the research. This can be done by posting an announcement online with a disorder support group, by developing a research home page or research Facebook page, among others (Scaffe et al. 2000; Lee et al. 2017; Hintz and Dean 2020; Langbaum et al. 2020).

Investigators must be especially mindful when using social media such as Facebook to promote their research. For privacy purposes, investigators should refrain from engaging in discussions about the research or the disorder on a social site. It is wise to quickly direct the conversation to "off-line" communication. When participating in an online forum, it is important to be sensitive and respectful. Many of these forums are composed of people with the diseases being studied or their family members and friends. Participation by an investigator, while often welcomed by the participants, has the potential to feel like an intrusion if not handled appropriately.

It is also crucial to remember that what makes social networks so great for recruitment can also pose a risk to the investigators. When successful, it is easy for potential participants to spread the word about your research or post positive comments on your site. However, the reverse is also true. It is hard to overcome bad internet press, even if claims are unfounded. Related to this, research representatives should be mindful of their personal online reputation. Participants may search Facebook for a research team member they met or "Google" the person. What will they find? Is there anything there that will discredit the reputation or image of the study?

Institution Databases

Many institutions maintain patient databases encompassing a variety of information that may be useful to the investigator, as well as biobanks with a variety of sample types (see Chapter 4). A database may be so large as to encompass all patients served by a particular hospital and the treatments they received or may be generated out of a clinic and contain only disorder or specialty specific information. Investigators may be able to utilize such databases to identify eligible study participants and should inquire about their existence at their institution. Due to privacy issues, use of the database may have certain restrictions. It is not acceptable for the investigator to contact members of the database regarding the research study without prior permission from the IRB. If the database is composed of a limited number of physicians, it is ideal for them to communicate with their patients about the study via letter. If they are willing and this method is used, it must be stipulated in the IRB protocol. The letter should contain a brief explanation of the study and the investigator's contact information should they be interested. This can be an extremely efficient method to introduce the study to a large number of eligible individuals in a short period of time (Estabrooks et al. 1999). If individuals are asked to send something back to indicate their interest in the study, they should be provided with an addressed envelope and not an addressed postcard to ensure that the patient's confidentiality is maintained while the letter is en route.

Medical Clinics

Investigators who ascertain potential participants using medical clinics must be especially careful to clearly define the difference between the patient's clinical care and his or her participation in a study. Patients may mistakenly believe that the research study is part of their medical care. Patients may also have the impression that their medical care is provided contingent upon participation in the proposed research study (which contradicts the principle of a voluntary study). In addition, potential study participants may confuse the research study with a clinical service and may anticipate test results or treatment. Therefore, the investigator should clearly state the difference between the clinical and research settings and emphasize their independence (e.g., a research lab does not provide clinical services). The investigator should also be sure that the participant fully understands this difference.

In addition to clearly delineating the line between research and clinical services, clinician investigators who also provide medical care to research participants should maintain study files separate from medical records (Wolf and Beskow 2019). Furthermore, medical practitioners should avoid referencing the research study in the patient's medical chart if it is stipulated in the IRB protocol that research information will not go in the record. In this way, information gathered by the study is not subject to disclosure requests by insurance companies, except in legal proceedings. As described previously, investigators may apply for a CoC that grants research data immunity from such as subpoenas.

Recruitment by Family Members

In genetic studies, it is often necessary to obtain samples from entire families for analysis. As mentioned previously, investigators and coordinators need to be particularly mindful of coercion when a study involves recruitment of more than one family member. In some instances, people may be tempted to pressure their own family members to join a research study and may in fact do so. An investigator should guard against coercion, albeit well-intentioned, by speaking with all family members individually and eliciting their opinions about participating in the research study. Logistically, it may be necessary for one family member to inform all the other family members about the study, but the investigator should ask for permission to communicate individually with each potential participant.

Informed Consent

Participation in a genetic research study involves the comprehension of complex information. This information must be presented to each individual participating in the study in a clear and concise manner. The process of communicating this information to potential study participants is the basis on which that individual can give his or her informed consent. The investigator must provide the potential participant with an appropriate explanation of the study, for only in this way can informed consent assure voluntary participation and informed decision-making on behalf of the individual. When individuals understand the research protocol, they will be able to set realistic expectations about the benefits and the limitations of participating in a genetic research study.

In response to documented human rights violations in the area of human studies research, regulations have evolved to ensure the protection of research participants. The importance of "voluntary consent" in research did not emerge until the end of World War II in accordance with the Nuremberg Code. In the United States, the required elements of the informed consent process are set forth by federal regulations (OHRP). In addition to federal regulations, state laws and case law precedents may apply to research involving human subjects.

Participants should be told the purpose of the study, as well as the risks, benefits, and alternatives to participation, the policy on disclosure of results (if any), and the financial and time commitment involved (Genetic Alliance 1997; Stoeklé et al. 2016; Eisenhauer et al. 2019). Potential research participants should understand that their participation is optional and that they may withdraw from the research at any time after enrolling in the study without harm or penalty to themselves or to their family members per several regulations including the U.S. Code of Federal Regulations, the Common Rule (45 CFR 46); the Declaration of Helsinki (Shrestha and Dunn 2019); the 2002 research guidelines of the Council For International Organizations of Medical Sciences; and the Belmont Report (45 CFR 46; National Commission for the Protection of Human Subjects of Biomedical and Behavioral Research 1979; CIOMS 2002; Schaefer and Wertheimer 2010; Shrestha and Dunn 2019).

Investigators should have a plan for handling participant withdrawal in their protocol. When a participant requests to withdraw from a research protocol, the investigator needs to determine the extent of their withdrawal. Most participants who withdraw do so because the study has become burdensome in some way or they simply do not want to be contacted further regarding the study. The investigator should cease interacting or intervening with the participant for research purposes and should not collect any additional private health information related to the research.

Documentation procedures need to be determined by the investigators and IRBs. When documentation is appropriated, the OHRP recommends that the participant's specific requests regarding withdrawal, and the reasons for withdrawal, if known, be documented in the research record (U.S. Department of Health and Human Services 2010). It may also be that the participant is only withdrawing from the primary intervention but is willing to continue with secondary aspects to the study. This may also be permissible and addressed in the research policy.

In rare instances, participants not only wish to withdraw from further research contact but wish to have all data and samples destroyed so they can no longer be used for research purposes. Investigators are permitted and, in some cases, required (i.e. when necessary to maintain the integrity of the research), to retain and analyze existing data and samples (U.S. Department of Health and Human Services 2003; U.S. Department of Health and Human Services 2008b; U.S. Department of Health and Human Services 2010). In addition, any information pertaining to the research that has been placed in the medical record cannot be removed.

The informed consent process should be documented in two forms. This is typically done by signing the consent and HIPAA form, if separate, and recording this process and the date of occurrence in the research database. Participants should be given a signed copy of the consent form for their files. While standard research practice requires a signed consent form before enrolling a research participant in a protocol, the existence of such a document does not guarantee that a participant fully understands all the ramifications of the research. To comply with the IRB approval process, consent forms and procedures must meet the facility's standards, as well as, federal and state regulations.

Vulnerable Populations

According to federal regulations, individuals belonging to vulnerable populations are afforded special protection with regard to their participation in research protocols. Such populations include children, individuals with a mental disability, prisoners, pregnant women, fetuses, and economically or educationally disadvantaged persons (U.S. Department of Health and Human Services 2009). In addition to federal laws, studies involving research participants who receive services from state-funded agencies or institutions may require additional approval by those state agencies (e.g., state department of mental retardation). Therefore, as part of the IRB review process, investigators must define the populations to be included and provide a rationale for their inclusion in genetic research protocols.

Minors

By nature of their design, genetic research studies often involve sampling entire families. While a parent or legal guardian ultimately gives permission for a child's participation in a research protocol, federal regulations suggest that children should receive an explanation of the study at an age-appropriate level and, whenever possible, should give their assent in which they affirm their agreement to participate in the study, especially if the study has no direct medical benefit to the child (U.S. Department of Health and Human Services 2009). While the validity of assent obtained from children continues to be debated, some investigators feel that children should have the opportunity to refuse to participate in a study (Callier et al. 2016). In addition, according to the specific protocol and the associated risks, limits may be placed on the amount of blood, if any, which may be drawn from a child for a research study (U.S. Department of Health and Human Services 2009). Notably, the age of consent, while generally considered 18, does vary within the US and its territories. For example, the age of consent in Puerto Rico is 21 and in Nebraska it is 19. Alabama and Massachusetts allow subjects as young as 14 years of age to consent to participate in clinical trials. The investigator should familiarize his/herself with the laws where enrollment is taking place.

The two most likely scenarios involving the sampling of minors for genetic studies include the sampling of an affected child for linkage or mutation studies, and the sampling of a child (regardless of affected status) to enable the reconstruction of parental genotypes. However, as with all research results, results should be disclosed only by CLIA-certified laboratories within the context of genetic counseling.

Persons with Cognitive Impairment

Many genetic research studies involve researching the genetic basis of disorders that affect mental and cognitive abilities in a congenital, progressive, or late-onset manner. Persons lacking mental capacity (e.g., persons with mental retardation or dementia) who are, thus, unable to give informed consent for participation in a study, are afforded special consideration. As discussed above with regard to children, in conjunction with obtaining consent from the legal guardian, assent should be sought from the potential participant whenever possible. Investigators should check with their IRB about federal and state laws for including persons with mental impairment in their research protocols.

Data and Sample Collection

Sample Collection

As required when designing a study protocol, investigators should define their methods for obtaining samples prior to initiating sample collection. See Chapter 4 for a discussion of collecting and storing samples. The study design will determine which samples should be collected. Depending on the requirements of the study, genetic research samples can be collected by the investigator at a medical clinic or obtained via field studies (visits to participants at their homes). Alternatively, sample collection kits can be mailed to research participants. In these instances, the investigator or the participant must arrange for the collection of blood or other tissue samples. The investigator should clarify in the protocol whether the participant will be reimbursed for any expenses associated with sample collection, such as blood-drawing fees, parking, and food. According to the OHRP, an institution that assists with sample collection or other services on behalf of a research project is not considered "engaged" in research as long as their staff are not enrolling participants (i.e. obtaining informed consent), the services offered are ones they typically perform for non-research purposes, the institution does not administer study interventions, and individuals providing these services will not obtain authorship privileges or professional recognition for doing so. For more details on what constitutes engagement in research, consult, OHRP Guidance on Engagement of Institutions in Human Subjects Research (http://www.dhhs.gov/ohrp/policy/engage08.html, October 16, 2008).

In addition, investigators should educate themselves about local, state, and federal laws that govern packing, marking, and labeling biological shipments. Packaging of biological samples sent through the mail or delivery service must meet requirements of the Federal Occupational Safety and Health Administration (OSHA). Biologic specimens sent via air must also be shipped according to applicable government and International Air Transport Association (IATA) regulations. Many research institutions have knowledgeable staff on hand who can be consulted regarding laws and protocols. In addition, air couriers, such as FedEx, typically have useful information and instructions on their websites.

Confirmation of Diagnosis

A good study design requires the confirmation of family history information and medical diagnoses. An accurate pedigree is critical to establishing the genetic relationships among family members necessary for genetic research. Likewise, diagnostic information on affected individuals is essential for proper analysis.

Diagnostic confirmation can be obtained in several ways depending on many factors including the type of disorder being investigated, proximity of the patients to the investigator, research funding for travel, and previous medical workup of research participants. Participants can be examined in a clinic or during a field study (see below, The Art of Field Studies). Alternatively, investigators may confirm diagnoses or traits by reviewing medical records.

For some genetic research, especially linkage studies, it is as important to rule out a diagnosis in an apparently unaffected family member as it is to confirm the diagnosis in an affected individual. This rigorous clinical evaluation may lead to the diagnosis of a genetic disorder in an unsuspecting family member who did not know that he or she was affected with the disorder being researched in the family. Before participating, research participants should be informed that there is a possibility that they will be found to have the disorder being investigated. Investigators should have a plan in advance for determining how they will handle such situations, and this should be communicated to the participant. The investigator may need to consult with the IRB to determine what is permissible. As mentioned in the section on the release of research results, the options include not informing of the status, providing an option on the consent that gives the participant a choice about being notified, and notifying regardless. If the investigator feels it is appropriate, patients interested in learning whether they have a disorder can be referred to a health care professional for clinical and follow-up care (see below, Referring for Additional Medical Services). As with all studies, unanticipated problems arise and should be dealt with on a case-by-case basis in conjunction with the investigator's IRB.

Accurate clinical diagnoses of affected individuals are crucial to performing genetic research aimed at identifying disease genes. Occasionally, a diagnosis cannot be confirmed unless additional clinical studies are done. In such instances, participants should be informed of the rationale for requesting additional tests and told how the costs of further clinical evaluations will be covered (e.g., by the investigator, by the participant, or by the participant's insurer). Again, people should be reminded of their right to refuse testing. Furthermore, participants should be informed of procedures related to reporting results of clinical testing and/or evaluations, and whether this information will be placed in the medical record. As discussed previously (under Medical Clinics), it is important for the investigator to clearly delineate the distinction between research participation and clinical care to ensure the family's continued informed and voluntary participation in the protocol. Any clinical diagnostic procedure performed explicitly for the research protocol must be approved by the IRB.

The Art of Field Studies

The field study has a long tradition in genetic research. While no formal assessment exists, many investigators attest to the success of field studies as the single most valuable practice to ensure success in family ascertainment for genetic research. Typically, investigators who do field studies travel to meet with study volunteers in their homes at a time convenient for them. This increases the efficiency of the research study by minimizing the toll on participants.

A field study takes place after a family has met the criteria for participating in the study, the diagnosis in the family has been confirmed, and all family members required for participation have agreed to participate. Detailed information about what a field study involves should be provided to each individual in the family who may participate. This information includes a meeting location (e.g., the research subject's home or another site, such as a clinical facility) and an estimate of how much time will be required for each individual and for the group.

A research team should consist of a clinician who can perform the directed physical examinations to confirm the diagnosis of the disorder being studied, an experienced phlebotomist, and an investigator who can obtain informed consent, coordinate the field visit, elicit a family history, explain the purpose of research, answer participants' questions, and administer other study-related tests. For field visits that involve working with a large family, more study personnel may be needed to facilitate the examination, sampling, and informed consent processes in a timely fashion.

In conducting a field study, it is customary to introduce the entire research team to the family and to explain what each team member will be doing, and to whom questions should be directed. The research team should try to complete the exams and sample collection in a timely fashion and allow time at the end of the visit for closing questions and comments by family members. Before leaving the family, one member of the research team (usually the genetic studies director) should define how contact and follow-up will be maintained with the family and research center. Finally, it is courteous to follow up a visit to a family with a thank-you letter that can be sent with a signed copy of their informed consent.

Interacting with families is qualitatively different from interacting with individuals and presents unique circumstances for both the families and the investigator. Visiting families in their homes allows for a more detailed discussion of family history and the family folklore surrounding a disorder. However, while some families are very open and accepting of their disease, others view it more as a stigma and treat the disorder as a taboo within the family. Thus, in addition to observing normal social conventions and being courteous when in someone's home, it is important for the investigator to anticipate and be sensitive to different attitudes of family members toward the disease in question.

Referring for Additional Medical Services

When appropriate, the study coordinator should refer individuals to other health care professionals or to clinics for additional medical services. Research participants with a family history of a genetic disorder may have specific questions about their personal reproductive risk and should be referred for genetic counseling. In addition, families with an affected child may seek community and federal services such as early intervention programs and financial compensation for medical care. The study coordinator can refer these families to a local genetic counselor, social worker, or support group to serve as a family advocate by exploring available programs. Finally, family members needing supportive counseling or additional medical or psychological care should be referred to the appropriate clinician when indicated.

Maintaining Contact with Participants

Maintaining contact with study participants provides both the investigators and the participants with several advantages. The participant is informed of pertinent information relating to a particular genetic disorder, including any breakthroughs that have resulted from the research study. The investigator maintains contact with the family, providing the potential for re-contact, re-examination, and/or re-sampling in the future. Ideally, given adequate funding and staffing, the study coordinator provides participants with yearly updates on the progress of the study as well as bulletins when research breakthroughs occur.

Contact can be maintained through a variety of ways: a follow-up letter to the families, periodic updates on the status of the research (e.g., a newsletter), and telephone calls. It is helpful to designate a contact person in each family and to direct all correspondence to that person. In addition, each person should receive the telephone number of the investigator (e.g., study coordinator) in the event that future questions arise. It is beneficial to inform families who wish to be re-contacted about their responsibility to notify the research team of any change in address or phone number so that they may receive timely reports on research progress and breakthroughs. In case contact is lost with a family, mailing a certified or registered ("return receipt requested") letter to the family at their last known address is suggested.

Future Considerations

As mentioned at the beginning of this chapter, many of the issues in genetic research studies discussed here are not completely resolved and the procedures currently in use are constantly evolving. Some of the strongest debates center around questions related to the return of research results, especially IFs, and what constitutes sufficient informed consent (and for which age groups), particularly with regard to future use of research samples. The resolution of these debates will have a major and lasting impact on the research community.

References

Angrist, M. (2011). You never call, you never write: why return of 'omic' results to research participants is both a good idea and a moral imperative. *Personalized Medicine* 8 (6): 651–657.

Bookman, E.B., Langehorne, A.A., Eckfeldt, J.H. et al. (2006). Reporting genetic results in research studies: summary and recommendations of an NHLBI working group. *American Journal of Medical Genetics. Part A* 140 (10): 1033–1040.

Callier, S.L., Abudu, R., Mehlman, M.J. et al. (2016). Ethical, legal, and social implications of personalized genomic medicine research: current literature and suggestions for the future. *Bioethics* 30 (9): 698–705.

CIOMS. Council For International Organizations of Medical Sciences (2002). International ethical guidelines for biomedical research involving human subjects. Geneva, Switzerland. *Bulletin of Medical Ethics* 182: 17–23.

Clayton, E.W., Halverson, C.M., Sathe, N.A., and Malin, B.A. (2018). A systematic literature review of individuals' perspectives on privacy and genetic information in the United States. *PLoS One* 13 (10): e0204417.

Council on Ethical and Judicial Affairs (1995). Testing children for genetic status. *Code of Medical Ethics Reports, American Medical Association* 6: 47–58.

Eisenhauer, E.R., Tait, A.R., Rieh, S.Y., and Arslanian-Engoren, C.M. (2019). Participants' understanding of informed consent for biobanking: a systematic review. *Clinical Nursing Research* 28 (1): 30–51.

Estabrooks, S., Dowdy, E., Granger, C. et al. (1999). Family ascertainment strategies in common complex genetic disorders: an example in early onset coronary artery disease. *Journal of Genetic Counseling* 8 (5 Abstract): 400.

Feliciano-Astacio, B.E., Celis, K., Ramos, J. et al. (2019). The Puerto Rico Alzheimer Disease Initiative (PRADI): a multisource ascertainment approach. *Frontiers in Genetics* 10: 538.

Genetic Alliance 1997: Informed consent: participation in genetic research studies. http://www.geneticalliance.org/sites/default/files/publicationsarchive/1997informedconsent.pdf (accessed September 2018).

GINA, Genetic Information Nondiscrimination Act of 2008. http://www.gpo.gov/fdsys/pkg/PLAW-110publ233/pdf/PLAW-110publ233.pdf (accessed September 2018).

Goodman, B. (2016). What's wrong with the right to genetic privacy: beyond exceptionalism, parochialism and adventitious ethics. In: *The Ethics of Biomedical Big Data*, 139–167. Cham: Springer.

Greely, H.T. (2007). The uneasy ethical and legal underpinnings of large-scale genomic biobanks. *Annual Review of Genomics and Human Genetics* 8: 343–364.

Green, R.C., Berg, J.S., Grody, W.W. et al. (2013 Jul). American College of Medical Genetics and Genomics. ACMG recommendations for reporting of incidental findings in clinical exome and genome sequencing. *Genetics in Medicine* 15 (7): 565–574.

Hall, J., Hamerton, J., Hoar, D. et al. (1991). Policy statement concerning DNA banking and molecular genetic diagnosis. *Clinical and Investigative Medicine* 14: 363–365.

Hann, K.E., Freeman, M., Fraser, L. et al. (2017). Awareness, knowledge, perceptions, and attitudes towards genetic testing for cancer risk among ethnic minority groups: a systematic review. *BMC Public Health* 17 (1): 503.

Hindorff, L.A., Bonham, V.L., Brody, L.C. et al. (2018). Prioritizing diversity in human genomics research. *Nature Reviews Genetics* 19 (3): 175.

Hintz, E.A. and Dean, M. (2020). Best practices for returning research findings to participants: methodological and ethical considerations for communication researchers. *Communication Methods and Measures* 14 (1): 38–54.

Hudson, K.L. (2007). Prohibiting genetic discrimination. *The New England Journal of Medicine* 356: 2021–2023.

Ibuki, T., Yamamoto, K., and Matsui, K. (2020). Differences in conceptual understanding of the "actionability" of incidental findings and the resultant difference in ethical responsibility: an empirical study in Japan. *AJOB Empirical Bioethics* 11 (3): 187–194.

Kalia, S.S., Adelman, K., Bale, S.J. et al. (2017 Feb). Recommendations for reporting of secondary findings in clinical exome and genome sequencing, 2016 update (ACMG SF v2.0): a policy statement of the American College of Medical Genetics and Genomics. *Genetics in Medicine* 19 (2): 249–255.

Kaufman, D., Murphy, J., Scott, J., and Hudson, K. (2008). Subjects matter: a survey of public opinions about a large genetic cohort study. *Genetics in Medicine* 10 (11): 831–839.

Langbaum, J.B., High, N., Nichols, J. et al. (2020). The Alzheimer's prevention registry: a large internet-based participant recruitment registry to accelerate referrals to Alzheimer's-focused studies. *The Journal of Prevention of Alzheimer's Disease* 7: 242–250.

Lee, Y.J., Arida, J.A., and Donovan, H.S. (2017). The application of crowdsourcing approaches to cancer research: a systematic review. *Cancer Medicine* 6 (11): 2595–2605.

Lowstuter, K.J., Sand, S., Blazer, K.R. et al. (2008). Influence of genetic discrimination perceptions and knowledge on cancer genetics referral practice among clinicians. *Genetics in Medicine* 10 (9): 691–698.

Majumder, M.A., Guerrini, C.J., and McGuire, A.L. (2020). Direct-to-consumer genetic testing: value and risk. *Annual Review of Medicine* 72: 151–166.

McConkie-Rosell, A. and Markel, D. (1995). Facilitating research: the many roles of the genetic counselor. Perspectives in genetic counseling (newsletter). *National Society of Genetic Counselors* 17 (1): 4.

Melvin, E., Estabrooks, S., Rotblatt, M. et al. (1999). Report of focus groups to identify the attitudes, motivations, and preferences of families participating in genetic research. *Journal of Genetic Counseling* 8 (6), Abstract: 419.

Melvin, E., Estabrooks, S., Wolpert, C. et al. (2000). Releasing research results for gene identification studies. *Journal of Genetic Counseling* 9 (6), Abstract: 524.

Moore v. Regents of University of California. (1990). (51 Cal. 3d 120; 271 Cal. Rptr. 146; 793 P.2d 479) (Sup. Ct. CA).

Murphy, J., Scott, J., Kaufman, D. et al. (2008). Public expectations for return of results from large-cohort genetic research. *The American Journal of Bioethics* 8 (11): 36–43.

National Commission For the Protection of Human Subjects of Biomedical and Behavioral Research. (1979). The Belmont Report: ethical principles and guidelines for the protection of human subjects research. http://www.hhs.gov/ohrp/humansubjects/guidance/belmont.html (accessed September 2018).

National Human Research Protections Advisory Committee (NHRPAC). *Clarification of the Status of Third Parties When Referenced by Human Subjects in Research.* 2002. http://wayback.archive-it.org/4657/20150930183542; http://www.hhs.gov/ohrp/archive/nhrpac/documents/third.pdf (accessed September 2018).

NHRPAC Working Group on Genetics. 2002. IRB Guidebook Chapter on Human Genetics Research, Draft 2 June 27, 2002. http://wayback.archive-it.org/4657/20150930183547; http://www.hhs.gov/ohrp/archive/nhrpac/documents/nhrpac13.pdf (accessed September 2018).

Palley, H.A., Bright, C.L., and Afkinich, J. (2017). DNA, privacy, and social justice: the policy-making process in the state of Maryland. *Journal of Social Welfare and Human Rights* 5 (1): 1–14.

Parker, L.S. and Lidz, C.W. (1994). Familial coercion to participate in genetic family studies: is there cause for IRB intervention? *IRB: Ethics and Human Research* 16 (1/2): 6–12.

Paul, J.L., Leslie, H., Trainer, A.H., and Gaff, C. (2018). A theory-informed systematic review of clinicians' genetic testing practices. *European Journal of Human Genetics* 26 (10): 1401–1416.

Payne, J., Williams, S., Maxwell, D. et al. (2018). Familial hypercholesterolaemia patient support groups and advocacy: a multinational perspective. *Atherosclerosis* 277: 377–382.

Plichta, J.K., Griffin, M., Thakuria, J., and Hughes, K.S. (2016). What's new in genetic testing for cancer susceptibility? *Oncology* 30 (9).

Presidential Commission for the Study of Bioethical Issues, Privacy and Progress in Whole Genome Sequencing. (2012). Washington, DC. Report Archived. https://bioethicsarchive.georgetown.edu/pcsbi/sites/default/files/PrivacyProgress508_1.pdf (accessed September 2018).

ASHG/ACMG Report (1995). Points to consider: ethical, legal, and psychosocial implications of genetic testing in children and adolescents. *American Journal of Human Genetics* 57: 1233–1241.

Riazi, S. and Kraeva, N. (2020). Mobilizing mutations: human genetics in the age of patient advocacy. *Anesthesia & Analgesia* 130 (3): e88–e89.

Roberts, M.C., Wood, E.M., Gaieski, J.B., and Bradbury, A.R. (2017). Possible barriers for genetic counselors returning actionable genetic research results across state lines. *Genetics in Medicine* 19 (11): 1202–1204.

Roden, D.M., Pulley, J.M., Basford, M.A. et al. (2008). Development of a largescale de-identified DNA biobank to enable personalized medicine. *Clinical Pharmacology & Therapeutics* 84 (3): 362–369.

Rothstein, M.A. (2010). Is deidentification sufficient to protect health privacy in research? *American Journal of Bioethics* 10 (9): 3–11.

Saelaert, M., Mertes, H., De Baere, E., and Devisch, I. (2018). Incidental or secondary findings: an integrative and patient-inclusive approach to the current debate. *European Journal of Human Genetics* 26 (10): 1424–1431.

Scaffe, A., Pollin, T., Xu, J., and Kittner, S. (2000). An evaluation of different recruitment strategies used in a family study. *Journal of Genetic Counseling* 9 (6), Abstract: 530.

Schaefer, G.O. and Wertheimer, A. (2010). The right to withdraw from research. *Kennedy Institute of Ethics Journal* 20 (4): 329–352.

Shalowitz, D.I. and Miller, F.G. (2008). Communicating the results of clinical research to participants: attitudes, practices, and future directions. *PLoS Medicine* 13;5 (5): e91.

Shrestha, B. and Dunn, L. (2019). The declaration of Helsinki on medical research involving human subjects: a review of seventh revision. *Journal of Nepal Health Research Council* 17 (4): 548–552.

Shuster, E., Markman, J.R., and Markman, M. (1997). Fifty years later: the significance of the Nuremberg Code. *The New England Journal of Medicine* 337: 1436–1440.

Soini, S. (2012). Genetic testing legislation in Western Europe—a fluctuating regulatory target. *Journal of Community Genetics* 3 (2): 143–153.

Stewart, S.H., Walitzer, K.S., Blanco, J. et al. (2019). Medication-enhanced behavior therapy for alcohol use disorder: Naltrexone, Alcoholics Anonymous Facilitation, and OPRM1 genetic variation. *Journal of Substance Abuse Treatment* 104: 7–14.

Stoeklé, H.C., Mamzer-Bruneel, M.F., Vogt, G., and Hervé, C. (2016). 23andMe: a new two-sided data-banking market model. *BMC Medical Ethics* 17 (1): 1–11.

Terry, S., Whittemore, V., Terry, P. et al. (2000). Partners in research. *American Journal of Human Genetics* 67 (4, Suppl): 243.

Title 45, Code of Federal Regulations, Part 46 (45CFR46), Protection of Human Subjects. https://www.ecfr.gov/cgi-bin/retrieveECFR?gp=&SID=83cd09e1c0f5c6937cd9d7513160fc3f&pitd=20180719&n=pt45.1.46&r=PART&ty=HTML (accessed September 2018).

U.S. Department of Health and Human Services, Food and Drug Administration, Office of the Commissioner (OC), Good Clinical Practice Program (GCPP). (2008a). Guidance for sponsors, clinical investigators, and IRBs: data retention when subjects withdraw from FDA-regulated clinical trials. http://www.fda.gov/downloads/regulatoryinforamtion/guidances/ucm126489.pdf (accessed September 2018).

U.S. Department of Health and Human Services, National Institutes of Health (NIH). 2003. Protecting personal health information in research: understanding the HIPAA privacy rule. http://privacyruleandresearch.nih.gov/pr_02.asp (accessed September 2018).

U.S. Department of Health and Human Services: Office of Human Research Protections (OHRP). 2008b. *Guidance on Research Involving Coded Private Information or Biological Specimens.* Washington, DC: Department of Health and Human Services. http://www.hhs.gov/ohrp/policy/cdebiol.html (accessed September 2018).

U.S. Department of Health and Human Services: Office of Human Research Protections (OHRP). (2010). Guidance on withdrawal of subjects from: data retention and other related issues. http://www.hhs.gov/ohrp/policy/subjectwithdrawal.html (accessed September 2018).

U.S. Department of Health and Human Services: Office of the Assistant Secretary for Health (ASH), Office of Human Research Protections (OHRP). (2018). Regulations: Human Subjects Research (45 CFR 46): Code of Federal Regulations: Title 45: Public Welfare: Department of Health and Human Services: Part 45: Protection of Human Subjects. https://www.hhs.gov/ohrp/regulations-and-policy/regulations/45-cfr-46/index.html (accessed September 2018).

Wauters, A. and Van Hoyweghen, I. (2016). Global trends on fears and concerns of genetic discrimination: a systematic literature review. *Journal of Human Genetics* 61 (4): 275–282.

Wertz, D.C., Fanos, J.H., and Reilly, P.R. (1994). Genetic testing for children and adolescents: who decides? *JAMA* 272: 875–881.

Wijmenga, C. and Zhernakova, A. (2018). The importance of cohort studies in the post-GWAS era. *Nature Genetics* 50 (3): 322–328.

Winkler, E.C. and Wiemann, S. (2016). Findings made in gene panel to whole genome sequencing: data, knowledge, ethics–and consequences? *Expert Review of Molecular Diagnostics* 16 (12): 1259–1270.

Wolf, L.E. and Beskow, L.M. (2019). Genomic databases, subpoenas, and certificates of confidentiality. *Genetics in Medicine* 21: 2681–2682.

Wolf, S.M. and Kohane, I.S. (2017). 16 The rise of patient-driven research on biospecimens and data: the second revolution. In: *Specimen Science: Ethics and Policy Implications* (eds. H.F. Lynch, B.E. Bierer, I.G. Cohen and S.M. Rivera), 317. The MIT Press.

Wong, C.A., Hernandez, A.F., and Califf, R.M. (2018). Return of research results to study participants: uncharted and untested. *JAMA* 320 (5): 435–436.

Wright, K.M., Rand, K.A., Kermany, A. et al. (2019). A prospective analysis of genetic variants associated with human lifespan. *G3: Genes, Genomes, Genetics* 9 (9): 2863–2878.

Wynn, J., Lewis, K., Amendola, L.M. et al. (2018). Clinical providers' experiences with returning results from genomic sequencing: an interview study. *BMC Medical Genomics* 11 (1): 45.

6

Linkage Analysis

Susan H. Blanton

Dr. John T. Macdonald Foundation Department of Human Genetics, University of Miami Miller School of Medicine, Miami, FL, USA

As discussed in Chapter 2, genes located near each other tend to be transmitted together to offspring. When there is variation at both of the genes, i.e. at least one parent is heterozygous at both loci (a double heterozygote), it is possible to track the cosegregation of the specific alleles through the pedigree. If we know the location of one of the genes, we can use this to determine the location (or map) the other gene; this latter gene is typically one that gives rise to a disease phenotype. This concept is known as linkage analysis. While disease gene mapping in humans is relatively recent, T. H. Morgan mapped several phenotypes in *Drosophila melanogaster* as early as 1911 (Morgan 1911). The possibility of a genetic map was first formally investigated by A. H. Sturtevant, who ordered five markers on the X chromosome in Drosophila and then estimated the relative spacing among them (Sturtevant 1921).

In its simplest form, linkage analysis involves the counting of the number of recombinant and nonrecombinant chromosomes and comparing the distribution to that expected under random assortment (Table 6.1). In experimental organisms, the mating types can be dictated, which allows for greater control. When studying diseases in humans, this is, of course, not possible. Therefore, maps of markers in humans have been developed by means of one of several statistical algorithms used in computer programs such as CRIMAP and MAPMAKER (Lander and Green 1987), CLINKAGE and MULTIMAP (Matise et al. 1994), and MAP-O-MAT (Matise and Gitlin 1999). Genetic maps can assume equal recombination between males and females or can allow for sex-specific differences in recombination since it has been well established that there are substantial differences in recombination frequencies between men and women; on average, the female autosomal map is 1.66 times the male autosomal map (Li et al. 1998; Nachman 2002; Cheung et al. 2007). These maps are generally produced utilizing a single set of reference pedigrees, such as those developed by the Centre d'Etude du Polymorphisme Humain (CEPH) (Dausset et al. 1990), which are mostly composed of three-generation pedigrees with a large number of offspring (average 8.5). Both sets of maternal and paternal grandparents are usually available, so linkage phase (which alleles are on the same chromosome) frequently can be established. The collection of CEPH pedigrees, in its entirety, consists of more than 60 pedigrees and includes more than 600 individuals; DNA from this valuable resource is available, however, through the Coriell Institute for Medical Research. The

This revised chapter contains material contributed by David Goldgar, Jonathan Haines, Elizabeth Hauser, Jason Moore, and Silke Schmidt, in the previous edition of this book.

Genetic Analysis of Complex Diseases, Third Edition. Edited by William K. Scott and Marylyn D. Ritchie.

Table 6.1 Example development of genetic map using four linked Loci *A*, *B*, *C*, and *D*, scored in 100 Offspring[a].

A. Score Recombination Events between All Pairwise Combinations of Four Loci

Loci scored	Number of recombinants	Frequency of recombination
A–B	10	0.10
A–C	3	0.03
A–D	15	0.15
B–C	7	0.07
B–D	5	0.05
C–D	12	0.12

B. Process

1) Determine which two loci have the highest frequency of recombination between them; these two loci are the farthest apart on the map.
2) Fit the other loci into the map like pieces of a puzzle.

C. Resulting Genetic Map

Percent recombination loci order		0.03		0.07		0.05	
	A		*C*		*B*		*D*

[a] This example demonstrates no evidence for interference. Positive interference would be manifested as a decrease in overall map length from what would be expected by adding the pairwise distances. In this example, if $(A–C)+(C–B)+(B–D)>A–D$, positive interference may be present.

complexity of the underlying statistical methods used to generate genetic maps renders them sensitive to marker genotyping errors, particularly in small intervals (Buetow 1991), and these maps are less useful in small regions of less than about 2 cM. While marker order is usually correct, genotyping errors can result in falsely inflated estimates of map distances.

The first maps consisted largely of blood protein polymorphisms. Morton and colleagues (Keats et al. 1979) assembled them into linkage groups on chromosomes. There were only approximately 80 linkage groups, and there was not much success in mapping with them as they were widely spaced and the polymorphisms were generally biallelic with a low heterozygosity. In addition, a number of the loci were either not mapped or placed on the incorrect chromosome. Restriction-fragment length polymorphisms, or RFLPs, were discovered in the 1970s (Kan and Dozy 1978; Wyman and White 1980). RFLPs are generated by treating DNA with bacterial enzymes that recognize specific DNA sequences (these are typically palindromic sequences) and cut the DNA if the sequence is an exact match. A change in the sequence within the palindromic region will cause the enzyme to not recognize it and thus not cut the DNA at that point. Examples of enzymes are EcoRI (from *Escherichia coli*) which recognizes $\frac{\text{GAATTC}}{\text{CTTAAG}}$ and HinfI (from Haemophilus influenza) which recognizes $\frac{\text{GANTC}}{\text{CTNAG}}$. While still biallelic, there were many more RFLPs than blood polymorphisms and they often had higher heterozygosities (proportion of individuals who are heterozygous for the bases, e.g. A/C or C/G) than the blood proteins. A disadvantage was the genotyping method required the use of radioactivity and x-ray exposure, which could sometimes take several weeks. VNTRs (variable number of tandem repeats) were the next class of markers developed

(Jeffreys et al. 1985). The VNTRs used in linkage analysis had repeat sizes in the range of 15–70 base pairs. There were many more of these than RFLPs and they were multiallelic; both characteristics made them much more useful for linkage studies, as discussed below. However, they were often difficult to interpret, e.g. deciding between patterns on electrophoresis gels corresponding to different repeat sizes, such as 10 repeats vs. 11 repeats. As with RFLPs, VNTRs also required the use of radioactivity. In the late 1980s, STRs (single tandem repeat), or microsatellite repeats, were discovered (Weber and May 1989). These were repeats of 2–4 base pairs with typically 5–10 alleles and could be detected with either radioactivity or silver staining. Lastly, maps of SNPs, or single nucleotide polymorphisms (some of these were the original RFLPs), were developed; SNPs occur, on average, every 100–300 base pairs so that there are millions throughout the genome (Sachidanandam et al. 2001). However, they, like their RFLP predecessor, are largely biallelic and thus fairly uninformative individually for linkage studies.

Disease Gene Discovery

Disease gene discovery is greatly facilitated by the availability of these dense genetic maps. While linkage analysis for the localization of disease genes boils down to the "simple" idea of counting recombinants and nonrecombinants, in humans this process is complicated for a variety of reasons. For example, the generation time is long in humans such that large, multigenerational pedigrees in which a disease or trait is segregating in living individuals are rare; scientists cannot dictate matings or exposures, nor can they require participation of specific individuals in a study. Thus, the process of linkage analysis in humans requires a statistical framework in which various hypotheses about the linkage of a trait locus and marker locus can be considered in an observational study design.

Linkage analysis has traditionally been performed using either a parametric approach or a nonparametric approach. Parametric approaches require the assumption of a genetic model and the specification of various genetic model parameters such as disease allele frequency and penetrance. In contrast, nonparametric methods assume that the disease has a genetic basis, but do not require specification of a genetic model and thus do not suffer from the potentially hazardous effects of model misspecification. When the genetic model for a disease is clearly known, parametric-based methods are preferred over nonparametric approaches. The four major advantages of the parametric method are (i) statistically, it is a more powerful approach than any nonparametric method if the assumed genetic model is approximately correct. (ii) It utilizes every family member's phenotypic and genotypic information. (iii) It provides an estimate of the recombination fraction between marker and disease locus. (iv) It provides a statistical test for linkage and for genetic (locus) heterogeneity.

Example 1 Parametric Linkage Analysis in Pedigree with Unlinked Marker

For illustrative purposes, consider the pedigree in Figure 6.1a. In this pedigree, shaded individuals are affected with disease and unshaded individuals are unaffected. Results for a genetic marker are shown underneath each individual. In pedigrees in which the genetic model can be deduced with a high degree of certainty, linkage analysis can be broken down into five steps.

It should be emphasized that this example is a simplified version of an extremely complicated process. Also, before beginning, it is imperative that careful attention be paid to phenotyping all individuals. This requires a thorough understanding of the disease and the range of its expression. Many linkage projects have been led astray by the misclassification of an individual.

(Continued)

Example 1 (Continued)

Step 1: State Components of Genetic Model. For parametric linkage analysis, the genetic model must be specified. Components of the genetic model include the inheritance pattern of the disease locus (autosomal or X-linked; dominant, recessive, or codominant), disease allele frequency and penetrance (probability of affection given the genotype), and frequency of phenocopies (affected individual does not have a risk genotype) and new mutations. Rough estimates of the disease allele frequency and penetrance can often be obtained from the literature or from computer databases such as Online Mendelian Inheritance in Man (https://www.ncbi.nlm.nih.gov/omim); estimates of the rate of phenocopies and new mutation are frequently guesses, included as a nuisance parameter in some cases to allow for the fact that these can exist. Linkage analysis using log of the odds (LOD) scores is relatively robust to modest misspecification of the disease allele frequency and penetrance, but misspecification of whether the disease is dominant or recessive can lead to incorrect conclusions of linkage or nonlinkage (Clerget-Darpoux et al. 1986). In this example, the disease allele will be assumed to be rare and to function in an autosomal dominant fashion with complete penetrance; the disease locus will be assumed to have two alleles, N (for normal or wild type) and A (for affected or disease). In addition, mutation and phenocopies are assumed to be absent. These assumptions allow substantial simplification of the problem, as outlined further below.

In addition to specifying the parameters of the disease locus, frequencies for the alleles at the marker locus are required. In pedigrees in which genotypes are missing in founding individuals (due either to an unsampled individual or to laboratory complications), the misspecification of allele frequencies can have substantial impact, leading to incorrect conclusions of linkage or nonlinkage and biased estimates of the recombination fraction (Knowles et al. 1992; Ott 1992). Other factors can also wreak havoc in a linkage analysis. Such factors include linkage heterogeneity and scoring errors in the pedigree structure, diagnostic status, or marker genotyping, which are discussed later in this chapter.

Step 2: Assign Underlying Disease Genotypes Given Information in Genetic Model (Figure 6.1b). In preparation for scoring recombinant and nonrecombinant individuals in a pedigree, the relationship between genotype and phenotype as defined by the genetic model can be used to assign the underlying genotype of pedigree members at the disease locus. The assumption of complete penetrance of the disease allele allows all unaffected individuals in the pedigree to be assigned a disease genotype of NN. Since the disease allele is assumed rare, the disease genotype for affected individuals can be assigned a disease genotype of NA. In other words, since the disease allele is rare, the chance that an affected individual is homozygous for the disease allele is so small that, for the purposes of this example, it can be considered to be zero.

Again, it is important to emphasize that this example is a simplified version; most linkage analysis is performed by computer analysis that allows the consideration of the small probability that a founder in such a pedigree (e.g. individual I-1) is homozygous for the disease allele. When a computer program performs this analysis, it assigns probabilities for genotypes AA and NA in individuals like I-1 by using user-specified information on disease allele frequencies.

Step 3: Determine Putative Linkage Phase. Individual II-1 has inherited the disease trait together with marker allele 2 from his affected mother. Thus, the A allele at the disease locus and the 2 allele at the marker locus were inherited in the gamete transmitted to II-1. There are two mutually exclusive hypotheses to consider: The null hypothesis is that the disease and marker loci are unlinked to one another. If the loci are genetically unlinked, there will be an approximately equal number of recombinant and nonrecombinant gametes among the offspring of II-1. The alternate hypothesis is that the disease and marker loci are linked to one another. If the loci are genetically linked, there will be more nonrecombinants than recombinants among the offspring ("meiotic events") of II-1. The basic idea is to consider these two competing hypotheses of linkage versus nonlinkage and determine which hypothesis best describes the available data from the pedigree.

Thus, the putative linkage phase (the disease allele "segregates" with marker allele 2) has been established, and this phase can be tested in subsequent generations.

Step 4: Score Meiotic Events as Recombinant or Nonrecombinant. For this mating type, there are four possible gametes from the affected parent II-1: N1, N2, A1, and A2. Based on the putative linkage phase assigned above in step 3, gametes A1 and N2 are recombinant. Thus, all affected offspring of II-1 and II-2 who have inherited marker allele 2 from their father will be scored as nonrecombinant for the disease and marker; affected offspring who have inherited the 1 allele will be scored as recombinant for the disease and marker. Similar reasoning applies to the unaffected offspring, except that the unaffected offspring who have inherited allele 1 are non-recombinant and those who have inherited allele 2 are recombinant.

In this pedigree (Figure 6.1c), five offspring of II-1 are recombinant and five are nonrecombinant. Thus, out of 10 scorable meiotic events, the number of recombinant gametes is equal to the number of nonrecombinant gametes. These data are consistent with the hypothesis of nonlinkage between the disease and marker loci.

One frequent question arises – why is the transmission of the 2 allele from the affected grandmother to the affected son not counted as a meiotic event? Or, why are there not 11 instead of 10 meioses in this pedigree? The answer is that we do not know the linkage phase in individual II-1; we are just using the transmission from his affected mother to him to determine our hypothesis about what the linkage phase would be if the disease and marker loci were linked to one another.

Step 5: Calculate and Interpret LOD Scores. The pedigree data are used to consider both the hypothesis of linkage between the disease and marker locus and nonlinkage between the disease and marker locus. Morton (1955) suggests a likelihood ratio approach in which the likelihood of the pedigree and marker data are calculated under the null hypothesis that assumes free recombination ($\theta = 0.5$) between the disease and marker loci, where θ represents the recombination fraction, and then is compared to the likelihood of the hypothesis of linkage between the disease and marker loci. This likelihood of the pedigree data under the hypothesis of linkage between the disease and marker loci is calculated at various increments of $\theta < 0.5$ within the range of allowable values (0.00–0.49), representing unique subhypotheses of linkage between the disease and marker loci. The likelihood ratio (LR) is constructed as $L(\text{pedigree} \mid \theta = x)/L(\text{pedigree} \mid \theta = 0.5)$, where x is some value of θ. The likelihood of observing the pedigree data is just $\theta^{R}(1-\theta)^{NR}$, where R and NR are the number of recombinants and nonrecombinants, respectively, for the two phases and N is the total number of scored offspring (R + NR = N); this likelihood is the numerator in the ratio of likelihoods. The denominator is the likelihood assuming that the two loci are unlinked (i.e. when $\theta = 0.5$). Thus, the ratio of the likelihoods of the two hypotheses is constructed as

$$LR = \frac{\theta^{R}(1-\theta)^{NR}}{0.5^{R}(0.5)^{NR}}$$

And reduces to

$$LR = \frac{\theta^{R}(1-\theta)^{NR}}{0.5^{N}}$$

(*Continued*)

Example 1 (Continued)

Typically, the base-10 logarithm of this ratio is taken to obtain an LOD (Log of the ODds) score [Note: $Z(\theta)$ is sometimes used to denote a LOD score; Z_{max} is often used as the maximum LOD score]. An LOD score of 3 indicates odds of 10^3:1 (1000:1) in favor of linkage and is considered to be conclusive evidence for linkage between two markers (or a marker locus and disease

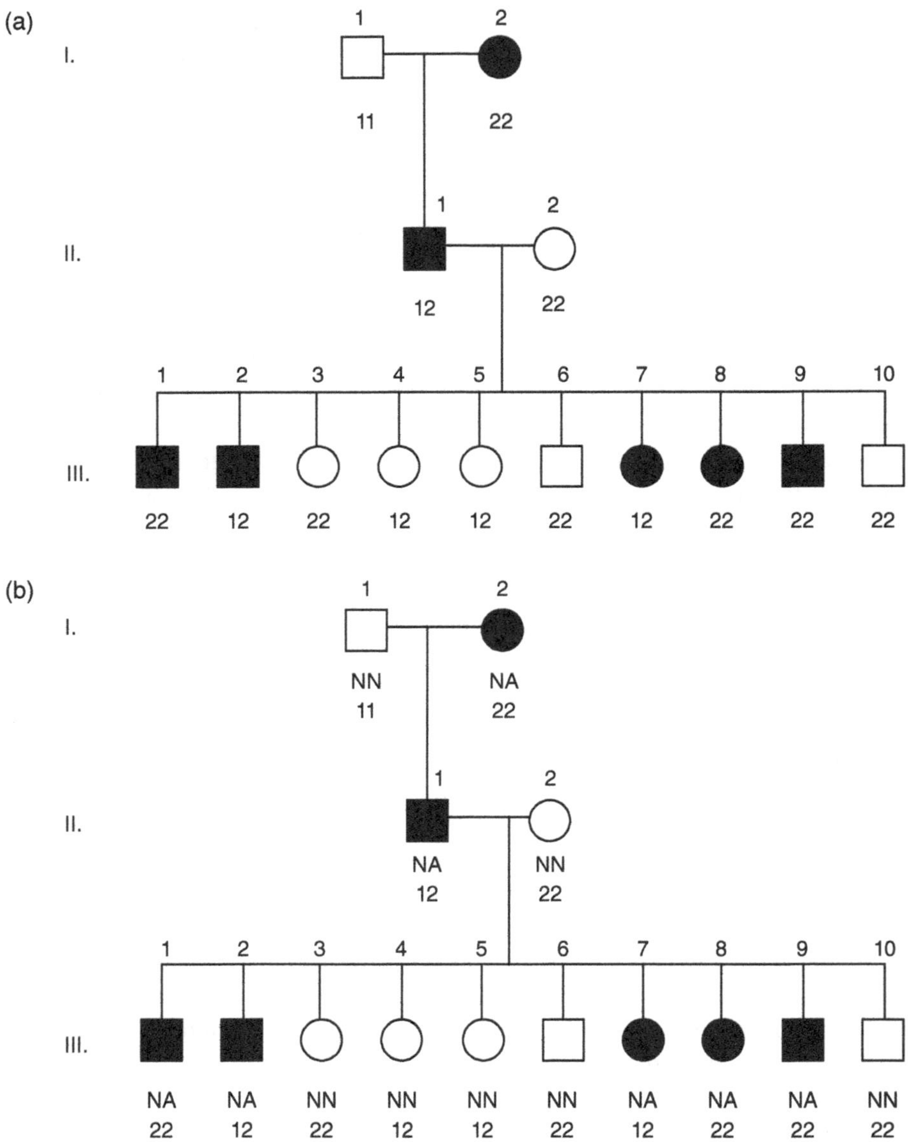

Figure 6.1 (a) Pedigree in which a rare, fully penetrant autosomal dominant disease is segregating. Genotypes are shown beneath pedigree symbols. (b) Putative disease genotype listed beneath pedigree figure. (c) Assignment of disease-associated haplotypes. Results on the left of the bar were transmitted from the male parent and those on the right of the bar were transmitted from the female parent. In the third generation, meioses are scored as recombinant (R) or nonrecombinant (NR). Since the number of recombinant meioses is equal to the number of nonrecombinant meioses, this pedigree is consistent with nonlinkage between the disease and marker loci.

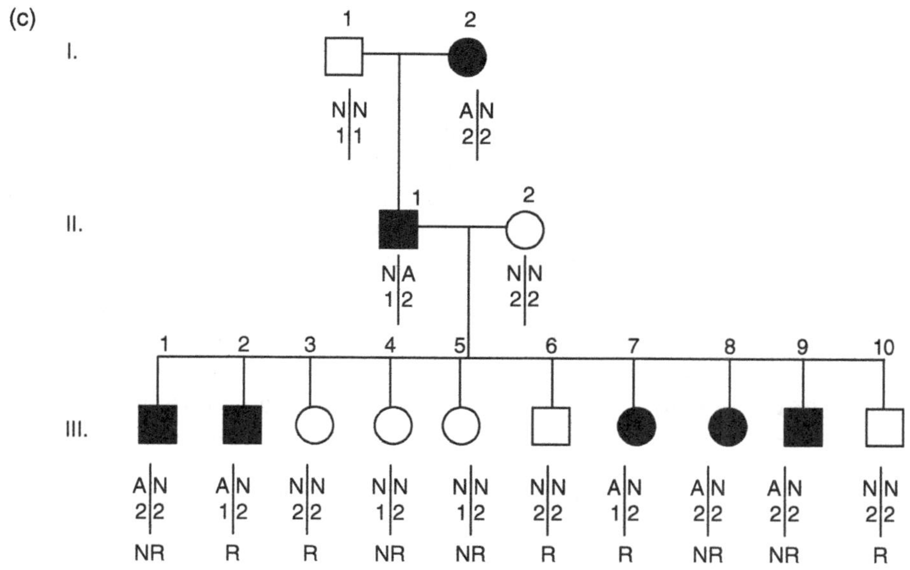

Figure 6.1 (Continued)

locus) in most cases. An LOD score of −2 indicates odds of 10^{-2}:1 (0.01:1) in favor of linkage, or more commonly, odds of 100:1 against linkage of the two markers. An LOD score of −2 or less is considered to be conclusive evidence that the two markers under study are unlinked *at the specified recombination fraction*. It is important to note that this "exclusion" of a disease gene from a region is only valid under the specific set of assumptions made for the analysis. For example, if an analysis was made assuming an autosomal dominant trait, but the trait was actually autosomal recessive, it could be falsely excluded from a region. LOD scores between the values of −2 and 3 are considered inconclusive and warrant additional study. A study is "expanded" by rendering the currently available family data more informative (i.e. testing the family with different or more informative markers), collecting and testing additional family members, or by increasing the number of families under study. An LOD score should always be considered in conjunction with its respective estimate of θ.

It is important to note that using an LOD score of 3.0 (odds 1000:1 in favor of linkage) as a test statistic does equal a type 1 error rate (α) of 0.001, which is much more stringent than the standardly used type I error rate of 0.05 in most statistical analyses. Because of the prior probability of linkage between two traits, the true p-value associated with an LOD score of 3.0 is approximately 0.04, consistent with a false-positive rate of 1 in 25. In other words, when evidence for linkage between two loci is declared significant at an LOD score of 3.0, there is a 4% chance it is a spurious positive result (for more detail, see Ott 1999).

Table 6.2, Example 1, shows the two-point LOD scores for the marker at a variety of hypotheses about the estimate of the recombination fraction between the disease and marker locus. In this example, the highest LOD score is −0.09 at $\theta = 0.40$ and at no value of θ is the LOD score positive, let alone ≥3.0, so this pedigree demonstrates no evidence in favor of linkage between the disease and marker loci. However, the pedigree does provide important information about where the disease locus is not located. Visual inspection of the LOD score data suggests that

(*Continued*)

Example 1 (Continued)

the value of θ at which the LOD score exceeds −2.0 is between 0.10 and 0.15, so approximately 13 cM on either side of the marker locus can be excluded as harboring the disease gene, for a total exclusion of 26 cM as a result of typing this marker!

There are a number of well-tested computer programs, including the LINKAGE computer package (Lathrop et al. 1984), FASTLINK (Schaffer et al. 1994; Cottingham et al. 1993), VITESSE (O'Connell and Weeks 1995), GENEHUNTER (Kruglyak et al. 1996), ALLEGRO (Gudbjartsson et al. 2000), and MERLIN (Abecasis et al. 2002), which are often utilized for calculation of LOD scores.

Table 6.2 LOD scores for pedigrees in Examples 1, 2, and 3.

	$\theta = 0.01$	$\theta = 0.05$	$\theta = 0.10$	$\theta = 0.15$	$\theta = 0.20$	$\theta = 0.30$	$\theta = 0.40$
Example 1	−7.01	− 3.61	−2.22	−1.46	−0.97	−0.39	−0.09
Example 2	0.97	1.51	1.60	1.55	1.44	1.09	0.62
Example 3	0.99	1.21	1.30	1.25	1.14	0.79	0.33

Example 2 Linkage Analysis in Pedigree with Linked Marker

Consider the pedigree in Figure 6.2a.

Step 1: State Components of Genetic Model. Assume the genetic model is the same as in Example 1 (rare, autosomal dominant disease allele with complete penetrance, no mutation, no phenocopies).

Step 2: Assign Underlying Disease Genotypes Given Information in Genetic Model. The assignment of disease genotypes to pedigree members is the same as in Example 1 (Figure 6.2b). Unaffected individuals have an underlying disease genotype of NN and affected individuals have a disease genotype of NA.

Step 3: Determine Putative Linkage Phase. The assignment of putative linkage phase is identical to that in Example 1, so that the disease allele is transmitted in the same gamete as marker allele 2.

Step 4: Score Meiotic Events as Recombinant or Nonrecombinant (Figure 6.2c). In this example, all five of the affected children have inherited marker allele 2 from their affected father and are thus nonrecombinant. In addition, four of the five unaffected children have inherited marker allele 1 from their affected father. These four offspring, too, are nonrecombinant with respect to the disease and marker loci. Individual III-6, however, is unaffected and has inherited marker allele 2 from his affected father. This individual is scored as a recombinant individual (a recombinant meiosis). This nonrandom segregation of the disease allele and a marker allele within a pedigree is consistent with linkage between the disease and marker locus.

Step 5: Calculate and Interpret LOD Scores. The LOD scores for this pedigree at a variety of hypotheses about the recombination fraction between the disease and marker loci are shown in Table 6.2, Example 2. No LOD score at any tested recombination fraction is higher than 3.0, so this pedigree does not provide significant evidence in favor of linkage between the disease and marker locus. The highest LOD score in this pedigree is 1.6, indicating that more data from additional pedigrees need to be generated in order to draw firm conclusions about linkage. However, this LOD score is probably "interesting" enough that it would indeed warrant such follow-up. Here, it is critical to note that the recombination event occurs in an unaffected individual. If the assumption of complete penetrance of the disease allele is incorrect (for example,

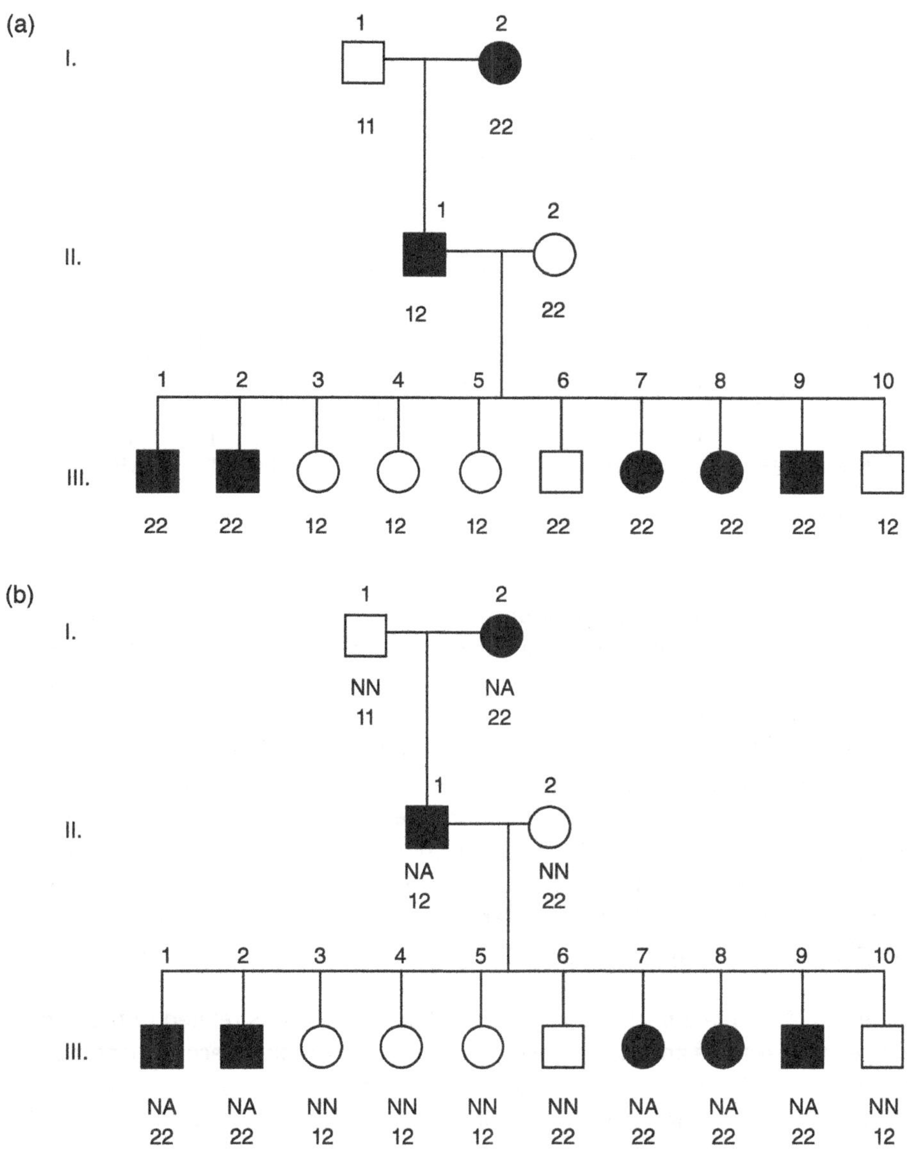

Figure 6.2 (a) Pedigree in which a rare, fully penetrant autosomal dominant disease is segregating. Genotypes are shown beneath pedigree symbols. (b) Putative disease genotype listed beneath pedigree figure. (c) Assignment of disease-associated haplotypes. Results on the left of the bar were transmitted from the male parent and those on the right of the bar were transmitted from the female parent. In the third generation, meioses are scored as recombinant (R) or nonrecombinant (NR). Since the number of nonrecombinant gametes is larger than the number of recombinant gametes, this pedigree is consistent with linkage between the disease and marker loci.

(*Continued*)

Example 2 (Continued)

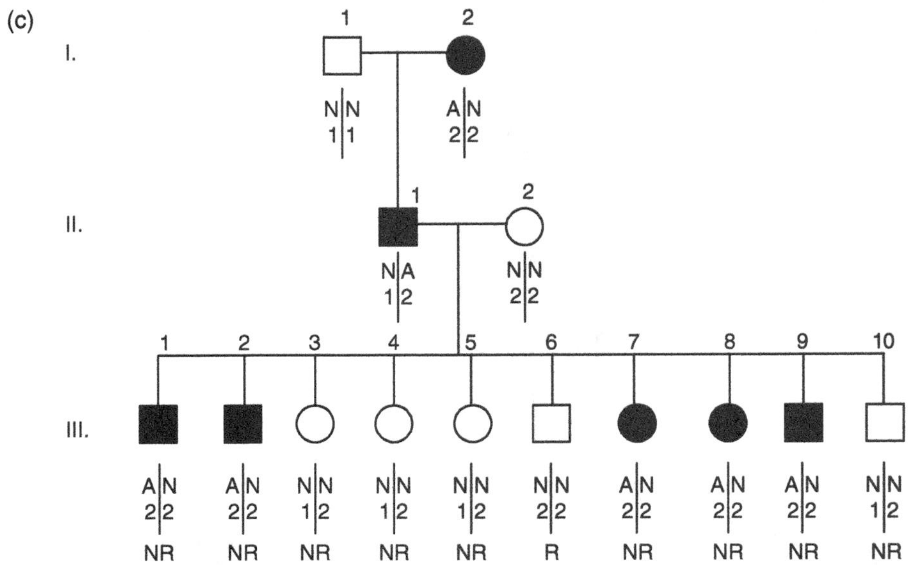

Figure 6.2 (Continued)

if the individual is younger than the age at onset for an age-dependent trait and thus is not yet expressing it), the assignment of this meiotic event as a "recombinant" could be in error. Thus, it is incumbent on the investigator to weigh carefully information about disease gene localization gleaned from unaffected family members.

The maximum-likelihood estimate (MLE) of the recombination fraction is that value of θ at which the LOD score is largest. It is this value of the LOD score (the highest LOD score) with which the determination of statistical significance is made. In this case, the maximum LOD score is 1.60, which occurs when $\theta = 0.10$, and so 0.10 is the MLE for θ in this example. This result does not meet the established criterion of odds ⩾1000 : 1 (a LOD score ⩾ 3) for concluding evidence for linkage, yet it is interesting in and of itself and perhaps merits further investigation by the genotyping of additional tightly linked markers or additional families.

An intuitive, direct estimate of the recombination fraction, which is related to the genetic distance between the disease and marker locus, is just the proportion of recombinant meioses counted among total meioses scored. In this example, the direct estimate of the recombination fraction, based on the observation of 1 recombinant out of 10 scorable meioses, is 1/10, or 10%. This direct approach is possible in this example because these are phase-known events so that all recombinants and nonrecombinants can be counted.

When significant evidence for linkage is found, a formal presentation of results of a linkage analysis is incomplete without an indication of a support interval for the MLE of θ. A description of the calculation of these support intervals, usually performed using the "one LOD score down" method, can be found in either Ott (1999) or Conneally et al. (1985). In this approach, all values of the recombination fraction that fall within one LOD score below the highest attained LOD score are considered within the support interval.

Example 3 Linkage Phase Unknown

Next consider an example in which the grandparents are unavailable (Figure 6.3). In this pedigree, the linkage phase cannot be established with certainty; however, only one of two possible linkage phases is the truth. Under phase 1, the disease allele is segregating with the 1 allele. Alternatively, the disease allele could be segregating with the 2 alleles (phase 2). Under each of these alternate scenarios, it can be determined whether or not the offspring – these 10 scorable meioses – are recombinant or nonrecombinant gametes. In this example, the LOD scores are calculated a little differently:

$$\text{LOD} = \log_{10}\left[\frac{1}{2}\left(\underbrace{\frac{\left(\theta^{R}(1-\theta)^{NR}\right)}{(\theta=0.5)^{N}}}_{\text{phase 1}} + \underbrace{\frac{\left(\theta^{R}(1-\theta)^{NR}\right)}{(\theta=0.5)^{N}}}_{\text{phase 2}}\right)\right]$$

where R and NR are the number of recombinants and nonrecombinants for the two phases. Note that the uncertainty associated with the unknown linkage phase reduces the overall linkage information. The highest LOD score obtained in this pedigree is 1.3 (Table 6.2), as opposed to the highest LOD score of 1.6 in the earlier pedigree. In effect, the likelihood ratio has been reduced by approximately half that of the phase known pedigree ($10^{1.6} = 39.81$; $10^{1.3} = 19.95$). Although this may not seem like a substantial loss of information, it does influence the results when working with pedigree samples that are large enough to give evidence in favor of linkage. If that much information is lost on a series of 10 pedigrees, a substantial amount of potential linkage information has been lost. The novice mapper is often tempted to disregard the case of the disease segregating with the 1 allele, since it results in many more recombinant chromosomes than nonrecombinant ones. However, this is a circuitous argument and while children may be used to infer parental genotypes, they may not be used to infer parental phase.

Additional Notes. Note that in the pedigree in Figure 6.2c, the 2 allele is segregating with all affected individuals and with only one of five unaffected individuals. One might characterize this observation as an "association" between the 2 allele and the disease allele within this family. By definition, two traits that are linked to one another will show an association within a family, but the associated allele may vary between families linked to the same locus.

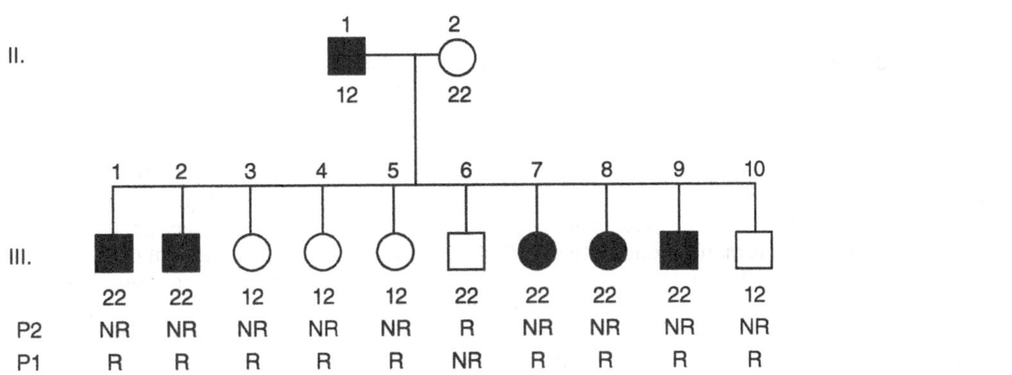

Figure 6.3 Pedigree for Example 3 for calculation of LOD score for linkagephase-unknown pedigree.

Ability to Detect Linkage

For the meiotic events of a parent to be scorable for linkage analysis, the parent must be heterozygous at both loci (disease and marker) under consideration, i.e. a "double heterozygote" (Figure 6.4). The family studied in the above examples is large and relatively atypical for a human pedigree since all family members are available for sampling. How much information is required to obtain significant results from a linkage analysis? The answer to this question can be complicated and highly dependent on the inheritance pattern of the disease (i.e. autosomal or X-linked, dominant or recessive), the penetrance of the disease, whether sporadic cases of the disease (phenocopies) may be present, and other factors. As a guideline, each scorable, phase-known meiotic event (with no recombination between the trait and marker locus) contributes about 0.30 to the LOD score. Table 6.3 gives some additional information about the contribution of each meiotic event to the LOD score at a variety of recombination fractions. Thus, at a minimum, 10 phase-known meiotic events demonstrating no recombination between the trait and marker locus are required to obtain an LOD score of 3.0. These can be from multiple families, although that may bring in other issues, or from a single large family. In general, human pedigrees have few phase-known meiotic events since some individuals may be unable or unwilling to participate in the study. Computer simulation studies can be performed to assess the power of an available data set to detect linkage under an assumed genetic model prior to initiating a screen of the entire genome to detect linkage.

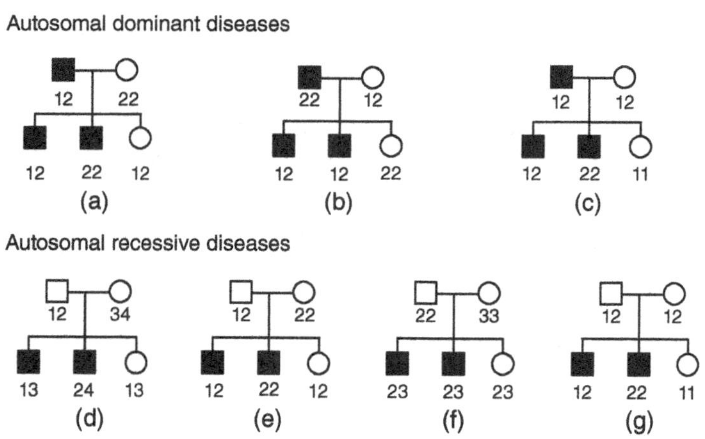

Figure 6.4 Pedigree examples demonstrating families that are informative and noninformative for linkage analysis. Both the autosomal dominant and autosomal recessive disease are assumed to be fully penetrant. Shaded individuals are affected with the disease, unshaded individuals are unaffected. Marker results are indicated beneath each individual. Autosomal dominant diseases: The affected parent, who is by definition homozygous at the disease locus, must also be heterozygous at the marker locus. (a) Fully informative pedigree: All three offspring can be scored for linkage analysis. (b) Uninformative pedigree: Cannot tell which 2 allele is transmitted with the disease allele. (c) Partially informative pedigree (an "intercross"): Homozygous offspring of this mating contribute significantly to linkage analysis. Autosomal recessive diseases: For pedigrees to be fully informative, both parents, who are heterozygous at the disease locus by virtue of the fact that they have at least one affected child with a recessive disease, must also be heterozygous for different alleles at the marker locus. (d) Fully informative pedigree: Both paternal and maternal gametes can be scored with respect to disease and marker locus. (e) Partially informative pedigree: Only paternally contributed gametes contribute to linkage analysis. (f) Uninformative pedigree: Both parents are homozygous at the marker locus. (g) Partially informative pedigree: Homozygous offspring contribute significantly to the linkage analysis.

Table 6.3 Number of phase-known, fully informative meioses needed to detect linkage at various values of recombination fraction θ.

θ	Number of informative Meioses needed for LOD = 3	Expected LOD per Meiosis
0.00	10	0.30
0.01	11	0.28
0.02	12	0.26
0.05	14	0.21
0.10	19	0.16
0.20	36	0.08
0.25	53	0.06

Several programs including SIMLINK and FASTSLINK are available for performing these power studies (Boehnke 1986; Ott 1989; Ploughman and Boehnke 1989; Weeks et al. 1990a). Also see more details in Chapter 12.

Real World Example of LOD Score Calculation and Interpretation

We will illustrate LOD score calculation with an example linkage study of Alzheimer disease (AD) and markers on chromosome 19*q*. We use a data set that includes pedigrees from the original paper reporting linkage of AD to this chromosomal region (Pericak-Vance et al. 1991), which subsequently led to the identification of the apolipoprotein E (*APOE*) gene as a susceptibility gene for late-onset AD (Corder et al. 1993). Although the mode of inheritance for AD is unknown, we assume an autosomal dominant model allowing for phenocopies and incomplete penetrance in this example. The disease allele frequency is assumed to be $q = 0.001$, which corresponds to a frequency of approximately 0.2% for people with one or two copies of the disease allele (i.e., $[2q(1-q)+q^2] = [(2\times0.001\times0.999)+0.001^2] = 0.002$, under Hardy–Weinberg equilibrium; see Chapter 2 for a discussion of Hardy–Weinberg equilibrium). The population allele frequencies for the marker, D19S246, were calculated from a representative set of unrelated non-Hispanic, European ancestry individuals (164 chromosomes). The penetrance for carriers of the disease allele (f_{DD}, f_{Dd}) is assumed to be 0.80, and the penetrance for individuals without the disease allele (phenocopy frequency, f_{dd}) is assumed to be 0.01. With these parameter choices, the proportion of nongenetic cases among all affected individuals is approximately 86% (i.e. $[0.01\times0.999^2]/[(0.01\times0.999^2)+(0.80\times0.002)] = 0.862$). For a single-gene disorder, the population prevalence should approximately match the prevalence φ derived from the assumed disease allele frequency and penetrance values (e.g. $\varphi = [(q^2\times f_{DD})+(2q[1-q]\times f_{Dd})+([1-q]^2\times f_{dd})]$). However, for a complex and common disease such as AD, with a prevalence of approximately 10% in adults over 65 years of age, it is not realistic to try to achieve this relationship. More appropriate ways to analyze such complex phenotypes with the genetic model-based LOD score approach are discussed later in this chapter.

Table 6.4 presents the two-point linkage analysis for AD and marker D19S246 under the above dominant model. The highest LOD score is 1.57 at $\theta = 0.15$. Using the classical criterion for linkage ($Z_{max}(\theta)\geq3.0$), one can easily conclude that there is no significant evidence for linkage under this genetic model.

Table 6.4 Two-point linkage analysis for Alzheimer disease and D19S246.

	Recombination fraction (θ)						
Pedigree	**0.00**	**0.05**	**0.10**	**0.15**	**0.20**	**0.30**	**0.40**
401	0.016	0.013	0.010	0.008	0.006	0.002	0.001
701	−0.972	−0.445	−0.242	−0.129	−0.062	−0.004	0.003
736	1.010	0.893	0.773	0.651	0.527	0.282	0.081
757	0.297	0.301	0.286	0.258	0.218	0.123	0.035
763	0.225	0.191	0.158	0.126	0.096	0.045	0.012
794	−0.473	−0.224	−0.092	−0.016	0.026	0.045	0.021
820	0.058	0.108	0.133	0.141	0.135	0.094	0.035
911	0.039	0.030	0.023	0.017	0.012	0.005	0.001
1086	0.163	0.133	0.105	0.081	0.059	0.027	0.007
1207	−0.568	−0.196	−0.063	−0.003	0.023	0.025	0.008
1229	−0.367	−0.217	−0.120	−0.058	−0.022	0.002	0.000
1396	0.230	0.194	0.160	0.128	0.097	0.046	0.012
1399	0.822	0.719	0.611	0.500	0.387	0.179	0.040
1491	0.046	0.056	0.069	0.081	0.086	0.080	0.053
1547	0.081	0.063	0.048	0.035	0.025	0.010	0.002
1592	0.029	0.022	0.017	0.012	0.008	0.003	0.001
1677	−0.325	−0.253	−0.192	−0.141	−0.100	−0.041	−0.010
1682	0.321	0.317	0.297	0.266	0.224	0.127	0.038
1685	0.071	0.066	0.057	0.046	0.034	0.014	0.003
1725	−0.201	−0.168	−0.135	−0.105	−0.078	−0.035	−0.009
1738	−0.297	−0.230	−0.171	−0.122	−0.082	−0.030	−0.006

1743	−0.272	−0.223	−0.180	−0.140	−0.105	−0.048	−0.012
1843	−1.675	−0.960	−0.655	−0.457	−0.315	−0.129	−0.031
1971	0.014	0.011	0.008	0.006	0.004	0.002	0.001
1999	−0.520	−0.351	−0.237	−0.155	−0.098	−0.031	−0.005
2043	0.219	0.190	0.160	0.131	0.103	0.051	0.014
2100	0.693	0.602	0.511	0.420	0.331	0.168	0.046
2120	−0.011	−0.009	−0.007	−0.006	−0.004	−0.002	−0.001
Total	−1.347	0.633	1.335	1.572	1.536	1.011	0.339

Disease Gene Localization

Once linkage between a disease locus and marker locus is established via two-point linkage analysis, the next step is to identify the smallest region of the genome that should contain the disease gene, the minimum-candidate region (MCR). Two approaches to disease gene localization are generally used: haplotype analysis and multipoint linkage analysis.

Haplotype analysis is a tool for ordering alleles on a chromosome. It complements multipoint linkage analysis in that it provides visual confirmation of statistical testing. Given the order of the genetic markers on a chromosome, we can establish with varying degrees of certainty the alleles that were passed from parent to offspring in any specific gamete by haplotype analysis and identify critical recombination events that flank the MCR. Haplotype analysis can be done by hand or with the use of a computer program such as SimWalk2 (Weeks et al. 1995). Through haplotype analysis in Mendelian disease, critical recombination events may allow us to identify the upper and lower bounds of the disease gene interval.

The process of haplotyping involves identifying which alleles were transmitted in the same gamete from each parent and identifying where recombination events between markers and/or between markers and the disease gene are located. The main goal is to identify markers that flank the disease gene. In the presence of reduced penetrance, recombination events that occur in affected individuals tend to be associated with a higher degree of certainty than recombination events that involve unaffected individuals because an unaffected individual could either be a nondisease allele carrier or a nonpenetrant carrier of the disease allele. The confidence with which a recombination event defines a boundary of the MCR then is heavily dependent on the penetrance.

An example of haplotype analysis is shown in Figure 6.5. In the pedigree on the left, the unordered genotypes for three markers are shown underneath the pedigree symbols. The markers are arranged linearly and are tightly linked to one another such that the intermarker distances are 2 cM. The task in haplotype analysis is to assign the allelic combinations that were transmitted in the

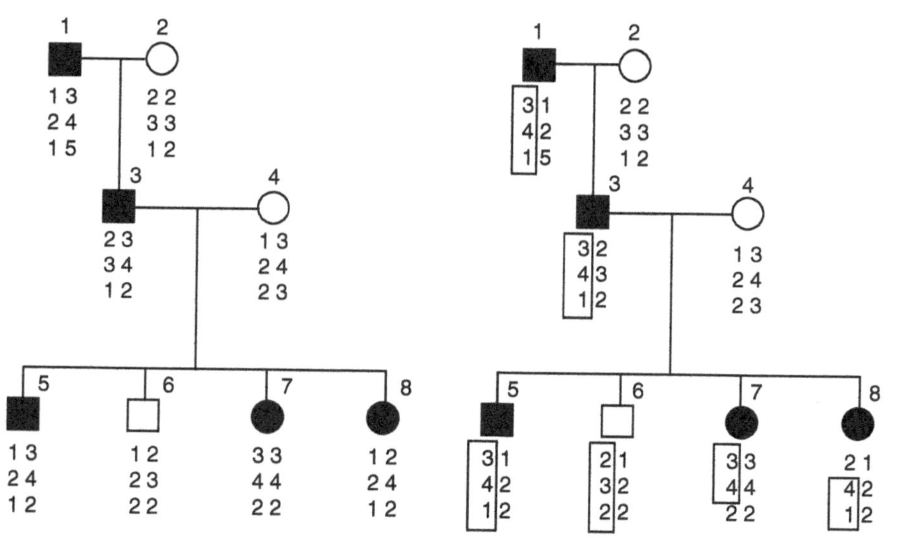

Figure 6.5 Pedigree on the left shows unordered genotypes underneath pedigree symbol for three linked markers. Pedigree on the right shows ordered genotypes, with alleles transmitted from father on the left. Alleles included in the disease-associated haplotype are in the box. See text for details.

gamete from parent to offspring. In the pedigree on the right one can see that from the affected grandfather the gamete with the combination of alleles 3/4/1 was transmitted to the affected son. The haplotype with the combination of alleles 2/3/2 was transmitted from the unaffected grandmother to the affected son. The haplotype 3/4/1 is now considered to represent the linkage phase that we test in the third generation. This haplotype was transmitted in its complete state from the affected father to his first affected son, individual five. The unaffected individual six inherited the grand maternal combination of alleles at the three loci. Both individuals five and six represent nonrecombinant gametes. Individual seven inherited the disease-associated haplotype from markers one and two but inherited the grand maternal derived allele one at marker two.

The combination of markers that individual seven inherited demonstrates that a recombination event between the disease gene and the marker loci occurred distal to marker two. The affected daughter, eight, inherited the disease-associated haplotype for markers two and three. She inherited the grand maternal allele from her affected father at marker one so that a recombination event occurred between the disease and marker loci proximal to marker two. These two recombination events suggest that the disease gene in this family is located between markers one and two.

Multipoint Analysis

Multipoint linkage analysis is a statistical technique using available genetic maps in which the order of markers and the distances between the markers are known. Multipoint linkage analysis allows the simultaneous consideration of genotypes from multiple linked loci. This technique is useful for localization of disease genes between two markers (the flanking markers) and for maximizing the informativeness of a series of markers. In a general sense, given a series of markers of known location, order, and spacing, the likelihoods of the pedigree data are sequentially calculated for the disease gene to be at any position within the known map of markers. The multipoint LOD scores are typically graphed as in Figure 6.6. The x-axis represents the genetic distance between the markers, and the y-axis represents the LOD score. In this example, the most likely location for the disease is in the 7-cM interval bounded by IL9 and D5S178. In this interval, the LOD score reaches

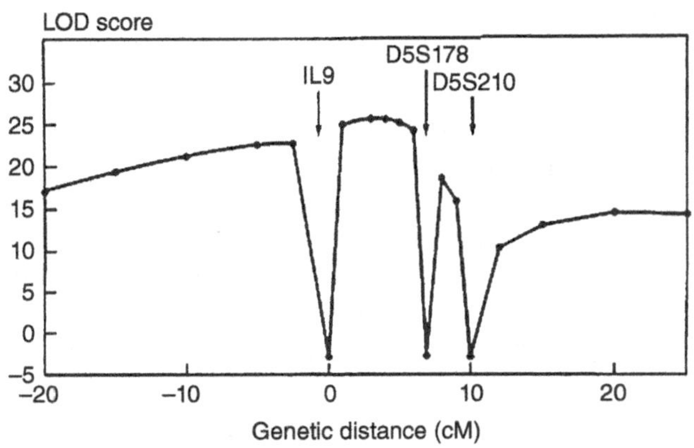

Figure 6.6 The most likely location for the disease gene is in the 7-cM interval bounded by IL9 and D5S178. The maximum LOD score is 23.50. Narrow support intervals for disease gene localization are calculated by subtracting 3 from the maximum LOD score; all locations within this three-LOD unit are within the approximate 99% confidence interval for disease gene localization.

its highest value of approximately 23.5. The next most likely interval is the interval outside IL9, where the maximum LOD score reaches approximately 21.0. Note that the interval between D5S178 and D5S210 has a LOD score that maximizes at approximately 18. This interval between D5S178 and D5S210 is outside the approximate 99% confidence interval for disease gene localization because it does not fall within a three-LOD unit support interval.

There are two major advantages of multipoint LOD score analysis. First, it provides an opportunity to recover linkage information at an originally uninformative locus via haplotype inference. Thus, the linkage results are less sensitive to uninformative or missing genotypes at any single marker. In essence, multipoint analysis can extract more of the total inheritance information from a pedigree. In one simulation study, when a rare dominant disease gene was simulated in the middle of an 18 cM map using a pedigree with two affected fourth cousins, an LOD score of 2.2 (91% of its theoretical maximum) was obtained when 20 markers were analyzed simultaneously. In contrast, the highest two-point LOD score was only 0.83 (34% of its theoretical maximum), and even simultaneous six-marker analysis yielded, at most, an LOD score of 1.74 (72% of theoretical maximum) (Kruglyak et al. 1996).

Second, the multipoint LOD score approach can be very useful to pinpoint a disease gene location in the fine mapping of a Mendelian disorder. Any true recombinant at some location x will contribute a strongly negative LOD score and decrease the possibility that this position contains a candidate locus. This significantly narrows the region in which the disease locus can exist, thus helping to better define the minimum candidate region. For a single pedigree, the information conveyed by a multipoint LOD score curve mirrors the cosegregation of multimarker haplotypes with the disease phenotypes of pedigree members. Therefore, the process of haplotyping, i.e. the ordering of genotypes by parental origin, can be considered a natural "by-product" of multipoint LOD score analysis. In a given pedigree with evidence for linkage, haplotyping can reveal the critical recombinants, i.e. those affected individuals that share only part of the disease-associated haplotype due to a recombination that shortens the region of interest.

Another use of multipoint LOD score analysis for Mendelian disorders is exclusion mapping, where "negative" information is used to exclude regions of the genome from possibly harboring a disease gene. The classical criterion is to exclude all regions with $Z(x) < -2.0$, but it is important to be aware of the effect of heterogeneity and misspecified model parameters on multipoint exclusion mapping. Some researchers believe that multipoint LOD score curves are not as helpful for narrowing the disease gene region for complex disorders as they are for Mendelian phenotypes. The curves often peak throughout extended regions of the genome, and exclusion mapping, using the classical criterion of $Z_{max} < -2.0$ may also be less meaningful. Therefore, it has been suggested to postpone parametric multipoint analysis of complex disorders altogether until a region with preliminary evidence for linkage has been found, i.e. to apply the method for follow-up analysis, but avoid its use with genome screen data (Ott 1999). However, it has been suggested that the sensitivity of parametric multipoint analysis may have been overestimated and that multipoint LOD scores combined with the admixture test (multipoint HLODs) can be a powerful tool even for linkage analysis of complex traits (Greenberg and Abreu 2001).

While multipoint LOD scores are potentially more informative than two-point LOD scores, they will not be of higher absolute value at any given marker locus if the marker itself is already fully informative for linkage. Note that performing a multipoint analysis requires additional information over two-point analysis, a map of ordered markers and estimated intermarker distances, which means that the linkage test is sensitive to misspecified maps (either in order or distance) in addition to the usual genetic model parameters. For many years, the Marshfield maps, based on eight three-generation CEPH families containing 188 meioses, had been considered the most comprehensive mapping resource. More recently, Matise and colleagues presented a combined physical/genetic map, which is consistently updated for use with modern SNP-based genotyping array data (available at: http://compgen.rutgers.edu/rutgers_maps.shtml) (Matise et al. 2007).

In multipoint linkage analysis, the computational complexity of pedigree likelihoods of the form presented above is greatly magnified, since the joint inheritance of many markers has to be traced through a pedigree to evaluate the LOD score at each point along the marker map, and the number of possible genotypes for untyped pedigree members increases exponentially. Two different algorithms were developed to recursively compute the likelihood terms: The Elston-Stewart "peeling" algorithm (Elston and Stewart 1971) was implemented in the first generally available software for pedigree likelihood computation (LIPED, (Ott 1974) and was the basis of the subsequent programs FASTLINK (Cottingham et al. 1993), a faster version of the original LINKAGE package (Lathrop et al. 1984), and VITESSE (O'Connell and Weeks 1995); (O'Connell 2001). The Elston–Stewart algorithm scales linearly in the number of pedigree members and exponentially in the number of markers and is thus typically used for large pedigrees and a moderate number of markers (three to six, depending on the number of untyped pedigree members). VITESSE is a faster option for simple pedigrees without loops that descend from one founder pair, while FASTLINK is able to handle inbred pedigrees and those with multiple founder pairs.

The Lander–Green algorithm (Lander and Green 1987) has been implemented in the packages GENEHUNTER (Kruglyak et al. 1996)) and ALLEGRO (Gudbjartsson et al. 2000) and MERLIN (Abecasis et al. 2002). MERLIN employs a modified version by replacing the traditional array organization of likelihoods with sparse binary trees. While each program has strengths and weaknesses, MERLIN is superior in terms of computational speed. The Lander–Green algorithm scales exponentially in the number of pedigree members and linearly in the number of markers and is thus ideally suited for analyzing all the markers for an entire chromosome on small to moderate-sized pedigrees (in GENEHUNTER, $2n-f \leq 16$, where n is the number of nonfounders and f the number of founders; ALLEGRO and MERLIN are able to handle somewhat larger pedigrees). These programs also implement various types of nonparametric linkage tests (see below) and provide other useful statistics of interest, such as the information content of a pedigree and the overall data set along the marker map, which highlights regions where additional marker genotyping would be most useful.

Another class of computational algorithms is of the Markov Chain Monte Carlo (MCMC) type, which is a stochastic rather than exact algorithm. SimWalk2 (Sobel and Lange 1996) can be used to compute a random-walk-based approximation to the exact parametric LOD score computed by FASTLINK and VITESSE and is able to simultaneously analyze a large number of markers on pedigrees that are too large for the Lander–Green algorithm.

Some packages implement so-called hybrid algorithms, where the algorithm employed is pedigree specific. For example, MENDEL implements either the Elston-Stewart algorithm or the Lander–Green algorithm depending on which is better suited for a particular pedigree (Lange et al. 2013). For others, pedigrees amenable to exact analysis with the Lander–Green algorithm are processed accordingly, while those too large for exact LOD score computation are analyzed with a SimWalk2-type MCMC algorithm (e.g. MERLIN, S.A.G.E.) (Elston and Gray-McGuire 2004). SUPERLINK belongs to this latter group and has the advantage of automatically parallelizing its calculations and can therefore handle dense SNP maps (Silberstein et al. 2013).

Effects of Misspecified Model Parameters in LOD Score Analysis

To illustrate the impact of misspecified model parameters on the power to detect linkage or the type I error (false-positive rate), computer simulations using the 28 AD families discussed earlier in the chapter were performed. Keeping the actual pedigree structure and observed disease phenotypes constant, the program SLINK (Weeks et al. 1990b) was used to simulate disease and marker genotypes in 1000 replicates assuming either complete linkage ($\theta = 0.0$) or no linkage ($\theta = 0.5$) between the disease and marker locus. The disease locus was simulated under the model assumed above for the analysis of the real data: autosomal dominant inheritance with a disease allele frequency $q = 0.001$, penetrance for gene carriers (f_{DD}, f_{Dd}) of 0.80 and for phenocopies (f_{dd}) of 0.01. The marker was simulated using the actually observed allele frequencies of marker D19S246 (12 alleles with frequencies 0.25, 0.02, 0.04, 0.02, 0.03, 0.18, 0.11, 0.06, 0.14, 0.11, 0.02, 0.02). Each replicate was analyzed using the simulated model as well as other models representing model misspecification: three models with different disease allele frequencies, three models with different penetrance ratios, three models with different mode of inheritance, and two models with different marker allele frequencies. When one parameter was evaluated, the other parameters were generally fixed at their "true" values used in the simulating model. Only when the mode of inheritance was varied, we included an analysis model where the disease allele frequency was adjusted to produce the same gene carrier frequency. The effect of misspecified model parameters on the LOD score analysis was measured by the mean maximum LOD score (Z_{max}) and mean estimated MLE (θ) obtained from the 1000 replicates. The results at $\theta = 0.0$ reflect the impact on the power to detect linkage, and those at $\theta = 0.5$ reflect the impact on the type I error.

Impact of Incorrect Disease Allele Frequency

With a true disease allele frequency of 0.001, the data were analyzed using modified frequencies of 0.01, 0.1, and 0.0001, corresponding to gene carrier frequencies of 2, 20, and 0.02%, respectively (Table 6.5). This represents a wide variation of disease allele frequencies. In the absence of linkage, there is little variation in the mean maximum LOD score (from 0.11 to 0.12). In the presence of complete linkage, the mean Z_{max} decreases slightly when the disease allele frequency is either underestimated or overestimated. In an autosomal dominant model, an increased disease allele

Table 6.5 Impact of misspecifying disease allele frequency on LOD score analysis.

Analysis model				Generating model: $q = 0.001, f_{DD} = f_{Dd} = 0.80, f_{dd} = 0.01$				
				$\theta = 0.00$			$\theta = 0.50$	
q	f_{DD}	f_{Dd}	f_{dd}	Mean Z_{max}	Mean MLE(θ)	Proportion Z_{max} with >3.0 (%)	Mean Z_{max}	Mean MLE(θ)
0.001	0.80	0.80	0.01	11.78	0.0	100	0.11	0.40
0.01	0.80	0.80	0.01	11.44	0.0	100	0.12	0.40
0.10	0.80	0.80	0.01	9.95	0.0	99.8	0.11	0.40
0.0001	0.80	0.80	0.01	11.55	0.0	100	0.11	0.40

frequency can either increase the probability of affected parents being homozygous rather than heterozygous or increase the probability that the disease allele is introduced into the pedigree through married-in individuals rather than a single founder. However, within a reasonable range of plausible values (0.0001–0.10), the impact of an incorrect disease allele frequency on the LOD score and on the estimated recombination fraction is generally small.

Impact of Incorrect Mode of Inheritance

In general, an incorrect mode of inheritance has a large impact on the LOD score (Clerget-Darpoux et al. 1986). This effect is particularly serious when a dominant disease locus is coded as recessive (Table 6.6). In the AD example, the mean maximum LOD score decreases from 11.78 to 1.64, the power to obtain $Z_{max} \geq 3$ decreases from 100 to 7.5%, and the MLE (θ) increases from 0.0 to 0.10 when the disease allele frequency remained at $q = 0.001$. When the disease allele frequency is adjusted ($q = 0.10$) to obtain approximately the same gene carrier frequency, the decrease in mean Z_{max} and power is less pronounced (from 11.78 to 3.94, and from 100 to 75.2%, respectively) but still quite drastic. An intuitive explanation for this observation is that, half the time, the random transmission of alleles at the disease locus from an unaffected parent to affected offspring is scored as a recombination between the disease and marker locus, because under a dominant model, an unaffected parent does not carry the susceptibility allele, but under a recessive model, the affected offspring must possess two susceptibility alleles. In addition, the affected parent will be considered homozygous at the disease locus and is thus uninformative for linkage. The impact of assuming an additive model when the true model is dominant is much less dramatic (decrease of Z_{max} from 11.78 to 9.75), illustrating that the additive model can be viewed as a good "compromise" between dominant and recessive models when the true mode of inheritance is unknown. In the absence of linkage, Table 6.6 shows that an incorrect mode of inheritance has little impact on the LOD score.

Impact of Incorrect Disease Penetrance

As long as incomplete penetrance is allowed for in the genetic model, the exact values for the penetrance have a relatively small impact on the LOD score when there is either linkage or no

Table 6.6 Impact of misspecifying mode of inheritance on LOD score analysis.

				Generating model: $q = 0.001$, $f_{DD} = f_{Dd} = 0.80$, $f_{dd} = 0.01$				
Analysis model				$\theta = 0.00$			$\theta = 0.50$	
q	f_{DD}	f_{Dd}	f_{dd}	Mean Z_{max}	Mean MLE(θ)	Proportion with $Z_{max} > 3.0$ (%)	Mean Z_{max}	Mean MLE(θ)
0.001	0.80	0.80	0.01	11.78	0.0	100	0.11	0.40
0.001	0.80	0.01	0.01	1.64	0.1	7.5	0.11	0.40
0.10	0.80	0.01	0.01	3.94	0.1	75.2	0.11	0.40
0.001	0.80	0.40	0.01	9.75	0.0	99.9	0.12	0.40

Table 6.7 Impact of misspecifying disease penetrance on LOD score analysis.

Analysis model				Generating model: q = 0.001, f_{DD} = f_{Dd} = 0.80, f_{dd} = 0.01				
				θ = 0.00			θ = 0.50	
q	f_{DD}	f_{Dd}	f_{dd}	Mean Z_{max}	Mean MLE(θ)	Proportion with Z_{max} > 3.0 (%)	Mean Z_{max}	Mean MLE(θ)
0.001	0.80	0.80	0.01	11.78	0.0	100	0.11	0.40
0.001	0.80	0.80	0.10	4.91	0.0	94.5	0.10	0.40
0.001	0.80	0.80	0.00	9.64	0.1	99.8	0.12	0.40
0.001	0.40	0.40	0.01	9.76	0.0	99.9	0.12	0.40

linkage (Table 6.7). As the ratio of penetrance between disease allele carriers and disease allele noncarriers decreases (from 80 to 40 to 8 in Table 6.7), the mean Z_{max} decreases (from 11.78 to 9.76 to 4.91, with a power decrease from 100 to 99.9 to 94.5%), since a low ratio decreases the certainty of whether an affected individual is a gene carrier and an unaffected individual a noncarrier. In the presence of linkage, most individuals are nonrecombinants, which is why the low ratio results in a decrease in power. When the penetrance of noncarriers is reduced to 0, Z_{max} decreases from 11.78 to 9.64. Thus, not allowing for the presence of phenocopies is not detrimental as long as the model allows for incomplete penetrance, which permits multiple possible disease genotypes to underlie the observed phenotype. When a Mendelian disorder is analyzed, a nonzero phenocopy rate is sometimes interpreted as a "misdiagnosis parameter", to account for the fact that some family members are diagnosed as affected with less certainty than others and thus may or may not be carriers of the gene that causes the familial aggregation. In general, in the presence of model uncertainty, choosing a lower penetrance ratio by specifying both a nonzero penetrance of phenocopies and incomplete penetrance for susceptible genotypes ("weaker model") is sometimes recommended to make LOD score analysis less sensitive to incorrect model parameters (Risch and Giuffra 1992).

Impact of Incorrect Marker Allele Frequency

In our example data set, incorrect marker allele frequencies do not have a large impact on the mean Z_{max} for the overall data set (Table 6.8). This can be explained partly by the relatively small degree of misspecification, where, in the worst case, the most common allele of a marker with a total of 12 alleles has a frequency of 0.23 and is coded as 0.02 (and vice versa). Unlike allelic association (Chapter 8), the presence of linkage alone is compatible with the cosegregation of different alleles of a marker in different pedigrees. Therefore, changing the frequency of a particular allele may reduce the LOD score for some families, but may increase the LOD score for others. When these effects balance each other, the LOD score for the entire data set will hardly change at all. However, there are situations when marker allele frequencies can have a substantial impact on the LOD score obtained from a single pedigree (Ott 1992). In general, a higher proportion of founders without genotype data leads to a greater impact of marker allele frequencies on LOD score analysis, since the frequencies are used to compute probabilities of

Table 6.8 Impact of misspecifying marker allele frequencies on LOD score analysis.

	Generating model: q = 0.001, f_{DD} = f_{Dd} = 0.80, f_{dd} = 0.01, marker allele frequencies 0.25, 0.02, 0.04, 0.02, 0.03, 0.18, 0.11, 0.06, 0.14, 0.11, 0.02, 0.02				
	θ = 0.00			θ = 0.50	
Analysis model	**Mean Z_{max}**	**Mean MLE(θ)**	**Proportion with Z_{max} > 3.0 (%)**	**Mean Z_{max}**	**Mean MLE(θ)**
Same as simulated	11.78	0.0	100	0.11	0.40
Equally frequent alleles	12.09	0.0	100	0.16	0.40
Common/rare alleles reversed	11.84	0.0	100	0.23	0.40

marker genotypes for untyped individuals. For example, if two affected cousins share a marker allele in common and not all their parents and grandparents are genotyped, the evidence for linkage increases with the rarity of the allele. If the allele is quite common, there is an increased probability that parents are homozygous for that allele, and thus uninformative for linkage, or that married-in individuals, rather than a single common ancestor, may have transmitted the allele to one or more affected relatives, thus diminishing the evidence for linkage. The greatest impact is observed when the marker has only two alleles with very different frequencies. In the presence of linkage between disease and marker, misspecification of the rare allele as common would reduce linkage evidence, i.e. produce false-negative evidence against linkage. In the absence of linkage, misspecification of the common allele as rare would lead to false-positive evidence for linkage (e.g. Spence et al. 1977). This probably explains why, in the example data set (Table 6.8), the mean Z_{max} score under the null hypothesis of no linkage increased from 0.11 (marker alleles as simulated) to 0.16 (equally frequent marker alleles) to 0.23 (the most common alleles coded as the most rare and vice versa).

Control of Scoring Errors

Apart from incorrect model parameters, there are three types of scoring errors that can affect the value of the LOD score: Phenotype errors, genotype errors, and incorrectly identified relationships within a pedigree. Phenotype errors occur when an affected individual is coded as unaffected or vice versa. Not much can be done from an analysis perspective to minimize the probability of phenotype errors, although their effect on the LOD score can be large. For example, when full penetrance is assumed and an unaffected individual, mistakenly classified as affected, shows an obligate recombination between marker and putative disease locus, the LOD score for a single pedigree might drop from, say, 1.5 to $-\infty$ at $\theta = 0$. Therefore, an accurate clinical diagnosis is of utmost importance for any linkage analysis. The program VARYPHEN allows the user to carry out a sensitivity analysis with respect to disease phenotypes (Xie and Ott 1990). By setting the disease phenotype of each family member in turn to "unknown" and repeating the linkage analysis, a table of resulting changes in LOD scores and estimated θ values is obtained. It is then easy to see which individuals produce the largest changes and are thus the most crucial in terms of diagnostic accuracy.

Genotype errors, just like missing data, are unavoidable in any real data set since a machine's or a technician's interpretation of experimental results is an imperfect way of measuring the true marker genotype. For most labs, an empirical value of about 1% genotype error is probably a realistic estimate for single SNP genotyping. Chip-based genotyping methods available from a variety of vendors are generally even more accurate, with error rates of approximately 0.1% (Saunders et al. 2007). The introduction of quality control procedures in the genotyping lab (Rimmler et al. 1999) is an extremely useful tool for keeping the actual error rate as low as possible. As with phenotype errors, genotype errors may have an appreciable impact on the LOD score by introducing false recombinants, thus decreasing the evidence for linkage. Several methods can be used to identify genotypes likely to be erroneous after they have been submitted for analysis. The first and most straightforward check is the identification of Mendelian inconsistencies, e.g. the occurrence of an allele in the offspring that is not found in the parents, the occurrence of more than four alleles in a sibship, etc. Software such as PEDCHECK (O'Connell and Weeks 1998) can be used to identify these inconsistencies, followed by rereading in the lab. For densely spaced marker genotypes obtained at the fine-mapping stage of a project, rather than the genome screen stage, some additional error checking approaches can be applied. The program SimWalk2 (Sobel and Lange 1996) is often used to estimate haplotype configurations of multiple linked markers in a pedigree. As an extension of this haplotyping capability, a "mistyping" option can be used to detect excess recombination in small genomic regions, which typically indicates likely genotype errors. The program SIBMED (Douglas et al. 2000) is particularly suitable for error detection in affected sib-pair data without parental genotypes, where haplotyping is more difficult to carry out. It compares the multipoint probability of the sib-pair data with the marker in question included vs. excluded and pinpoints marker-sib-pair combinations with probability ratios that exceed a simulation-based marker-specific threshold. For a comparison of the various error checking methods, the reader is referred to Mukhopadhyay et al. 2004.

Seemingly erroneous genotypes in a pedigree may also be due to misspecified relationships between pairs of individuals. A loss of power to detect linkage may be caused by the inclusion of pairs who are less closely related than assumed. For example, two putative full sibs may be unrelated in the case of unknown adoption or sample switches, or they may be half sibs in the case of misassigned paternity. False-positive evidence for linkage may be created by sample duplication or incorrect assignment of monozygotic twins as full sibs. These types of problems may not be detectable with Mendelian inconsistency checks, particularly if a sample includes only sib-pair data without parental genotypes. Once a large number of markers (≥50) have been genotyped, software such as RELPAIR (Boehnke and Cox 1997); (Epstein et al. 2000)), RELATIVE (Goring and Ott 1997), PREST (McPeek and Sun 2000), or GRR (Abecasis et al. 2001) can be used to compute the multipoint likelihood of the marker data conditional on certain possible relationships (e.g. MZ twins, full sibs, half sibs, unrelated pair for putative full sib-pairs). By comparing the probability of the marker data under the specified vs. a different assumed genetic relationship of various relative pairs, the most likely relationship of the pair can be inferred. RELPAIR is able to consider all possible pairs of individuals in the sample, not just those within families, and even allows for the presence of genotyping error. PREST can analyze inbred as well as outbred pedigrees.

Genetic Heterogeneity

Every LOD score analysis that involves more than one pedigree should include an examination of the scores for each pedigree. In the AD example, the LOD scores ranged from −1.67 to 1.01 (at $\theta = 0.0$), raising the possibility that only some of the pedigrees may be linked to the chromosome

19q13 locus. A complex disorder such as AD is very likely to be genetically heterogeneous. In general, genetic locus heterogeneity is said to exist when two or more genes act independently to cause the same clinical disease phenotype, or to increase disease susceptibility. The power of a study to detect linkage can be greatly improved either by increasing sample homogeneity to start with by examining only a clinical subtype of the disorder or only patients with a particularly early age of onset or by the ability to tease out more homogeneous subgroups that may exhibit a stronger linkage signal, which is obscured by the heterogeneity in the overall data set. This type of heterogeneity is likely to be a major reason for the many failures to replicate linkage findings in complex disease. However, there are many ways in which heterogeneity can be allowed for in the statistical analysis of pedigree data, and several approaches are described below.

The null hypothesis of homogeneity can be formally tested using two different approaches: the M-test (Morton 1956; Ott 1999), or its extension known as the β-test (Risch 1988), and the admixture test (Smith 1963). When families can be preassigned to several different groups based on some disease characteristics, such as age of onset, severity of disease, clinical features, or transmission pattern, the M-test can be used to test for different recombination fractions in these k different groups of families, where $\theta_1 \neq \theta_2 \neq \ldots \theta_k$. The first hypothesis ($H_1$) of linkage and homogeneity specifies $\theta_1 = \theta_2 = \ldots \theta_k < 0.5$. Under the second hypothesis of heterogeneity (H_2): $\theta_1 \neq \theta_2 \neq \ldots \theta_k$, the recombination fractions are potentially different in the different groups of families and are estimated separately. Note that H_1 is obtained from H_2 by restricting $(k-1)$ parameters ($\theta_2, \ldots, \theta_k$) to be the same as θ_1. To test H_1 against H_2, one computes

$$X^2 = 2\, x\, \ln(10)\left(\sum_{i=1}^{k} Z_i\left(\hat{\theta}_i\right) - Z\left(\hat{\theta}\right)\right)$$

where $Z\left(\hat{\theta}\right)$ is the maximum LOD score in each of the family groups, and $Z\left(\hat{\theta}\right)$ is the maximum LOD score over all the pedigrees. Asymptotically, under the assumption of homogeneity (H_1), this statistic follows a χ^2 distribution with $(k-1)$ df. The M-test requires the estimation of a separate θ_i parameter in each of the k groups of families. Alternatively, a Bayesian approach can be taken, in which the θ values are assumed to follow a β-distribution with two parameters, which are estimated from the posterior distribution of θ given the observed data. This likelihood-based test is known as the β-test and was shown to be more powerful than the M-test in many situations (Risch 1988).

The admixture test (Smith 1963) is used when families cannot be preassigned to different groups based on clinical criteria. In the most commonly employed version of this test, two types of families are assumed, with α denoting the proportion of type I families and $(1-\alpha)$ the proportion of type II families. The recombination fraction in type I families is equal to θ_1 and that in type II families is $\theta_2 = 0.5$. The null hypothesis of no linkage and homogeneity is H_0: $\alpha = 0, \theta_1 = \theta_2 = 0.5$. The hypothesis of linkage and heterogeneity is H_2: $\alpha < 1, \theta_1 < 0.5$, and the hypothesis of linkage and homogeneity is obtained from H_2 by a single restriction, H_1: $\alpha = 1, \theta_1 < 0.5$. Under H_2, the likelihood is maximized over α and θ_1. Under H_1, the restricted hypothesis, θ_r is obtained assuming homogeneity ($\alpha = 1$). The test of the hypothesis of homogeneity (H_1 vs. H_2) is carried out by calculating $\chi^2 = 2\,[\ln L(\alpha, \theta_1) - \ln L(\alpha = 1, \theta_r)]$, which asymptotically has a χ^2 distribution with 1 df with probability ½ and is equal to 0 with probability ½. Thus, the p-value from a two-sided test has to be halved. Both the M-test and the admixture test can be carried out with the HOMOG program package (Ott 1986). In practice, the admixture test is used more frequently than the M-test and β-test.

A problem with the admixture test concerns the question of significance when the total LOD score (under homogeneity) is less than 3.0, yet a test of homogeneity yields a p-value below the conventional level of 0.05 or 0.01. Declaring the heterogeneity significant would imply the presence of linkage in a subset of the families, even though the usual criterion for significant linkage ($Z \geq 3$) is not met. The more conservative school of thought argues that an overall $Z \geq 3$ ($p \leq 0.0001$) is necessary to reject the null hypothesis of no linkage and that the additional hypothesis of heterogeneity should only be tested if that hypothesis is rejected. A less stringent approach tests linkage and homogeneity simultaneously. In this situation, an overall log-likelihood ratio difference of 3.0 or more between homogeneity with no linkage and heterogeneity with linkage is considered sufficient to declare significant heterogeneity and linkage in a subset of the families. A heterogeneity LOD score, where the numerator is maximized over both α and θ, and the denominator is the likelihood under H_0: $\alpha = 0, \theta = 0.5$, is often called HLOD. An asymptotic null distribution of the test statistic $\chi^2 = 4.6 \times \text{HLOD}$, which tests for linkage while allowing for heterogeneity, has been derived (Faraway 1993). The critical value is slighter higher (3.3 vs 3.0) than for a LOD score (Hodge et al. 1997). For the Alzheimer data, the admixture test does not indicate the presence of locus heterogeneity. The HLOD is the same as the previously obtained $Z_{max} = 1.57$ at $\theta = 0.15$ ($\alpha = 1.0$).

The power of these statistical tests of homogeneity is influenced by a number of factors, including the distance between the linked marker and the disease locus and the amount of linkage information provided by the individual families. A larger recombination fraction between marker and disease gene reduces the power to detect heterogeneity. Larger, more informative families will yield more power than a series of smaller families in discriminating linked and unlinked groups. The main advantage of the admixture test is that it does not require any prior knowledge about possible sources of heterogeneity. If there is a valid prior reason to divide the data set into more homogeneous subgroups, however, the admixture test is less powerful than the M-test.

Various extensions of the simple admixture model described above have also been implemented in the HOMOG package. For example, HOMOG2 assumes that a disease is caused by either of two loci, which are within measurable distance of each other on the same chromosome and thus linked to the same marker. Two recombination fractions are then estimated together with the proportion of families linked to the first locus α, assuming a proportion 1-α to be linked to the second locus. HOMOG3R can be used for a model assuming linkage to one disease locus in some families, linkage to another locus, unlinked to the first (e.g. on a different chromosome), in other families, and a third subset of families that are unlinked to either of the two disease loci. It has been shown (Schork et al. 1993) that this approach is a good approximation for two-locus linkage analysis, where the simultaneous gene mapping of two disease loci is attempted. HOMOGM is the extension of this type of model to an arbitrary number m of disease loci. MERLIN can also be used to calculate parametric HLODs.

As in two-point analysis, multipoint heterogeneity LOD scores can be computed with the admixture test (e.g. using HOMOG). However, the interpretation of these tests in the multipoint situation is less straightforward because the asymptotic properties of multipoint LOD scores, which are generally multimodal, are not well characterized and the χ^2 approximations are thus questionable. Rather than using p-values, it may be preferable to simply report the observed likelihood ratio between the hypotheses of no linkage and homogeneity vs. linkage and heterogeneity. As in two-point analysis, heterogeneity has a major impact on the power to correctly detect linkage using multipoint analysis. With two-point LOD scores, heterogeneity, if not taken into account in the analysis, or other types of model error leading to false apparent recombinations, can be partially absorbed by an inflated estimate of the recombination fraction (Risch and Giuffra 1992). However, in multipoint analysis, the flanking markers prevent the recombination fraction from floating very far, which makes multipoint LOD scores somewhat less robust with respect to genotype and model errors than two-point LOD scores. A typical multipoint LOD score curve in the presence of

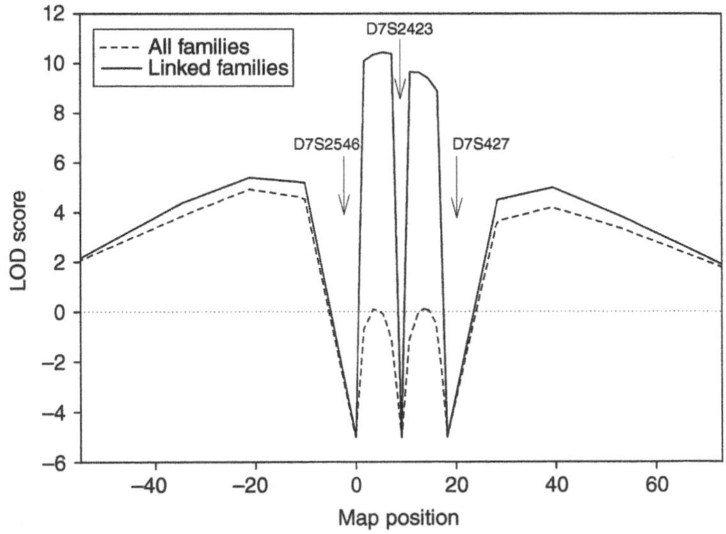

Figure 6.7 Example of multipoint linkage analysis in the presence of genetic heterogeneity for limb-girdle muscular dystrophy. When all families are analyzed, no evidence for linkage to this region is obtained. Examination of linked families after formal heterogeneity testing reveals strong evidence for linkage. *Source:* Adapted from Speer et al. (1999).

heterogeneity is shown in Figure 6.7. When heterogeneity is ignored, there is no evidence for linkage to the region holding the gene of interest (all families). When heterogeneity is taken into account (linked families), the region is clearly identified. Figure 6.7 also illustrates the typical shape of multipoint LOD score curves, which peak in regions between two markers and drop to $-\infty$ at the position of the markers themselves. This is due to the presence of obligate recombinants, which make it impossible for the disease locus to be located exactly at the position of the marker (corresponding to $\theta = 0$ in two-point analysis).

In summary, heterogeneity can be a major confounding factor in the gene mapping process. Once heterogeneity has been confirmed, one must ask what criteria should be used to classify a family as "linked" for molecular follow-up purposes. The parameter estimates from the HOMOG admixture test can be used to calculate the estimated conditional (given marker observations) probability that a family is of the linked type. Even though this is just a statistical estimate derived from a particular set of data, such a probability may be useful to focus laboratory time and effort on the families most likely to harbor a disease gene.

Practical Approach for Model-Based Linkage Analysis of Complex Traits

For complex traits, susceptibility genes are presumably more common and of more modest effect than Mendelian disease genes, and they are likely to act in concert with multiple other genes and/or environmental factors. Thus, any assumed single-gene mechanism (genetic model) for parametric LOD score analysis is almost certainly incorrect. It was shown that analysis under a single wrong model (i.e. calculation of "wrod" [wrong lod] scores) does not inflate the type I error rate of the LOD score method (Williamson and Amos 1990). However, analyzing the data under multiple models and estimating unknown parameters by maximizing the LOD score over model parameters does lead to grossly inflated significance levels (Weeks et al. 1990a). The question then remains as to how one should choose the model parameters if a parametric LOD score analysis is to be applied to complex trait data.

Traditionally, genetic model parameters have been estimated via segregation analysis (Chapter 3). This is a statistical technique for estimating the most likely mode of inheritance for a trait from pedigree data that have been collected in a systematic fashion, ideally in such a way that a correction for ascertainment bias can be applied (Khoury et al. 1993). A properly performed segregation analysis of a large data set can be quite useful for the linkage analyst since it provides MLEs of the mode of inheritance (MOI), disease allele frequency and penetrance for susceptible and nonsusceptible genotypes. As mentioned above, due to the unlikely existence of single-gene mechanisms, this type of analysis is only rarely attempted for complex traits, although it has been carried out successfully, for example, to inform the search to locate highly penetrant genes for early-onset breast cancer (Claus et al. 1991; Easton et al. 1993). Here, the segregation analysis also included the estimation of an age-dependent penetrance function for the gene of interest. Since the ratio of genetic and nongenetic cases is often variable across different age groups, with younger cases typically more likely to be "genetic" than older cases, the concept of liability classes can be used to flexibly model age-of-onset distributions in parametric linkage analysis, as illustrated by the breast cancer example (Easton et al. 1993; Terwilliger and Ott 1994). For Mendelian diseases, phenocopies are typically assumed to be absent and age-dependent penetrance is often modeled by assuming a normal age-of-onset distribution with mean and variance estimated from the pedigree data at hand (see (Ott 1999), Chapter 6.5 for more detail).

A much simpler alternative to segregation analysis is to estimate model parameters by making use of epidemiologic information (Terwilliger and Ott 1994). If estimates of population prevalence and proportion of nongenetic cases among all affected individuals in the population are available, the penetrance for a single-locus model can be determined for an assumed disease allele frequency, because the following relationship between prevalence φ, penetrance of susceptible genotypes f, penetrance of phenocopies f_p, and disease allele frequency q should hold for a dominant disease:

$$\varphi = \left[q^2 + 2q(1-q)\right]f + (1-q)^2 f_p$$

(an analogous expression holds for recessive traits). While it may seem appealing to use prior information about disease epidemiology, some guesswork clearly remains and, again, any single-gene mechanism is unlikely to accurately capture the true complex trait etiology.

While there are several approaches to using parametric methods to analyze complex diseases, the simplest and most commonly used is the "affected-only" approach. The idea of the affected-only ("very low-penetrance") analysis is to discard the phenotypic information from unaffected individuals since they may be unaffected due to incomplete penetrance of a single gene, a missing copy of a second susceptibility gene, absence of an environmental trigger, or similar complicating factors. However, their genotypic information may still be very useful. The simplest way to set up an affected-only analysis is to code all individuals not known to have the disease as "unknown". Having phenotype "unknown" essentially means that all three possible disease genotypes (*DD, Dd, dd*) are equally likely to produce the observed phenotype. While allowance for phenocopies is often necessary for Mendelian disorders/traits, phenocopies typically are assumed to be absent in this type of analysis. Note that the occurrence of phenocopies is essentially a form of heterogeneity (Durner et al. 1996). For the AD example, an affected-only analysis under a dominant model produces $Z_{max} = 3.83$ at $\theta = 0.10$ and under a recessive model $Z_{max} = 1.93$ at $\theta = 0.05$.

Multipoint model-based analysis of complex disorders at the initial genome screen stage appears to be applied somewhat less frequently than the corresponding nonparametric approaches popularized by GENEHUNTER and MERLIN (see below). For follow-up analysis of regions with positive linkage signals at the genome screen stage, a multipoint HLOD analysis, assuming both dominant and recessive models and using either "simple" genetic models that take into account the population prevalence or the affected-only approach, can be recommended. However, possible statistical problems associated with the unknown distribution of multipoint LOD scores should be kept in mind. It is important to evaluate multipoint LOD scores at locations outside the fixed map, which is equivalent to allowing for overestimation of the recombination fraction in the two-point analysis. Multipoint analysis is then essentially the same as testing for linkage to a (potentially more informative) haplotype of markers. A large collaborative study of prostate cancer provides a good example for a parametric multipoint follow-up analysis based on models derived from segregation analysis as well as epidemiologic information and implemented through the concept of liability classes (Xu 2000).

Nonparametric Linkage Analysis

In contrast to the parametric methods discussed above, nonparametric linkage (NPL) analysis makes no assumptions about the underlying genetic model of the trait/disease being studied, only that the trait has some genetic component. It also assumes that the genes involved follow Mendelian laws of inheritance. NPL analysis can be employed in mapping genes for both Mendelian and complex diseases. However, since the power to detect linkage using NPL methods in the Mendelian case is significantly less than the power of parametric analysis, most investigators do not use NPL when the inheritance model is clearly established. Thus, NPL analysis is generally restricted to complex diseases, where the inheritance model may be completely unknown. When we refer to complex traits, we typically divide them into two major categories: qualitative or dichotomous traits and quantitative or continuous ones. The qualitative traits most commonly studied are, of course, human diseases in which a person is either clinically affected or unaffected with the disease, such as birth defects or AD. Quantitative traits studied are generally normally distributed throughout the population but which may result in a disease phenotype at one or both ends of the distribution, e.g. blood pressure and hypertension. The genetic complexity may occur for one or more of the following reasons:

1) The disease may be etiologically heterogeneous, with only a subset of families or individuals affected due to genes conferring high risk.
2) The disease may involve many different genetic loci that act together to cause disease.
3) A gene for the disease may predispose only in the presence of a particular environmental exposure.

Although the common use of NPL methods is not as old as the use of parametric methods for reasons described below, it is important to remember that the concepts that underlie this methodology date back to the early part of the last century. Indeed, the first nonparametric method in humans was a sib-pair linkage approach in humans described in a seminal paper by Penrose published in 1935 in Annals of Eugenics, the predecessor of Annals of Human Genetics (Penrose 1935). Aside from a second paper on the sib-pair approach by Penrose (1953), the next major developments in sib-pair analysis took place in the 1970s. The first such work was the seminal paper by Haseman and Elston (1972), which described an elegant regression-based approach to search for loci influencing quantitative traits, followed by the important work of Suarez et al. (1978), who considered the distribution of identity-by-descent (IBD) status among sibs for a variety of genetic models and applied this method to the elucidation of the role of the major histocompatibility complex (MHC) in type I (juvenile onset) diabetes using a set of affected sib-pairs. The application of NPL methods has become standard practice in human genetic analysis of diseases of complex inheritance as well as quantitative traits.

Identity by State and Identity by Descent

All methods that have been developed for the mapping of complex traits in human relative pairs depend in some way on quantification of the degree to which related individuals "share" alleles at the marker locus or loci under investigation. All such methods can be divided into two basic classes: those that depend on marker alleles shared identical by state (IBS) and those that rely on alleles shared identical by descent (IBD). The simpler of the two to define and understand is identity by state. Two alleles are said to be IBS if they are the same variant of some polymorphic system; that is, two alleles are IBS if they cannot be distinguished by means of a particular method of detection. For example, for a SNP, if two individuals both have the same base pair, these two alleles are IBS. Any two individuals, whether related or not, can share 0, 1, or 2 alleles IBS; the two alleles possessed by any one individual can also be IBS, in which case the individual is said to be homozygous at this locus.

Identity by descent, on the other hand, depends not only on whether the alleles appear the same by their detection method but also on whether these alleles are derived or inherited from a common ancestor. Clearly, alleles that are shared IBD must also be IBS; however, the converse is not true. That is, alleles that are IBS are not necessarily IBD. Any two unrelated individuals who have the same genotype illustrate this; since they are unrelated, they by definition share no alleles IBD. Figure 6.8 shows two hypothetical examples to illustrate the differences between IBS and IBD.

In Figure 6.8a, in the first genotype configuration (i.e. both sibs are type "1 3"), it is clear from the parents that each affected child inherited the "1" allele from the father and the "3" allele from the mother. Thus, both these alleles are shared IBD as well as IBS. Contrast this with the situation in genotype configuration (3), in which both sibs have a "1" allele and thus share this allele IBS. Inspection of the parental genotypes, however, indicates that in one case this allele was inherited from the father, while for the other sib it was inherited from the mother. Accordingly, this

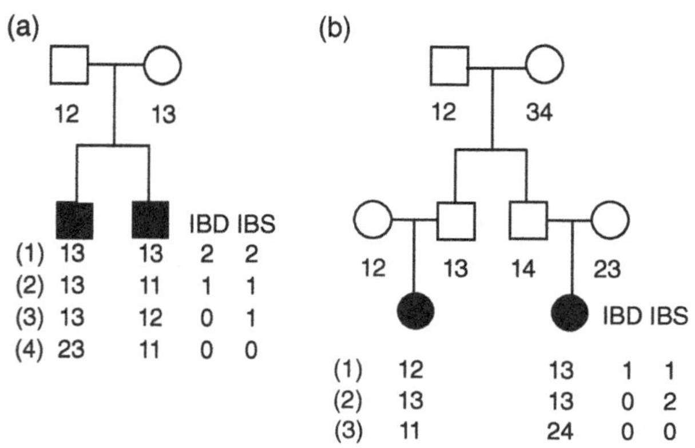

Figure 6.8 Examples of identity by state and identity by descent. See text for discussion.

allele, while shared IBS, is not shared IBD. It is important to point out that in this situation we were only able to distinguish with certainty the origin of each allele because the parents were both genotyped and because the mating type was informative (i.e. the parents were of different heterozygous genotypes). In practice, there will be cases in which one can only estimate the probability that the sibs share a particular number of alleles IBD. In many cases in which the parental genotypes are unknown, these estimates will depend on the frequency of the observed (and unobserved) alleles at the marker locus in question. The more polymorphic the marker, the more precisely we are able to estimate the IBD status.

For example, consider the case of two siblings who are identical homozygotes at a marker locus (i.e. they share two alleles IBS) and the parents are unavailable. If this allele is relatively rare in the population, then it is most likely that both unobserved parents are heterozygous at this locus and thus the two sibs share both alleles IBD. Conversely, if the allele is common, it is likely that at least one parent is homozygous at this locus and thus IBD status is more uncertain. A similar example for cousin pairs is shown in Figure 6.8b. In the second configuration, we see that although the two cousins have identical genotypes, they do not share either allele inherited from their common grandparents. Examination of the pedigree structure points out an important principle, namely, that individuals who are related through only one parent (unilineal relationships) can share at most only a single allele IBD. Examples of these are parent–offspring, cousins, half-siblings, and uncle–nephew. Individuals related through both parents (bilineal relationships) can share both alleles IBD. Other than siblings (and monozygotic twins), the most common bilineal relationship is double first cousins, which occur, for example, when two brothers marry two sisters; the offspring of these unions are double first cousins. Other bilineal relationships occur in complex genealogies in which there are regular patterns of intermarriage between related individuals. Table 6.9 lists the expected proportion of alleles shared IBD for a number of common relationships. These expected proportions are based on the principle of random segregation of gametes during meioses along with the number of meioses separating the two relatives of a pair.

As we saw above, unambiguously determining the IBD status of an affected relative pair (ARP) is not always possible, even when intervening relatives are genotyped. This is even more likely when intervening relatives are not available for genotyping. There are two additional sources of

Table 6.9 Expected percentage of affected pairs showing 0, 1, or 2 alleles IBD at a marker locus if no linkage is present.

	Alleles IBD		
Pair type	**0**	**1**	**2**
Monozygotic twins	0	0	100
Siblings	25	50	25
Parent–child	0	100	0
Grandparent–grandchild	50	50	0
Half-siblings	50	50	0
Uncle–nephew, etc.	50	50	0
Cousins	75	25	0
Double first cousins	56	38	6

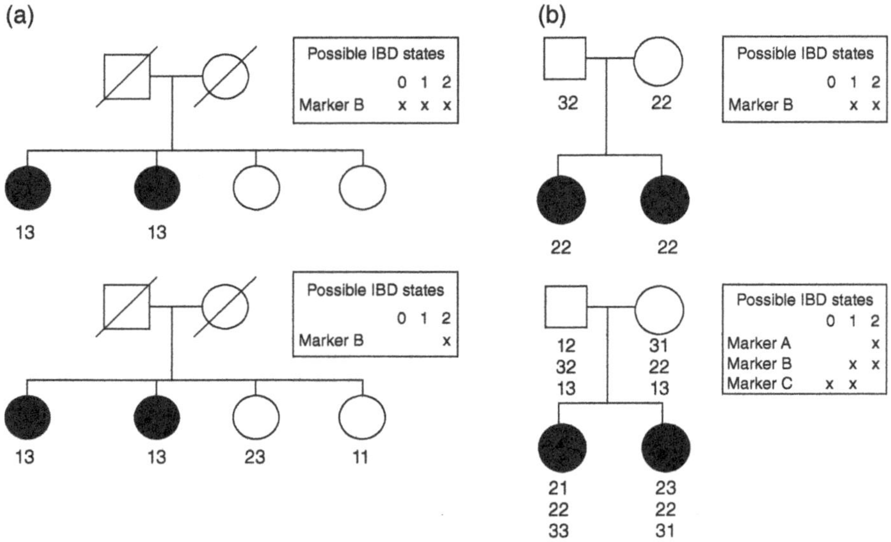

Figure 6.9 Example showing the inclusion of additional family members (here genotypes of unaffected siblings) aiding in the resolution of IBD.

information to help determine relative-pair IBD status at a given marker. The first source of information is from genotypes in additional family members to reduce the number of possible genotypes to consider for untyped or uninformative individuals. As Figure 6.9a illustrates, when the genotypes from the affected sib-pair alone are considered all IBD states are possible due to the large number of parental genotypes that could have resulted in those offspring genotypes. However, when the genotypes from the two unaffected siblings are included, the parental genotypes are clear, 12 × 13, and the affected sib-pair share 2 alleles IBD. The second source of information for IBD status is contained in the IBD states in nearby markers. As Figure 6.9b illustrates when we consider only Marker B, the affected sib-pair are equally likely to share 1 or 2 alleles IBD, since the mother is homozygous. However, when we add genotypes for closely spaced Markers A and C, the most likely IBD configuration for Marker B is that the affected sib-pair shares 1 allele IBD. The confidence with which we are able to choose one IBD configuration over another depends on the spacing of the markers. As the distance between the markers increases, the probability of recombination increases, decreasing the amount of information that the flanking markers contribute to determination of the IBD state at Marker B. IBD sharing can be calculated by a number of the programs discussed earlier in this chapter (e.g. GENEHUNTER, ALLEGRO, and MERLIN).

Methods for Nonparametric Linkage Analysis

Methods for NPL analysis can be distinguished based on a four key features: (i) which members of the family are included in the analysis; (ii) how phenotypic similarity and estimates of trait locus genetic effect are incorporated; (iii) how genotypic similarity at the marker locus is measured and estimated, i.e. by IBS or IBD; and (iv) whether the analysis includes IBD estimates at a single marker or at multiple markers. The most variability occurs in which family members are included in the analysis and how many markers can be analyzed, as evidenced by the differences in algorithms to estimate IBD. In general, the implementation of several different methods employing different constellations of family members can be quite useful in the analysis of complex traits, since different methods and different statistics highlight different features of the data (Anderson et al. 1999; Schmidt et al. 2001; Sengul et al. 2001; Sobel et al. 2001).

Tests for Linkage Using Affected Sibling Pairs (ASP)

Test Based on Identity by State

Identity-by-state relationships in sib-pairs compares the observed distribution of pairs sharing 0, 1, or 2 alleles IBS to that expected under the hypothesis of free recombination between disease and marker loci. The test for linkage therefore is a simple chi-square goodness-of-fit test comparing the observed IBS distribution to the expected. The expectations for IBS sharing under the null hypothesis are, of course, a function of the allele frequencies at the marker locus that take into account the distributions of mating types compatible with the observed pair genotypes. Table 6.10 shows the expected distribution of sharing 0, 1, and 2 alleles IBS as a function of the informativeness of the marker (indexed as the number of equally frequent alleles).

Because alleles can be IBS without being IBD, there are more pairs than expected sharing 1 and 2 alleles than the 1/2 and 1/4 expected for IBD sharing; however, as the number of marker alleles increases, it can be seen that the IBS distribution (at least in nuclear families) approaches the 1/4, 1/2, 1/4 expected for IBD relationships. Intuitively, as discussed earlier, this is because for highly polymorphic markers, the probability that one or both parents are homozygous, or both are identical heterozygotes (which makes IBD status ambiguous), becomes increasingly small. The power of these methods for sib and other relative pairs is considered in a paper by Bishop and Williamson (1990), who examined the effect on power of the strength of the genetic component, marker

Table 6.10 Analysis of IBS sharing probabilities for two siblings at a marker with *n* equally frequent alleles.

	Probability of siblings sharing alleles IBS		
Number of alleles	**0**	**1**	**2**
2	0.03	0.38	0.59
3	0.06	0.48	0.46
4	0.08	0.52	0.40
5	0.10	0.53	0.37
6	0.12	0.53	0.35
8	0.15	0.53	0.32
10	0.16	0.53	0.30
20	0.20	0.52	0.28
Infinite	0.25	0.50	0.25

informativeness, and type of relative pair. Although we have denoted these methods "identity by state," it is important to recognize that here one is simply using IBS as a surrogate for the unobservable IBD status. Many of the more sophisticated IBD methods presented in the next section are, in effect, estimating the unknown IBD distribution and thus are similar to the approach presented in this section.

Tests Based on Identity by Descent in ASPs

Simple Tests

In the simplest case, where the numbers of pairs sharing 0, 1, or 2 alleles IBD can be unambiguously determined (i.e. parents are typed and informative), several tests have been developed to test for linkage between a marker locus and a disease. The first such proposed test was a chi-square goodness-of-fit test comparing the observed proportion of sib-pairs sharing 0, 1, or 2 alleles IBD with the expected proportions of ¼, ½, and ¼ under the hypothesis of no linkage between the marker and disease locus. A related test proposed by Suarez et al. (1978) is to compare only the proportion of sib-pairs who share exactly 2 alleles IBD with the expected proportion of ¼. Only differences in which the observed proportion of shared alleles exceeds the expected are considered significant; thus, a one-sided test is used. The third test, called the means test, compares the average IBD sharing in the sample of pairs with the expected value of $0.5 = [(0\times ¼+1\times ½+2\times ¼)/2$ chromosomes]. Each of these tests can be written in terms of the observed numbers of pairs with each IBD sharing status:

0	1	2	Total
n_0	n_1	n_2	N

Statistics

1) *Goodness of Fit*

$$\chi^2 = \frac{4\left(n_0 - \frac{N}{4}\right)^2 + 2\left(n_1 - \frac{N}{2}\right)^2 + 4\left(n_2 - \frac{N}{4}\right)^2}{N}$$

Reject H_0 if χ^2 is greater than $\chi^2_{1-\alpha}$ with 2° of freedom for size α test.

2) *Proportion*

$$T_1 = 4\left(\frac{n_2}{N} - \frac{1}{4}\right)\left(\frac{N}{3}\right)^{\frac{1}{2}}$$ one-sided t test with $N-1$° of freedom

3) *Means*

$$T_2 = (2n_2 + n_1 - N)\left(\frac{2}{N}\right)^{\frac{1}{2}}$$ one-sided t test with $N-1$° of freedom

Table 6.11 Results of simple Sibpair tests on IDDM data.

Test	Statistic[a]	Degrees of freedom
Goodness of fit	$\chi^2 = 88.4$	2
Proportions	$T_1 = 9.22$	138
Means	$T_2 = 8.58$	136

[a] T_1 is the proportions test statistic; T_2 is the means test statistic.

As an example of the application of these three methods, we use data taken from Cox and Spielman (1989) for HLA sharing in 137 sib-pairs affected with type I diabetes (IDDM).

	Alleles or Haplotypes IBD		
	0	1	2
Observed	10	46	81
Expected	34	69	34

The results of these analyses are shown in Table 6.11.

Given the overwhelming evidence for the involvement of HLA in IDDM as shown by the data above, all three of the tests described give a highly significant p-value ($p < 10^{-15}$), but in other situations, these tests may provide different conclusions regarding the significance of observed IBD data. The choice of the optimal test from the set presented depends on the true underlying genetic model, which is usually unknown. However, studies have shown that the means test performs best under a fairly wide range of genetic models (Blackwelder and Elston 1985).

Tests Applicable When IBD Status Cannot Be Determined

The tests described above are useful when the IBD status of the entire data set is known with certainty, or when there is a sufficient subset of the entire data set to permit the subset, where IBD status is unknown, to simply be discarded. In many situations, however, marker genotype data is unavailable on one or both parents of the affected sib-pair. This is frequently the case, for example, for disorders with late age of onset, such as AD or prostate cancer. In these situations, however, information on additional family members often provides some information about the sharing status of the sibs. For this reason, methods that can use all the available data to make inferences about the IBD status of the affected pairs are desirable. The first such method was proposed in a seminal series of papers published by Risch (1990b,c) in which the author considered the basic models for general sib-pair analysis and the power of these methods. The method, denoted MLS (for maximum LOD score), compares the probability of the observed genotypic data in the affected sib (or other relative) pairs as a function of the proportion sharing 0, 1, or 2 alleles IBD (Z_0, Z_1, Z_2) with the corresponding probability of the observed marker data under the null hypothesis of no linkage ($Z_0 = 0.25$, $Z_1 = 0.5$, $Z_2 = 0.25$). Numerical search methods are used to find the values of Z_0, Z_1, Z_2 which make the numerator of this ratio (hence the ratio itself, since the denominator is constant) the largest. Put another way, we are finding the values of (Z_0, Z_1, Z_2) that are most consistent with the marker genotypes of the affected pairs. This method is designed to use information from affected sibs or other relatives in determining the optimal IBD proportions. To be analogous to parametric linkage analysis, the $\log_{10}$ of the likelihood ratio is taken to produce a LOD score. For example, for the HLA data above, the optimal values of the sharing probabilities (Z_0, Z_1, Z_2) are easily shown to simply be the observed proportions 10/137 = 0.073, 46/137 = 0.336, and 81/137 = 0.591, respectively, since we have complete information in this sample regarding IBD status. This produces a LOD score of:

$$\mathrm{LOD} = \log_{10}\left(\frac{0.073^{10}\,0.0336^{46}\,0.591^{81}}{0.25^{10}\,0.50^{46}\,0.25^{81}}\right) = 16.98$$

equivalent to odds of $10^{17}:1$ in favor of involvement of HLA in this disease.

However, not all values of the Z_i that may be considered in the numerical search are compatible with genetic models. For example, the case of none of the pairs sharing any alleles ($Z_0 = 1$) is not compatible with any Mendelian model of inheritance (in large samples). Holmans (1993) showed that the $\{Z_i\}$ for possible genetic models must satisfy

$$2Z_0 \le Z_1 \le ½$$

which he called the "genetically possible triangle." Note that this implies that $Z_0 \le 0.25$ and that $Z_2 \ge 0.25$. In this same paper, Holmans showed that restricting the estimation of the $\{Z_i\}$ to satisfy this criterion resulted in a more powerful test of linkage than the test based on unrestricted maximization.

The maximum likelihood estimates of (Z_0, Z_1, Z_2) can be used to make inferences about the contribution of this locus to the overall increased familial risk, as the expected sharing at a marker locus linked to a disease locus can be written in terms of the λ_S value for the disease locus and the distance between this locus and the marker locus (Suarez et al. 1978; Hodge 1984; Risch 1990a). These equations can then be solved to estimate the contribution of this locus to the overall familial risk. When there is no recombination between the marker locus and the disease locus – that is, when the marker locus is a candidate locus for the disease – the following equations relate the sharing probabilities and the familial relative risks attributable to a given locus.

$$Z_0 = \frac{1}{4\lambda_s}; \lambda_s = \frac{1}{4Z_0}$$

$$Z_1 = \frac{\lambda_0}{2\lambda_s}; \lambda_0 = \frac{Z_1}{2Z_0}$$

$$Z_2 = \frac{\lambda_M}{4\lambda_s}; \lambda_M = \frac{Z_2}{Z_0}$$

When the hypothesized disease gene is located some distance from the marker locus, more complex relationships (see, e.g. (Risch 1990b) hold, although they can be somewhat simplified by assuming that the recombination fraction between the disease locus and marker locus, θ, is not too large (e.g. $\theta < 0.10$) and that the offspring and sibling risks due to the locus are the same (i.e. $\lambda_O = \lambda_S$). If these assumptions can be justified, then the following holds:

$$Z_0 = \frac{1}{4} - \frac{(2\omega - 1)(\lambda_s - 1)}{4\lambda_s}$$

$$Z_1 = \frac{1}{2}$$

$$Z_2 = \frac{1}{4} + \frac{(2\omega - 1)(\lambda_s - 1)}{4\lambda_s}$$

where

$$\omega = \theta^2 + (1 - \theta)^2$$

In our HLA example, assuming zero recombination between HLA and IDDM susceptibility, we would estimate the sibling risk attributable to this locus to be $\lambda_{S\text{-HLA}} = 1/(4 \times 0.076) = 3.3$. If, however, the disease gene is not HLA but is located a distance of $\theta = 0.05$ from the HLA complex, the estimate of $\lambda_{S\text{-HLA}}$ increases to 7.1, using the estimate of Z_0. This example illustrates one of the problems of using single point analysis; the strength of the effect of a given locus and the distance of this locus from the tested marker is confounded. That is, in a genomic search with anonymous markers, it will be difficult to distinguish between a relatively weak disease locus located very close to a marker and a locus of larger effect located some distance away. There is no unique estimate of the disease gene location when using one marker at a time.

If we assume that HLA is, in fact, the predisposing locus (or located very close to it), we can use the estimate of 3.3 obtained above to get some idea of how much of the overall familial risk is likely to be accounted for by the HLA system. Since observed values from a variety of studies have estimated the recurrence risk to siblings for IDDM to be 0.06 and the disease has a population prevalence of about 0.004 (Spielman et al. 1980), this yields an overall λ_S value of $0.06/0.004 = 15$. Thus, based on the data, it is clear that other susceptibility loci must be involved in the determination of IDDM. The estimated contribution of the HLA locus is log(3.3)/log(15) = 44% for a multiplicative model and (3.3 − 1)/(15 − 1) = 16% for an additive multilocus model. The results of a complete genomic search for IDDM susceptibility loci are discussed later in this chapter.

Multipoint Affected Sib-Pair Methods

As we have seen, the methods discussed in this chapter depend on our ability to distinguish identity by descent from identity by state in affected sib-pairs. In addition, it is easy to see that the ability to detect linkage to a disease locus will depend on the distance between the marker locus at which IBD sharing is assessed and a disease locus influencing the disease under study. The use of multiple markers in a defined genetic region addresses both these issues by maximizing the probability that complete or partial IBD information is available in the region spanned by the markers. The first person to examine IBD sharing in nuclear families as a function of marker genotyping at multiple linked loci was Goldgar (1990), who examined the estimation of IBD sharing in a region spanned by genetic markers in order to map quantitative trait loci. For qualitative traits such as disease phenotypes, the theoretical advantages of using multiple markers (albeit perfectly informative ones) was first discussed by Risch (1990b). Hauser and Boehnke (1998) extended the general MLS method for ASPs to multipoint mapping in which the evidence for linkage to a disease locus on a map of markers is assessed. Risch's method is implemented in the computer program ASPEX.

Handling Sibships with More Than 2 Affected Siblings

Although the methods described earlier are oriented toward analyzing a set of affected sib-pairs, often other affected siblings are available for study. For methods designed to analyze pairs only, this requires the formation of all possible sib-pairs from the sibship. If the sibship contains n affecteds, there are $n(n-1)/2$ distinct pairs that can be formed for the analysis. The difficulty is that these pairs are not independent, and therefore significance levels based on such data will be inaccurate, resulting in increased type I error rates. For example, using the fact that there are four possible distinct parental alleles in a sibship, if sibs A and B share 0 alleles IBD, and sibs B and C also share 0 alleles IBD, we know that sibs A and C must share 2 alleles IBD. One solution proposed by Hodge (1984) to circumvent this problem is to weight the contribution of a sibship by $n-1$ (the number of independent pairs in a sibship of size n). For certain traits, consideration of discordant pairs (one affected, one unaffected) also may provide information for linkage (Olson 1995). In this case, one would expect there to be more pairs who share 0 alleles IBD at the expense of pairs sharing 2 alleles IBD. However, the power of the discordant pairs under almost all circumstances will be small compared to that for affected pairs. For diseases with variable (or late) age of onset and genes of relatively small effect, discordant pairs will be almost totally uninformative. However, if they are collected and genotyped to contribute additional information to the overall estimation of IBD status in the family, including them in the analysis may be worthwhile (Hauser et al. 1996).

Methods Incorporating Affected Relative Pairs

While studies of ASP have some practical advantages – the methods are conceptually simple, and in many cases, affected siblings are more abundant than other relative pairs – it is often the case that a large variety of ARPs are available for study. Families collected for study often contain both ASPs as well as affected avuncular and cousin pairs, sometimes extending across several generations and more distant relationships. The major disadvantage of sib-pair analysis is its inability to use additional information that can be extracted from other affected relatives, such as aunts, uncles, cousins, grandparents, or more distantly related individuals. Methods that use information from such affected relatives have been developed to take greater advantage of these extended families, such as APM and SimIBD (Weeks and Lange 1992; Davis et al. 1996). While these ARP methods have proven very useful for linkage studies, other methods have superseded them (St. George-Hyslop et al. 1990; Pericak-Vance et al. 1991; Haines et al. 1996).

NPL Analysis

In 1996, Kruglyak et al. introduced an ARP approach they termed NPL (nonparametric linkage) as part of the GENEHUNTER computer program (Kruglyak et al. 1996). Like the other ARP methods, the NPL statistic measures allele sharing among affected individuals within a pedigree and is based on allele sharing methods first described by Whittemore and Halpern (1994). The NPL method can analyze the data using a pairwise approach (using the $\text{NPL}_{\text{pairs}}$ statistic), but it has been extended to provide a simultaneous comparison of alleles in all affected individuals in a pedigree (the NPL_{all} statistic). The NPL approach is inherently multipoint, since it calculates the IBD probability for any given point along the chromosome (the "inheritance distribution"), using all available marker data on that chromosome.

The calculation of the NPL statistic can be broken into two parts. The first part, the calculation of the inheritance distribution, will be used to estimate the allele sharing among a set of affected individuals, be they just pairs or complete family groupings. This is done as outlined in Lander and Green (1987) and Kruglyak et al. (1996), and the reader is referred to these sources for a detailed description. The second part of the calculation is the individual evaluation of the scoring function that determines whether the inheritance information is indicative of linkage. As mentioned above, two scoring functions are described, the $\text{NPL}_{\text{pairs}}$ and the NPL_{all}.

If the inheritance pattern can be determined unambiguously in a pedigree, the resulting $\text{NPL}_{\text{pairs}}$ statistic is

$$S_{\text{pairs}} = \Sigma S_{ij}$$

where i and j are the two individuals being compared and $S_{ij} = 0, 1,$ or 2, depending on how many alleles are shared IBD.

In most genetically complex traits, however, the inheritance pattern cannot be determined unambiguously. In this case, the S_{pairs} statistic is an average taken across all possibilities.

To compare S_{pairs} to a statistical distribution, it is normalized:

$$Z_{\text{pairs}} = \frac{S_{\text{pairs}} - E\left(S_{\text{pairs}}\right)}{\left[\text{var}\left(S_{\text{pairs}}\right)\right]^{1/2}}$$

where $E(S_{\text{pairs}})$ and var.(S_{pairs}) are the expectation of the mean and variance of S_{pairs} under the null hypothesis.

The overall statistic across pedigrees is

$$Z = \Sigma\left(\frac{1}{\sqrt{m}}\right)\left(Z_{\text{pairs}}\right)$$

where $(Z_{\text{pairs}})_i$ = normalized score for pedigree i
m = number of pedigrees

The NPL_{all} statistic is defined as

$$S_{all} = 2^{-a}\sum_{h}\left[\prod_{j=1}^{2f} b_j(h)!\right]$$

where a = number of affected individuals in the pedigree
h = collection of alleles generated by taking one allele from each affected individual (there are 2a possible collections)
$2f$ = total number of founder alleles in the pedigree (i.e. the total number of different alleles of distinct origin)
$b_i(h)$ = total number of a specific founder allele (j) in the collection (h)

This statistic is averaged over all feasible inheritance patterns, normalized, and weighted across pedigrees in the same manner as the NPL_{pairs} statistic.

Statistical significance is determined by comparing the Z score to the standard normal distribution. The use of the standard normal distribution is an approximation, usually a conservative one. In other words, the true p-value is often smaller than the p-value obtained by using a standard normal table. To fix the conservative bias inherent in the NPL statistics based on the standard normal distribution Kong and Cox (Kong and Cox 1997) adapted the GENEHUNTER algorithm to provide a likelihood-based statistic presented as a LOD score. The LOD score is less conservative and more powerful.

The NPL_{all} statistic is the only ARP method that can consider all affected relatives simultaneously, rather than as a combination of all possible comparisons of pairs. This implements the intuitively attractive idea that if, for example, five affected individuals in a pedigree all share the same allele IBD, this information should carry more weight than if each of the 15 possible pairs in the same pedigree shared some allele IBD, not necessarily the same allele. This increased weighting is seen in the NPL_{all} equation by including the factorial of $b_j(h)$, rather than $b_j(h)$ itself. However, the NPL_{all} statistic does not perform as well as the NPL_{pairs} statistic in some situations, particularly those involving very common, recessive-acting alleles.

There is one potential disadvantage to the NPL statistic, related to its inherently multipoint approach. As discussed, the Lander–Green algorithm can incorporate many markers because computer time increases only linearly with the addition of new markers. However, computer time increases exponentially with the number of individuals in the pedigree. Thus, NPL is limited to smaller pedigrees. This is dictated by the rule that $2n-f\leq 16$, where n is the number of nonfounders and f is the number of founders. More complicated or extended pedigrees cannot be handled with any confidence, even on very fast workstations. Since the initial release of the GENEHUNTER software, there has been steady improvement in the speed of the algorithm and complexity of the pedigrees that can be analyzed (Gudbjartsson et al. 2000; Markianos et al. 2001). For example, MERLIN can analyze pedigrees with $2n-f\leq 32$ by using sparse binary trees and by taking advantage of symmetries in pedigree data for untyped individuals (Abecasis et al. 2002). SIMWALK2 is another option for generating NPL statistics.

In general, NPL analysis can be used in several situations:

1) When pedigrees of moderate size are available. Even pedigrees with only a single affected sib-pair can be useful, since a large number of markers can be examined simultaneously. However, even with the algorithmic improvements, large pedigrees cannot be examined.

2) When many relatives other than siblings are available for a genetically complex trait. Each pedigree needs only a single ARP.
3) When a large number of linked markers are being (or could be) examined. This is particularly useful if pedigrees have somewhat "sparse" genotyping (i.e. many individuals providing the biological links between affected individuals are not available for genotyping). This situation generally leads to difficulty in generating the IBD status for a single marker, and the problem may be reduced substantially if many linked markers are genotyped.

Fitting Population Parameters

To implement a genetic model that fits the prevalence and recurrence risk ratios requires knowledge of the λ_R value and an idea of the population prevalence. The following approximate equations can be used to construct a set of "reasonable" models that fit the epidemiological data:

$\lambda_S D^2 = qP^2$ for dominantly acting loci and $4\lambda_S D^2 = q^2P^2$ for recessively acting loci where λ_S is the sibling familial risk, D is the disease prevalence, q is the frequency of the allele associated with disease susceptibility, and P is the penetrance (probability of disease) associated with the risk allele. The risk of disease in noncarriers, t, is given by:

$$t = \frac{D - 2qP}{1 - 2q}$$

for dominant loci, and

$$t = \frac{D - q^2P}{1 - q^2}$$

for recessive loci.

For any known (or assumed) values of λ_S and D, values of q, P, and t can be determined from the equations above. Then traditional LOD score analysis (either two-point or multipoint) can be performed under this limited set of models, with appropriate correction of the significance threshold for the number of models tested (Hodge and Elston 1994; Abreu et al. 2002).

For example, consider a disease in which the sibling risk due to a hypothesized locus is 4.0 and the disease population prevalence is 1%. This set or parameters would be compatible with the following genetic models:

1) A rare autosomal dominant locus with an allele frequency of 0.0005 for the high-risk allele, and penetrance of 89% in individuals who carry at least one copy of the high-risk allele and 0.9% in noncarriers.
2) A common autosomal dominant locus with an allele frequency of 0.01 and a penetrance of 20% in carriers and 0.6% in noncarriers.
3) An autosomal recessive locus with a disease allele frequency of 0.05. Individuals who are homozygous for this allele have an 80% chance of disease, compared with 0.8% for heterozygous and normal homozygous individuals.

While these models all produce a familial relative risk of four and a disease prevalence of 1%, they differ in several respects. For example, the proportion of all cases due to the proposed risk allele will vary from 39% for model 2 to 20% for model 3 to 9% for model 1. These models will also differ with respect to the parent–offspring risk and in the proportion of families with many cases of disease. Thus, if one also has an idea of the offspring risk, it may be possible to rule out either a dominant or recessive model, and the frequency of extended families with high incidence of disease may help distinguish a rare high-penetrant allele from a more common, but lower penetrant gene. In any case, the accurate knowledge of the basic epidemiological profile of the disease in question can aid in choosing a small number of consistent models to use in parametric LOD score analysis of affected sib-pairs or other affected relative sets.

The major advantage of the likelihood-based method is that marker genotype information is used from all individuals, and the maximum likelihood approach captures all the possible marker information and thus generates the best estimates of IBD status. These estimates are used only in conjunction with the most certain of the trait information (i.e. phenotypes of affected individuals). This method can also be used on large and arbitrarily complex pedigrees, and limited (three to four markers) multipoint analysis can be performed. It also considers all affected individuals simultaneously, not just in a pairwise fashion. Finally, standard programs such as LINKAGE, VITESSE, and MENDEL can be used for these calculations.

The major disadvantage of the likelihood-based analysis is that the parameters of a genetic model must still be specified. Even if penetrance is set to zero, the inheritance pattern (e.g. autosomal dominant, autosomal recessive, or X-linked) remains to be specified, and misspecification will still generate misleading results. Furthermore, power to detect a locus can be extremely low if the genetic model is grossly misspecified. Thus, the results must be interpreted with caution. In addition, the amount of computer time needed for the calculations can become excessive if the pedigrees are large or have marriage or consanguinity loops, or if multipoint analysis is undertaken.

Although widely used, affected only analysis only compensates for poor knowledge of penetrance; it is not a satisfying solution to the other problems extant in complex traits. Nevertheless, it can be used in several situations when other methods cannot be implemented:

1) In single-gene disorders when the age at onset, or other penetrance functions, cannot be well specified.
2) When a major gene is suspected, but cannot be proven.
3) When large or complex pedigrees are being studied, even if the mode of inheritance is unknown.
4) When multipoint analysis is being contemplated, since other ARP methods (above) may be limited to two-point analysis.

The LOD score approach has the advantages of producing an estimate of the recombination fraction as well as permitting the estimation of other parameters of the genetic model. If the model is even approximately correct, the use of the LOD score method should provide higher power than nonparametric methods (Goldin and Weeks 1993). Also, under a wrong model, the type I error rate, (i.e. the chance of finding significant evidence of linkage when it is not present) is not inflated (Williamson and Amos 1990). Given that the LOD score method has been shown to be reasonably robust to deviations from the true model, has considerable flexibility in analyzing data from any pedigree structure with any number of affected and unaffected individuals (i.e. does not need to reduce the data to pairs), and makes complete use of available marker information, the use of this method should not be ruled out, even for studies dealing with sib-pair data. For further discussion of these issues, the reader is advised to consult Greenberg et al. (1996).

Power Analysis and Experimental Design Considerations for Qualitative Traits

Factors Influencing Power of Sib-pair Methods

The power of the affected sib-pair method – that is, the probability that a true disease-predisposing locus will be detected by examining a linked marker locus – was first considered in detail in the papers by Risch (1990b,c) for the MLS identity-by-descent method and by Bishop and Williamson (1990) for methods based on identity by state. As demonstrated by Risch (1990b), the power to detect linkage of a trait locus and a marker locus can be expressed solely in terms of the single parameter of sibling relative risk, λ_S, regardless of the underlying genetic model. The power is calculated as a function of the difference in the expected values of the IBD sharing probabilities (Z_0, Z_1, Z_2) if the disease is linked to the marker locus and the values expected under the null hypothesis of no linkage ($Z_0 = 0.25$, $Z_1 = 0.5$, $Z_2 = 0.25$). As we saw above, the expected distribution of (Z_0, Z_1, Z_2) can be written as simple functions of λ_S and the disease–marker locus recombination fraction θ.

For reference, Figure 6.10, reproduced from Risch's work (Risch 1990b), shows the power to detect linkage between a perfectly informative marker locus and a single disease locus that accounts for all the observed familial risk, as a function of sibling relative risk λ_S and the recombination fraction between the trait and marker locus for three sample sizes. The dramatic effect of recombination on power is easily seen, particularly for moderate ($N = 100$) and small ($N = 40$) sample sizes, even when λ_S is quite large.

Holmans (1993) showed higher power for the restricted-triangle method described above compared with the unrestricted MLS test for both dominant and recessive models. This gain in power was especially marked for cases of low marker informativeness and/or high recombination fraction between the marker and disease locus ($\theta = 0.1$). Holmans also examined the utility of genotyping parents in sib-pair analyses as a function of marker heterozygosity. The results indicated a ratio of required sample size of as much as 1.5 (i.e. not genotyping parents required 50% more pairs) when the marker polymorphism is low. If efficiency is measured by the number of individuals

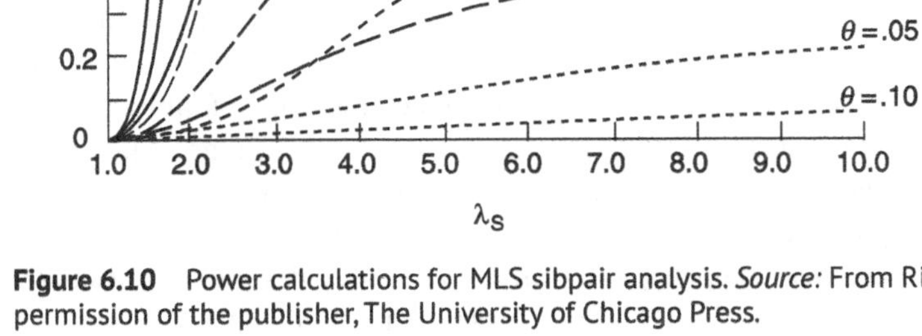

Figure 6.10 Power calculations for MLS sibpair analysis. *Source:* From Risch (1990b). Reproduced with permission of the publisher, The University of Chicago Press.

needing to be genotyped, then genotyping parents (which requires twice as much genotyping) could be considered to be inefficient, particularly with markers with heterozygosities of over 0.5. This of course assumes that the desired number of sib-pairs needed without typing parents can be easily ascertained and that they are genotyped accurately. Power for the case of interval mapping of a disease locus has been considered by Hauser et al. (1996), who derived conclusions regarding the genotyping of parents and other relatives similar to those of Holmans but with the increased power resulting from a multipoint analysis. It is clear that in designing a study to map a complex disease gene, one should consider the estimated magnitude of the effect of interest (as measured by λ_S), the marker density, the availability of parents or other relatives for genotyping, and the relative importance of type I and type II errors. To aid in the design of such studies, the DESPAIR (design of sib-pair) program, a part of the SAGE package, can be used.

In some situations, however, sib-pairs may not be the most efficient sampling strategy for mapping genes involved in complex disease. For example, Goldgar and Easton (1997) compared a variety of nuclear family sampling strategies in an analytical study of a number of two-locus disease models. For the majority of situations examined, affected sib trios were found to be a more efficient design for mapping disease loci than sib-pairs. This was especially true when one of the two loci had been localized/identified and linkage or mutation data were available and incorporated in the analysis. Even accounting for the increased difficulty in ascertaining affected sib trios, this design (or one more extreme) still proved superior to sib-pairs and should be considered for any gene localization effort. Sengul et al. (2001) review and compare a variety of ASP statistics applied to sibship sizes of five affected siblings for a variety of different models. They conclude that the S_{all} statistic performs very well in a variety of situations; however, there can be differences in performance based on the level of informativeness of the marker as well as on the distance between the marker and the disease locus.

Determining the power of a proposed study is clearly important in making decisions about the ability of the study to result in a useful conclusion, and these studies must be done prior to the start of the study. However, it is extraordinarily difficult to perform these power studies. Such studies require knowledge about the overall genetic effect to be observed as well about the role of individual genes, their interactions with each other, and interactions with environmental factors. In addition, particular family structures must be assumed, even before the families are collected, thus the number of ASPs, ARPs, and extended families achieved often is not easy to estimate in advance. Chapter 12 provides a detailed examination of power considerations for genetic studies of complex traits. The example below provides an example of power considerations for a simple example employing nonparametric linkage analysis.

The Example of Testicular Cancer

As part of an international effort to map susceptibility loci for testicular cancer, Easton (unpublished) calculated the expected LOD scores for the available sample under a variety of different models for the number and type of loci influencing the trait. All models were constrained to fit the observed sibling risk (λ_S) of 8.0 (e.g. (Goldgar et al. 1994)) and included one, two, or four loci acting additively or multiplicatively. These models were chosen to span a reasonable range of possibilities, although there are few data that a priori favor one model over another. The effects of intermarker distance and heterozygosity were also examined. The data consisted of 77 affected sib-pairs, 2 affected half-sibs, 15 affected uncle–nephew pairs, and 6 affected cousin pairs. Table 6.12 shows the expected LOD scores for each situation. While these analyses were done using a parametric

Table 6.12 Expected LOD scores for mapping testicular cancer susceptibility Loci using affected pairs under a variety of genetic models and genome search conditions.

			Heterogeneity[c]		Multiplicative[d]	
Marker density (cM)[a]	**Marker Heterozygosity (%)[b]**	**Single Gene**	**Two Genes**	**Four Genes**	**Two Genes**	**Four Genes**
0	100	14.4	3.3	0.8	5.8	1.9
10	90	7.8	2.1	0.4	3.3	1.2
(Multipoint)	80	5.8	1.4	0.3	2.6	0.9
20	90	5.0	1.2	0.3	2.2	0.8
(Multipoint)	80	3.6	0.9	0.2	1.7	0.6
20	90	3.1	0.7	0.2	1.4	0.5
(Two-point)	80	2.1	0.5	0.1	1.0	0.4

[a] Spacing between genetic markers: 0 cM refers to the situation of a disease gene that is a known candidate locus. For the other situations, the disease locus is assumed to lie midway between the two markers. Either a multipoint analysis using the two flanking markers simultaneously or a two-point analysis between the disease locus and each marker is performed.

[b] Informativeness of the marker (i.e. the proportion of individuals heterozygous at a given marker locus).

[c] In the heterogeneity models, most affected pairs are caused by only one of the two loci. The locus specific risks are 4.5 and 2.75 for the two- and four-locus models, respectively.

[d] In the multiplicative models, most pairs are the result of high-risk alleles at two or more loci.

analysis that assumed a rare autosomal dominant model with low penetrance, similar conclusions would apply with any of the methods discussed so far in this chapter.

Table 6.12 shows immediately that there is a dramatic decrease in the expected LOD score as a function of the number of loci involved. This is to be expected, since the effect of any single locus is reduced. We note however that in general, if the loci are acting multiplicatively, there is higher power (as measured by the expected LOD score) than if they are acting in an additive fashion, even though in the former case, the sibling relative risk due to an individual locus will be reduced (although the relative contribution will be the same). The reason for this was alluded to earlier, when it was observed that in multiplicative models, a higher fraction of pairs would be expected to be segregating risk alleles at multiple disease loci, while in additive models each pair may be due to only one of the set of disease susceptibility loci.

It is also apparent that moving from a completely informative marker completely linked to the disease locus ($\theta = 0$) to the more realistic genomic search situation of less than perfect markers with a marker density of 10–20 cM also reduces the expected LOD score considerably. Assuming the usual criteria of a LOD score of 3.0 for detection of linkage (note that an expected LOD score of 3.0 then corresponds to 50% power), one can see that under a realistic genomic search situation, for two or more loci contributing to disease susceptibility, it is unlikely that this sample is of sufficient size to detect linkage to any locus.

Examples of Sib-Pair Methods for Mapping Complex Traits

So far in discussing nonparametric linkage analysis, we have concentrated on presenting an overview of the methods developed to utilize nonparametric linkage methods in mapping genes contributing to complex human qualitative traits. However, the reader might well ask whether these methods and sampling schemes, for all their theoretical advantages, have successfully identified genes for any complex human traits through screening of random genomic markers. There has been great progress in several disorders using this approach, primarily in insulin-dependent (juvenile) diabetes, for which in addition to the candidate loci of HLA and insulin, a genomic search for additional IDDM susceptibility locus revealed evidence for at least three other genes (Davies et al. 1994). All these loci were characterized by relatively small effects (locus-specific λ_S values 1.3–1.5). Given that IDDM is characterized by an overall sibling relative risk of 15, even under a multiplicative model these loci account for only about 60–65% of the overall familial aggregation. Subsequent studies employing linkage or genome-wide association (Chapter 9) approaches have implicated over 40 genes, many with small effects not detectible using linkage approaches (Noble 2015).

An interesting example of the use of affected sib-pair methodology for a disease with a smaller overall familial component, but a higher prevalence in the population is provided by the analysis of type 2 (non-insulin-dependent) diabetes (NIDDM) by Hanis et al. (1996). This study utilized a primary sample of 330 Mexican-American affected sib-pairs typed for 490 genetic markers distributed throughout the genome and found evidence (maximum LOD score [MLS)] = 3.2) of a susceptibility locus on the terminal portion of the long arm of chromosome 2, designated *NIDDM1*. The maximum effect of this locus (assuming no recombination between the linked marker and the disease) was a locus-specific λ_S value of 1.37. Since the overall sibling relative risk for NIDDM is 2.8, the authors estimate that this locus accounts for 30% of the overall familial risk, assuming a multiplicative model for the interaction between all the loci involved. The original result resulted in a large region of linkage, not amenable to fine mapping. Using a method that conditions the linkage result at *NIDDM1* on linkage evidence on chromosome 15, the investigators were able to substantially narrow the *NIDDM1* region (Cox et al. 1999) such that association methods (Chapter 8) could be applied. Finally, *CALPAIN10*, a gene appearing to account for both the linkage and association evidence, was identified as the *NIDDM1* gene (Horikawa et al. 2000).

Another example of a genome-wide search for susceptibility loci for a complex disease involves several parallel genomic searches for susceptibility loci for multiple sclerosis (MS). A series of studies conducted in the mid-1990s were published in Nature Genetics (Ebers et al. 1996; Haines et al. 1996; Sawcer et al. 1996). Although all three studies found evidence for a susceptibility locus in (or near) the *HLA* complex (which had been suggested by earlier association studies and biological plausibility), no additional locus was found by more than one of the three studies. Subsequent studies, including a more recent study by Beecham et al. (2013) involving many of the same participants from these studies, have brought the total to at least 110 loci related to MS susceptibility.

Mapping Quantitative Traits

The previous sections have focused largely on qualitative or discrete traits, wherein an individual either has or does not have the trait of interest. In the following sections, we will now consider traits whose outcomes can take on a wide range of different values (i.e. continuously distributed); these are called quantitative traits (Moore 2003). Examples of quantitative traits include height, weight, blood pressure, and plasma enzyme levels. Quantitative traits are inherently more difficult to study than discrete traits and thus require different statistical methods.

Why study quantitative traits? The answer to this question is obvious when the clinical endpoint of interest is quantitative in nature. For example, clinical pharmacologists are interested in identifying genetic variations that are associated with interindividual variation in the metabolism or pharmacokinetics of specific drugs. Here, the clinical end point might be the activity of cardiovascular drug-metabolizing enzymes such as those from the cytochrome P450 family (Nakagawa and Ishizaki 2000). There are, however, compelling reasons to study quantitative traits even when the clinical end point is qualitative. Factors such as the uncertainty of diagnosis due to measurement error and lack of standard objective criteria for diagnosis can make it difficult to discern who, in fact, has disease. This is evident for common chronic diseases such as psychiatric disorders (Rice 1993; Tsuang et al. 1993) and asthma (Britton 1998). A strategy to facilitate identification of susceptibility genes for these diseases is to use intermediate risk factors as the trait of interest. Intermediate risk factors or endophenotypes (Ott 1995) may be closer to the underlying biological etiology of the disease and thus may make it easier to identify the underlying genes. For example, Williams et al. (1999) demonstrated that joint consideration of alcoholism diagnosis and quantitative event-related potentials increases the evidence for linkage of markers on chromosome 4 to an alcoholism susceptibility gene. Thus, for some diseases, quantitative traits may improve the power to identify susceptibility genes. Genetic architecture has been defined as (i) the number of genes that influence a trait, (ii) the number of alleles each gene has, (iii) the frequencies of the alleles in the population, and (iv) the influence of each gene and its alleles on the trait (Weiss 1993). While the first three points fall to some degree under the umbrella of population genetics, the last point lies at the heart of the genetics of quantitative traits. Although there are many different ways that genes can influence quantitative traits, most studies have focused on the influence of particular polymorphisms on the mean value of a trait.

Measuring Genetic Effects in Quantitative Traits

Individual differences in quantitative traits such as height, blood pressure, or triglyceride levels are typically the result of several genes together with both specific and random environmental components. These confounding factors in both quantitative and qualitative traits require the use of methods in which the precise mechanisms of disease causation are not required to be known; this has led to the recent interest in the methods described below.

For quantitative traits, we no longer characterize an individual as affected or not affected, but by the specific value of some quantitative measurement made on that individual. The values of this measurement will usually vary from individual to individual. Our goal in analyzing such traits is to "explain" this variation in terms of specific components, both genetic and environmental. Before discussing the role of specific genes, or specific environmental effects, it is useful to get an overall idea of how much of the observed variation in the trait is due to genetic factors, how much is due to nongenetic familial factors that are shared among family members (common environment), and how much is due to random environmental effects, which are unique to each individual. To do this, we look at the degree to which different categories of related and unrelated individuals have similar values for the quantitative trait of interest. If there is significant evidence of a genetic component, we can begin the process of localizing specific genes that may be responsible for the genetic variation. The methods available for both these tasks are described in the next several sections.

The strength of the genetic component is traditionally measured by a quantity called heritability. Usually, we distinguish between two different types of genetic variation, additive effects and dominance effects. As discussed in Chapter 3, broad sense heritability (H^2) is defined as the proportion of the total variation in the trait that is due to genetic factors ($[V_A + V_D]/V_P$), where V_P is the total phenotypic variance of the trait), while narrow sense heritability (h^2) is the proportion of the total variation due to additive genetic variation (V_A/V_P). For quantitative traits, the heritability can be estimated from the observed correlations between relatives. For nuclear family data, these consist of the intraclass correlation between siblings ρ_{SS} (estimated from the between-and within-sibship mean squares from analysis of variance) and the interclass correlation (ordinary Pearson product-moment correlation) between parents and offspring, ρ_{PO}. Thus, V_A/V_P is estimated as $2\rho_{PO}$, while V_D/V_P is estimated as $4(\rho_{SS} - \rho_{PO})$. Table 6.13 shows the sib–sib and parent–offspring correlations and corresponding estimates of the proportion of total variance due to additive and dominance genetic factors for systolic and diastolic blood pressure.

The estimate of V_D assumes that the increased similarity of sib-pairs compared to parent–child pairs is due to dominance variation; but, depending on the phenotype, it could be the result of

Table 6.13 Familial Correlations in Blood Pressure.

Trait	σ_{PO}	σ_{SS}	V_A/V_T	V_D/V_T	h^2 (Broad Sense)
Blood pressure					
Systolic	0.237	0.333	0.47	0.38	0.85
Diastolic	0.183	0.265	0.37	0.33	0.70

Source: Cavalli-Sforza and Bodmer (1971), based on work of Miall and Oldham (1963).

higher shared environment among sibs than parent–child pairs. Without detailed studies of other relatives, particularly, half-sibs, or adoption studies, it is impossible to distinguish true genetic dominance from common environment.

In general, studies to estimate variance components are most easily performed on a random sample of families. One example of such a study is the San Antonio Family Heart Study (Mitchell et al. 1996) that collected a random sample of extended Mexican-American families to examine genetic and environmental determinants of cardiovascular risk factors. This study has resulted in several publications reporting the heritability and linkage results for a variety of quantitative traits, e.g. lipid traits LDL and HDL (Almasy et al. 1999; Rainwater et al. 1999).

To illustrate how a variance components model works, let us assume that we have a candidate gene polymorphism with two alleles (*B* and *b*) and three genotypes (*BB*, *Bb*, and *bb*). In the population, we represent the overall trait mean by μ and each genotype-specific mean by μ_{BB}, μ_{Bb}, and μ_{bb}. Differences between the genotype-specific means can be due to the additive effects of the alleles, represented by a, and the dominant effect of one of the alleles, represented by d. If we define a constant, c, as $(\mu_{BB}+\mu_{bb})/2$ then each genotype-specific mean can be written as $\mu_{BB} = c-a$, $\mu_{Bb} = c+d$, and $\mu_{bb} = c+a$. If we assume $d = 0$, then the allelic effects are purely additive and the trait mean for the *Bb* genotype is midway between the two homozygotes. However, if $d = -a$, then the *B* allele has a purely dominant effect. If we assume Hardy–Weinberg equilibrium, the overall quantitative trait mean in the population can be expressed as a function of each genotype-specific mean multiplied by each genotype frequency. Thus, $\mu = p^2(c+a)+2pq(c+d)+q2(c-a)$.

In addition to this difference in genotype-specific means, the magnitude of these variance components is influenced by the allele frequencies (hence the genotype frequencies) at the locus of interest. These concepts can be extended to any number of loci. One general case that is often considered is that of an infinite number of loci, each with small effect, which form the basis of the polygenic inheritance model of quantitative traits.

Just as in the case with multilocus disease models, the number of loci will not, in general, be known. However, the magnitude of these effects as a proportion of the total phenotypic variance can be estimated from similarities among relatives, analogous to recurrence risks in discrete traits. For our purpose here, the important consideration is that two siblings contribute to half the additive variance and one-fourth of the dominance variance, while parents and offspring contribute to only the additive component of variance. Only bilineal relatives who can potentially share two alleles IBD will contribute to dominance variation; in fact, the proportion of this shared variance is equal to their probability of sharing two alleles IBD.

Clearly, there are several ways in which genes can influence quantitative trait levels (i.e. level genes). However, it is important to keep in mind that variances, co-variances, and dynamics of quantitative traits may also be genotype-specific. One of the earliest studies demonstrating that variability in a quantitative trait can be genotype-specific was carried out by Berg et al. (1981). This study, and a confirmatory study by Berg (1988), showed that variability of serum cholesterol is dependent on Kidd blood group locus genotype. Subsequent studies by Reilly et al. (1991), Haviland et al. (1995), and Nelson et al. (1999) reinforced the idea that variability genes may play an important role in the genetic architecture of quantitative traits. These studies demonstrated that variability in multiple lipid and apolipoprotein traits were *apolipoprotein E* genotype-specific. Why are variability genes important? What are the etiological inferences? Reilly et al. (1991) suggest that intragenotypic variability may be due to gene–gene and gene–environment interactions. Thus, the magnitude of the effects of a particular environmental of genetic factor may differ across genotypes. All three of the above studies demonstrated that pairwise correlations of various lipids and apolipoproteins were *apolipoprotein E* genotype-specific suggesting that genes can influence trait covariability.

In addition to considering different types of gene effects, it is also important to realize that each type of gene effect may be dependent on a particular genetic or environmental context. Thus, the influence of a particular polymorphism on quantitative trait levels, variability, co-variability, or dynamics may depend on one or more other polymorphisms (i.e. epistasis or gene–gene interaction) and/or one or more environmental factors (i.e. plastic reaction norm or gene–environment interaction). In fact, epistasis may be much more common than previously believed (Templeton 2000). For example, Nelson et al. (2001) demonstrated strong evidence for epistasis for most of the lipid-related quantitative traits examined using an exploratory data analysis approach called the combinatorial partitioning method or CPM. These epistasis effects were detected in the absence of any independent main effects of the candidate gene polymorphisms. Studies by Moore et al. (2002a,b) have also documented evidence that levels of PAI-1 and co-variability of t-PA and PAI-1 are dependent on interactions among polymorphisms in the *ACE* and *PAI-1* genes. This has been validated by protein–protein studies, as well as other genetic studies (Asselbergs et al. 2007; Pattin and Moore 2008).

Although context-dependent gene effects play an important role in genetic architecture, our ability to detect and characterize context-dependent effects is limited in several important ways. First, most parametric statistical methods suffer from the curse of dimensionality (Bellman 1961). For example, most linkage and association methods for quantitative traits use linear regression. As each additional genetic locus is added to the linear regression model, the number of gene–gene interaction terms grows exponentially. Few data sets are large enough to provide the degrees of freedom necessary to estimate the number of parameters required to model the epistasis effects. Second, as the total number of markers grows, it quickly becomes computationally infeasible to evaluate all possible combinations. One solution is to use stepwise linear regression to first condition on those markers that might have an independent main effect. However, gene effects that are purely interactive will be missed. The data dimensionality problem and the computability problem will need to be addressed before we can fully characterize the genetic architecture of quantitative traits.

Study Design for Quantitative Trait Linkage Analysis

There are several general approaches to determining the nature of the genetic component of a quantitative trait. The first approach starts with the trait of interest and attempts to draw inferences about the underlying genetics from looking at the degree of trait resemblance among relatives. This approach is sometimes referred to as a top-down or unmeasured genotype strategy because the inheritance pattern of the trait is the focus and no genetic polymorphisms are actually measured. The top-down approach is often the first step taken to determine whether there is evidence for a genetic component. Chapter 3 introduces heritability and segregation analysis as two top-down approaches.

With the bottom-up or measured genotype approach, either candidate gene polymorphisms or anonymous genomic markers are measured and then used to draw inferences about which genes might play a role in determining interindividual variability in a quantitative trait. The focus of the remaining sections in this chapter is the use of a bottom-up approach to identify QTLs using genetic linkage analysis.

Haseman–Elston Regression

In 1972, Haseman and Elston published a relatively simple linear regression method for identifying QTLs using full sibling pairs. This approach is based on the idea that if a particular marker locus is linked to a QTL, the squared difference in the trait values between sibling pairs should decrease as the proportion of alleles the sibs share IBD increases. In a randomly mating population, the mean proportion of alleles shared IBD is expected to be one half. Thus, sibs are expected to share greater than one half of their alleles IBD if the marker is linked to the QTL. Using linear regression, the dependent variable (Y) in the model is the squared difference in trait values and the independent or predictor variable (π) is the proportion of marker alleles shared IBD ($Y = \alpha + \beta\pi$). When the marker and the QTL are the same, the slope or β is equivalent to the additive genetic variance multiplied by −2. When the marker and the QTL are not the same and are separated by some distance, β is equivalent to the additive genetic variance multiplied by $-2(1-2\theta)^2$ where θ is the recombination fraction. Assuming linkage equilibrium between alleles at the marker locus and alleles at the QTL, the null hypothesis that there is no relationship between the squared difference in trait values and the proportion of marker alleles shared IBD can be tested using a one-sided t-test where the test statistic is the estimate of the slope divided by the standard error of the slope.

The Haseman–Elston approach is appealing because of its simplicity. However, it is limited in several important ways. First, for diallelic markers the power is fairly low because there are few informative matings for estimating allele sharing (Robertson 1973). This can be avoided by using markers with four or more alleles (Amos et al. 1989). Second, the distance of the QTL from the marker locus is confounded with the additive genetic variance component such that, when a statistically significant regression is observed, it is difficult to know whether there is a QTL with a large effect far away from the marker locus or a QTL with a small effect close to the marker locus. Fulker and Cardon (1994) address this issue by developing an interval mapping extension of Haseman–Elston regression. In their approach, two flanking markers separated by a known map distance are used to locate the QTL and estimate the size of its effect. This is carried out using the proportion of alleles shared IBD at each of the marker loci to estimate the proportion of alleles shared IBD at the QTL assuming a range of recombination values in the marker interval. The independent variable in this regression is the proportion of alleles shared IBD at the QTL. As with the original Haseman–Elston regression, the null hypothesis is tested using a one-sided t-test where the test statistic is the estimate of the slope divided by the standard error of the slope. It should be noted that multipoint extensions of this approach may provide more power and better estimates of QTL location when a dense marker map is available or when marker information is variable (Fulker et al. 1995; Kruglyak and Lander 1995a). However, multipoint methods assume that there is no linkage interference and that the distances between markers are accurately estimated.

Elston et al. (2000) described a revision of the popular regression approach described above. Here, the dependent variable (Y) becomes the mean corrected product of the sibling trait values ($[x_{1j}-\mu][x_{2j}-\mu]$) instead of the squared differences. Using simulation studies, this change was found to improve both the type I and type II error rates over that associated with the original Haseman–Elston regression approach described above. In addition, Xu et al. (2000) provide a modification of this revised approach that provides greater power when the sibling trait values are correlated. A study by Allison et al. (2000) show that this new Haseman–Elston approach is robust to deviations from trait normality. Even so, power may be improved by winsorization (i.e. adjusting extreme values) of nonnormal distributions (Fernandez et al. 2002). A detailed comparison of these recent developments with Haseman–Elston regression and others is presented by Feingold (2002).

An example application of Haseman–Elston regression is the study by Saccone et al. (2000). The goal of this study was to carry out a genome scan to identify QTLs for an intermediate trait or endophenotype for alcoholism. Here, the trait of interest was the maximum number of alcoholic drinks consumed in a 24-hour period. A total of 1105 full sib-pairs were analyzed using Haseman–Elston regression. Saccone et al. found very good evidence (LOD = 3.5) for a region on chromosome 4 that might harbor an alcoholism QTL. This region is very close to the alcohol dehydrogenase gene cluster on chromosome 4 which may prove to include several candidate level genes. The authors conclude that studying a quantitative trait for alcoholism improved their power to identify the candidate region and that similar measures of substance consumption may benefit studies of other substance abuse disorders.

What are the advantages of the Haseman–Elston regression approach? The most important advantage is the simplicity. These methods are relatively easy to understand and easy to interpret. Another important advantage is that this approach is model-free. That is, no particular genetic model is assumed. In general, model-free methods have good statistical properties (i.e. robust), are easier to extend to analyzing multiple loci simultaneously, and have more power than model-based approaches when the inheritance model has not been correctly specified. An important disadvantage in the original implementation is that this approach was limited to the analysis of siblings. However, Sham et al. (2002) implemented a multivariate regression approach to Haseman–Elston in MERLIN for the analysis of extended pedigrees and several groups have successfully used this in identifying QTLs in large families (Yang et al. 2012; Greenwood et al. 2013). For example, Greenwood et al. (2013) employed this method in the analysis of different aspects of personality as surrogate quantitative phenotypes for bipolar disorder in a group of 51 extended families with bipolar disorder. While they were unable to achieve a significant LOD score for any of the phenotypes, they did identify several promising regions for novelty seeking, persistence, and self-directedness.

Variance Components Linkage Analysis

The variance components approach to linkage analysis was extended by Amos (1994) to use relative pairs other than sibs (Amos 1994). Table 1 of Amos (1994) gives the additive major gene variance components for different relative pairs. For example, the additive major-gene variance component for full sibs is $[1/2 + (1-2\theta)^2(\pi_{ij} - ½)]\sigma^2_a$ where θ is the recombination fraction between the marker locus and the QTL, π_{ij} is the IBD sharing for the marker locus for the ith and jth relatives, and σ^2_a is proportion of variance due to the additive effects of the alleles at the major gene. Amos (1994) also outlines how the variance components for polygenes or unmeasured QTLs (σ^2_G) and unexplained environmental or chance factors (σ^2_e) can also be modeled. Maximum likelihood or estimating equation methods can both be used to estimate the various parameters in the model. Here, σ^2_a is the primary parameter of interest for hypothesis testing. Under the null hypothesis that a particular QTL is not associated with a quantitative trait, σ^2_a is equal to zero. This null hypothesis can be tested using chi-square statistics that are formed from −2 multiplied times the log-likelihood ratio comparing the unrestricted model in which σ^2_a, σ^2_G, σ^2_e, and θ are all estimated to the restricted model where σ^2_a is set to zero.

Amos (1994) carried out a number of simulation studies to validate the power and robustness of the variance components approach. The simulations were based on a major-QTL influencing a normally distributed quantitative trait. These studies confirm that this quantitative trait linkage approach has reasonable power and good statistical properties for identifying QTLs that have a relatively large independent main effect. This is further confirmed by Amos et al. (1996) with additional simulation studies and an application of the method to identifying QTLs for several intermediate risk factor traits for cardiovascular disease. Extensions to the variance component approach include the ability to model longitudinal data in pedigrees (de Andrade et al. 2002) and the ability to test for linkage in the presence of imprinting (Shete and Amos 2002).

Barnholtz et al. (1999) have applied the variance components approach to identifying QTLs associated with monoamine oxidase B, an intermediate risk factor trait for alcoholism, using families and markers from the Collaborative Study on the Genetics of Alcoholism (COGA). The authors carried out single-point and multipoint linkage analysis with covariates in the model and identified 24 markers on four different chromosomes that showed evidence for linkage with major-gene QTLs. However, when three monoamine oxidase B outliers were removed and the analysis repeated, no evidence for linkage of any markers to a QTL was found. This study illustrates the importance of exploratory data analysis prior to conducting a genetic study and the potential impact of violations of trait distribution assumptions on variance component models.

The variance component linkage analysis methods of Goldgar (1990) and Amos (1994) are more powerful and more flexible than the Haseman–Elston approaches but are limited to relative pairs. Blangero and Almasy (1997) introduced an extension of the variance components approach that considers the complete multivariate phenotypic vector of a pedigree thus allowing application of the method to full pedigree data; this is implemented in SOLAR. Blangero and Almasy (1997) suggest that for a simple model in which some number of quantitative trait loci (QTL) and residual polygenes play a role in determining quantitative trait levels, the jth individual's phenotype (y_j) can be assumed to be a function of the sum of $i = 1$ to n QTL effects (q_{ji}), an additive polygenic effect (g_j), a covariate effect (x_j), and a random environmental deviation (e_j). With these assumptions, the jth individual's phenotype can be modeled by the linear function: $y_j = \mu + \Sigma\, q_{ji} + \beta x_j + g_j + e_j$ where μ is the overall trait mean and β is the regression coefficient for the covariate effect. The expected covariance (Ω) of a quantitative trait between a pair of relatives in a general pedigree can be modeled as $\Omega = \Sigma \Pi_i \sigma^2_{qi} + 2\Phi\sigma^2_g + \mathrm{I}\sigma^2_e$ where σ^2_{qi} is the additive genetic variance due to the ith QTL, Π_i is a matrix whose elements provide the predicted proportion of genes that individuals share IBD at a QTL that is linked to a genetic marker locus, σ^2_g is the genetic variance due to residual additive genetic factors, Φ is the kinship matrix, σ^2_e is the variance due to individual-specific environmental effects, and I is an identity matrix. Parameters in this model can be estimated using maximum likelihood and the null hypothesis that the additive genetic variance of the ith QTL is equal to zero can be tested using chi-square statistics derived from likelihood ratio test statistics (Blangero and Almasy 1997). SOLAR is the commonly used implementation of variance component analysis, but variance component analysis can also be carried out using MERLIN.

Blangero et al. (2000, 2001) provide a detailed review of the assumptions underlying this approach. The primary assumption is that the phenotypes in the pedigrees have a multivariate normal distribution. Allison et al. (1999) studied violations of the normality assumption and found the variance component approach to be robust to some types of nonnormality but not to others. For example, leptokurtosis results in higher than nominal type I error rates (i.e. more false-positives than expected for a particular significance level). In general, the new Haseman–Elston approach (Elston et al. 2000) is more robust to deviations from normality than the variance components approach (Allison et al. 2000). Feingold (2002) has reviewed the mixed results of attempts to make the variance components approach more robust to deviations from normality.

Comuzzie et al. (1997) applied this approach to identifying QTLs for leptin levels in 10 Mexican-American families ranging in size from 35 to 71 individuals. A single marker locus was identified with strong evidence for linkage to a QTL that explained about 47% of the variation in serum leptin levels (LOD = 4.95). Thus, the variance components approach was able to detect evidence for a major gene on chromosome 2 that influences human obesity. More recently, Sacco and colleagues have conducted QTL mapping studies that utilize both family linkage studies (with SOLAR) and independent association studies in a large cohort to identify genes for a variety of quantitative traits, including carotid plaque and obesity (Dong et al. 2010, 2011). In these studies, the investigators were able to identify suggestive linkage regions (LOD > 2.0) for several phenotypes, including carotid plaque and BMI, in the family studies and then identify variants from these regions which were associated with the trait in the cohort.

It is well known that QTLs that influence quantitative traits primarily through epistatic interactions can be difficult to detect using traditional linkage methods such as variance components (Eaves 1994; Cheverud and Routman 1995; Tiwari and Elston 1998; Cheverud 2000). Mitchell et al. (1997) carried out a simulation study to evaluate the power of the approach to identify epistatic QTLs. Using a linear model, an individual's phenotypic value (y) can be written as $y = \mu + a_1 + a_2 + aa + g + e$ where μ is the overall mean of the trait, a_1 and a_2 are the additive effects of two QTLs, aa is the additive–additive interaction affect, g is the residual polygenic effect, and e is the residual random environmental effect. When applied to simulated data, the variance component approach had poor power to detect epistatic QTLs.

Kissebah et al. (2000) used the variance component linkage approach to carry out a genome scan for QTLs influencing intermediate quantitative risk factor traits for metabolic syndrome. A 10-cM scan was carried out in 2209 individuals from 507 nuclear families. A single marker on chromosome 3q27 was strongly linked to a QTL influencing six different intermediate quantitative traits (LODs = 2.4 to 3.5). Interestingly, a second marker on chromosome 17p12 was strongly linked to a QTL influencing plasma leptin levels (LOD = 5.0) through an epistatic interaction with the QTL from chromosome 3q27. Here, tests for epistasis were limited to the two chromosomal regions that showed the most evidence for independent main effects and were limited to additive–additive interactions. This study illustrates the usefulness of the variance component approach for testing hypotheses about epistasis. However, this limited set of analyses is likely to overlook additional QTLs whose effects are primarily through epistatic interactions.

The advantages of the variance components approach include the ability to handle complex pedigree structures and the ability to test for gene–gene and gene–environment interactions. While most linkage studies have traditionally employed a defined set of markers with relatively high heterozygosities, the advent of whole genome/exome sequencing allows for the identification of rare, often private, variants. Importantly, individual variants within the same gene may have very different effects. Consequently, Blangero et al. (2013) have extended SOLAR to include association analyses of cohorts combined with families to capture this information.

Nonparametric Methods

All of the quantitative trait linkage methods described above use the linear model and thus assume that the quantitative trait is normally distributed. However, in the event that this assumption is violated there are two alternatives. The first alternative is to apply an appropriate mathematical transformation to restore the distribution to normality. The second alternative is to use a nonparametric approach that is not sensitive to distributional properties. Kruglyak and Lander (1995b) provide a nonparametric approach to mapping QTLs that is based on a generalization of the nonparametric Wilcoxon rank-sum statistic. The advantage of this test is that quantitative traits with any distribution can be studied. The disadvantage is that this statistic does not provide a direct estimate of the genetic effect, rather, it simply tests for the presence of a QTL.

Each of the methods presented has reasonable power for identifying genes that have a relatively large independent main effect on interindividual variability in quantitative trait levels. However, identifying major genes that influence quantitative trait levels is only one small part of the genetic architecture of most quantitative traits. None of the methods presented are designed to identify variability or covariability genes much beyond the two gene level. Further, the power of these methods to identify genes whose effects are context-dependent is very limited. Why are more powerful methods for detecting interactions not available? The primary reason for this is largely historical in nature. Much of quantitative genetics as we know it today (Falconer and Mackay 1996; Lynch and Walsh 1998) can be attributed to the early work of Fisher (1918) and Wright (1932). Fisher's view of genetic architecture was that interindividual variability in quantitative traits was due to many genes each with independent effects that could be added together. This idea became very popular because it fit nicely with Mendelian genetics, and linear models could be used to partition the additive genetic variance. Epistasis or the nonadditive genetic variance was considered a nuisance and a statistical artifact (Chapter 11). Wright, on the other hand, emphasized that the relationship between genotype and phenotype is dependent on dynamic interactive networks of genes and environmental factors. This idea holds true today, as discussed above. Gibson (1996) stressed that gene–gene and gene–environment interactions must be ubiquitous given the complexities of intermolecular interactions that are necessary to regulate gene expression and the hierarchical complexity of metabolic networks. Thus, Fisher viewed epistasis as a statistical phenomenon in populations that depends on allele frequencies while Wright viewed it as a physiological phenomenon resulting from gene effects in individuals that are not dependent on allele frequencies. The debate as to whether interindividual variation in quantitative traits is due to the independent additive effects of genes or the interactive effects of gene networks continues today. For the most part, the Fisherian view has been more popular since the theory is relatively easy to understand and apply. However, there is accumulating evidence that gene–gene and gene–environment interactions play an important role in determining interindividual variability in quantitative traits (Schlichting and Pigliucci 1998; Templeton 2000; Pigliucci 2001; Li M et al. 2014; Buil et al. 2015). The quantitative trait linkage analysis methods presented here all have reasonable power to identify major-QTLs and can be implemented using freely available computer programs. It is critical that we continue to extend current methodologies, or develop novel ones, to expand both the number gene interactions as well as environmental effects that can be examined in our search to identify and characterize QTLs.

The Future

During much of the 1990s–2000s, linkage analysis and other family-based approaches fell out of favor and were replaced by association methods, largely deployed in case–control or cohort studies. However, the advent of next-generation sequencing has created a renewed interest in the family design. Variants can be filtered based on sharing within a family, considerably narrowing the number of variants which need to be followed up. "Cosegregation", i.e. linkage, with the phenotype can be determined. The phasing of multiple variants simultaneously is greatly simplified and, more importantly, is more reliable. As software is extended/developed that can truly capture the underlying architecture of complex diseases, we can truly make inroads into our understanding of disease and, hopefully, therefore, develop better treatments and preventions.

Software Available

An excellent reference for all of the linkage analysis software mentioned in this chapter is the list maintained by Dr. Jurg Ott (https://www.jurgott.org/linkage/home.html). This website provides direct links or information to obtain most of the statistical analysis software available.

References

Abecasis, G.R., Cherny, S.S., Cookson, W.O., and Cardon, L.R. (2001). GRR: graphical representation of relationship errors. *Bioinformatics* 17 (8): 742–743.

Abecasis, G.R., Cherny, S.S., Cookson, W.O., and Cardon, L.R. (2002). Merlin – rapid analysis of dense genetic maps using sparse gene flow trees. *Nat. Genet.* 30: 97–101.

Abreu, P.C., Hodge, S.E., and Greenberg, D.A. (2002). Quantification of type I error probabilities for heterogeneity LOD scores. *Genet. Epidemiol.* 22: 156–169.

Allison, D.B., Neale, M.C., Zannolli, R. et al. (1999). Testing the robustness of the likelihood-ratio test in a variance-component quantitative-trait loci-mapping procedure. *Am. J. Hum. Genet.* 65: 531–544.

Allison, D.B., Fernandez, J.R., Heo, M., and Beasley, T.M. (2000). Testing the robustness of the new Haseman-Elston quantitative-trait loci-mapping procedure. *Am. J. Hum. Genet.* 67: 249–252.

Almasy, L., Hixson, J.E., Rainwater, D.L. et al. (1999). Human pedigree-based quantitative-trait-locus mapping: localization of two genes influencing HDL-cholesterol metabolism. *Am. J. Hum. Genet.* 64: 1686–1693.

Amos, C.I. (1994). Robust variance-components approach for assessing genetic linkage in pedigrees. *Am. J. Hum. Genet.* 54: 535–543.

Amos, C.I., Elston, R.C., Wilson, A.F., and Bailey-Wilson, J.E. (1989). A more powerful robust sib-pair test of linkage for quantitative traits. *Genet. Epidemiol.* 6: 435–449.

Amos, C.I., Zhu, D.K., and Boerwinkle, E. (1996). Assessing genetic linkage and association with robust components of variance approaches. *Ann. Hum. Genet.* 60: 143–160.

Anderson, J.L., Hauser, E.R., Martin, E.R. et al. (1999). Complete genomic screen for disease susceptibility loci in nuclear families. *Genet. Epidemiol.* 17: S467–S472.

de Andrade, M., Gueguen, R., Visvikis, S. et al. (2002). Extension of variance components approach to incorporate temporal trends and longitudinal pedigree data analysis. *Genet. Epidemiol.* 22: 221–232.

Asselbergs, F.W., Williams, S.M., Hebert, P.R. et al. (2007). Epistatic effects of polymorphisms in genes from the renin-angiotensin, bradykinin, and fibrinolytic systems on plasma t-PA and PAI-1 levels. *Genomics* 89 (3): 362–369.

Barnholtz, J.S., de Andrade, M., Page, G.P. et al. (1999). Assessing linkage of monoamine oxidase B in a genome-wide scan using a univariate variance components approach. *Genet. Epidemiol.* 17 (Suppl 1): S49–S54.

Beecham, A.H., Patsopoulos, N.A., Xifara, D.K. et al. (2013). Analysis of immune-related loci identifies 48 new susceptibility variants for multiple sclerosis. *Nat. Genet.* 45 (11): 1353–1360.

Bellman, R. (1961). *Adaptive Control Processes*. Princeton, NJ: Princeton University Press.

Berg, K. (1988). Variability gene effect on cholesterol at the Kidd blood group locus. *Clin. Genet.* 33: 102–107.

Berg, K., Borresen, A.L., and Nance, W.E. (1981). Apparent influence of marker genotypes on variation in serum cholesterol in monozygotic twins. *Clin. Genet.* 19: 67–70.

Bishop, D.T. and Williamson, J.A. (1990). The power of identity-by-state methods for linkage analysis. *Am. J. Hum. Genet.* 46: 254–265.

Blackwelder, W.C. and Elston, R.C. (1985). A comparison of sib-pair linkage tests for disease susceptibility loci. *Genet. Epidemiol.* 2: 85–98.

Blangero, J. and Almasy, L. (1997). Multipoint oligogenic linkage analysis of quantitative traits. *Genet. Epidemiol.* 14: 959–964.

Blangero, J., Williams, J.T., and Almasy, L. (2000). Quantitative trait locus mapping using human pedigrees. *Hum. Biol.* 72: 35–62.

Blangero, J., Williams, J.T., and Almasy, L. (2001). Variance component methods for detecting complex trait loci. *Adv. Genet.* 42: 151–181.

Blangero, J., Diego, V.P., Dyer, T.D. et al. (2013). A kernel of truth: statistical advances in polygenic variance component models for complex human pedigrees. *Adv. Genet.* 81: 1–31.

Boehnke, M. (1986). Estimating the power of a proposed linkage study: a practical computer simulation approach. *Am. J. Hum. Genet.* 39: 513–527.

Boehnke, M. and Cox, N.J. (1997). Accurate inference of relationships in sib-pair linkage studies. *Am. J. Hum. Genet.* 61: 423–429.

Britton, J. (1998). Symptoms and objective measures to define the asthma phenotype. *Clin. Exp. Allergy* 28 (Suppl 1): 2–7.

Buetow, K.H. (1991). Influence of aberrant observations on high re-solution linkage analysis and reporting. *Am. J. Hum. Genet.* 49: 985–994.

Buil, A., Brown, A.A., Lappalainen, T. et al. (2015). Gene-gene and gene-environment interactions detected by transcriptome sequence analysis in twins. *Nat. Genet.* 47 (1): 88–91.

Cardon, L.R. and Fulker, D.W. (1994). The power of interval mapping of quantitative trait loci, using selected sib pairs. *Am. J. Hum. Genet.* 55: 825–833.

Cheung, V.G., Burdick, J.T., Hirschmann, D., and Morley, M. (2007). Polymorphic variation in human meiotic recombination. *Am. J. Hum. Genet.* 80: 526–530.

Cheverud, J.M. (2000). Detecting epistasis among quantitative trait loci. In: *Epistasis and the Evolutionary Process* (eds. M. Wade, B. Brodie III and J. Wolf), 58–81. New York, NY: Oxford University Press.

Cheverud, J.M. and Routman, E.J. (1995). Epistasis and its contribution to genetic variance components. *Genetics* 139: 1455–1461.

Claus, E.B., Risch, N., and Thompson, W.D. (1991). Genetic analysis of breast cancer in the cancer and steroid hormone study. *Am. J. Hum. Genet.* 48: 232–242.

Clerget-Darpoux, F., Bonaiti-Pellie, C., and Hochez, J. (1986). Effects of misspecifying genetic parameters in LOD score analysis. *Biometrics* 42: 393–399.

Comuzzie, A.G., Hixson, J.E., Almasy, L. et al. (1997). A major quantitative trait locus determining serum leptin levels and fat mass is located on human chromosome 2. *Nat. Genet.* 15: 273–276.

Conneally, P.M., Edwards, J.H., Kidd, K.K. et al. (1985). Report of the committee on methods of linkage analysis and reporting. *Cytogenet. Cell Genet.* 40: 356–359.

Corder, E.H., Saunders, A.M., Strittmatter, W.J. et al. (1993). Gene dose of apolipoprotein E type 4 allele and the risk of Alzheimer's disease in late onset families. *Science* 261: 921–923.

Cottingham, R.W., Idury, R.M., and Schaffer, A.A. (1993). Faster sequential genetic linkage computations. *Am. J. Hum. Genet.* 53: 252–263.

Cox, D.W. and Spielman, R.S. (1989). The insulin gene and susceptibility to IDDM. *Genet. Epidemiol.* 6: 65–69.

Cox, N.J., Frigge, M., Nicolae, D.L. et al. (1999). Loci on chromosomes 2 (NIDDM1) and 15 interact to increase susceptibility to diabetes in Mexican Americans. *Nat. Genet.* 21: 213–215.

Dausset, J., Cann, H., Cohen, D. et al. (1990). Centre d'Etude du Polymorphisme Humain (CEPH): collaborative genetic mapping of the human genome. *Genomics* 6: 575–577.

Davies, J.L., Kawaguchi, Y., Bennett, S.T. et al. (1994). A genome-wide search for human type 1 diabetes susceptibility genes. *Nature* 371: 130–136.

Davis, S., Schroeder, M., Goldin, L.R., and Weeks, D.E. (1996). Nonparametric simulation-based statistics for detecting linkage in general pedigrees. *Am. J. Hum. Genet.* 58: 867–880.

Dong, C., Beecham, A., Slifer, S. et al. (2010). Genomewide linkage and Peakwide association analyses of carotid plaque in Caribbean Hispanics. *Stroke* 41 (12): 2750–2756.

Dong, C., Beecham, A., Slifer, S. et al. (2011). Genome-wide linkage and peak-wide association study of obesity-related quantitative traits in Caribbean Hispanics. *Hum. Genet.* 129 (2): 209–219.

Douglas, J.A., Boehnke, M., and Lange, K. (2000). A multipoint method for detecting genotyping errors and mutations in sibling-pair linkage data. *Am. J. Hum. Genet.* 66: 1287–1297.

Durner, M., Greenberg, D.A., and Hodge, S.E. (1996). Phenocopies versus genetic heterogeneity: can we use phenocopy frequencies in linkage analysis to compensate for heterogeneity? *Hum. Hered.* 46: 265–273.

Easton, D.F., Ponder, M.A., Huson, S.M., and Ponder, B.A.J. (1993). An analysis of variation in expression of neurofibromatosis (NF) type 1 (NF1): evidence for modifying genes. *Am. J. Hum. Genet.* 53: 305–313.

Eaves, L.J. (1994). Effect of genetic architecture on the power of human linkage studies to resolve the contribution of quantitative trait loci. *Heredity* 72: 175–192.

Ebers, G.C., Kukay, K., Bulman, D.E. et al. (1996). A full genome search in multiple sclerosis. *Nat. Genet.* 13: 472–476.

Elston, R.C. and Gray-McGuire, C. (2004). A review of the 'Statistical analysis for genetic Epidemiology' (S.A.G.E.) software package. *Hum Genomics* 1 (6): 456–459.

Elston, R.C. and Stewart, J. (1971). A general model for the genetic analysis of pedigree data. *Hum. Hered.* 21: 523–542.

Elston, R.C., Buxbaum, S., Jacobs, K.B., and Olson, J.M. (2000). Haseman and Elston revisited. *Genet. Epidemiol.* 19: 1–17.

Epstein, M.P., Duren, W.L., and Boehnke, M. (2000). Improved inference of relationship for pairs of individuals. *Am. J. Hum. Genet.* 67: 1219–1231.

Falconer, D.S. and Mackay, T.F.C. (1996). *Introduction to Quantitative Genetics*, 4e. Harlow, Essex, UK: Longmans Green.

Faraway, J.J. (1993). Distribution of the admixture test for the detection of linkage under heterogeneity. *Genet. Epidemiol.* 10: 75–83.

Feingold, E. (2002). Regression-based quantitative trait-locus mapping in the 21st century. *Am. J. Hum. Genet.* 71: 217–222.

Fernandez, J.R., Etzel, C., Beasley, T.M. et al. (2002). Improving the power of sib pair quantitative trait loci detection by phenotype winsorization. *Hum. Hered.* 53: 59–67.

Fisher, R.A. (1918). The correlations between relatives on the supposition of Mendelian inheritance. *Trans. R. Soc. Edinburgh* 52: 399–433.

Fulker, D.W. and Cardon, L.R. (1994). A sib-pair approach to interval mapping of quantitative trait loci. *Am. J. Hum. Genet.* 54 (6): 1092–1103.

Fulker, D.W., Cherny, S.S., and Cardon, L.R. (1995). Multipoint interval mapping of quantitative trait loci, using sib pairs. *Am. J. Hum. Genet.* 56: 1224–1233.

Gibson, G. (1996). Epistasis and pleiotropy as natural properties of transcriptional regulation. *Theor. Popul. Biol.* 49: 58–89.

Goldgar, D.E. (1990). Multipoint analysis of human quantitative genetic variation. *Am. J. Hum. Genet.* 47: 957–967.

Goldgar, D.E. and Easton, D.F. (1997). Optimal strategies for mapping complex diseases in the presence of multiple loci. *Am. J. Hum. Genet.* 60: 1222–1232.

Goldgar, D.E., Easton, D.F., Cannon-Albright, L.A., and Skolnick, M.H. (1994). Systematic population-based assessment of cancer risk in first-degree relatives of cancer probands. *J. Natl. Cancer Inst.* 86: 1600–1608.

Goldin, L.R. and Weeks, D.E. (1993). Two-locus models of disease: comparison of likelihood and nonparametric linkage methods. *Am. J. Hum. Genet.* 53: 908–915.

Goring, H.H. and Ott, J. (1997). Relationship estimation in affected sib pair analysis of late-onset diseases. *Eur. J. Hum. Genet.* 5: 69–77.

Greenberg, D.A. and Abreu, P.C. (2001). Determining trait locus position from multipoint analysis: accuracy and power of three different statistics. *Genet. Epidemiol.* 21 (4): 299–314.

Greenberg, D.A., Hodge, S.E., Vieland, V.J., and Spence, M.A. (1996). Affecteds-only linkage methods are not a panacea. *Am. J. Hum. Genet.* 58: 892–895.

Greenwood, T.A., Badner, J.A., Byerley, W. et al. (2013). Heritability and linkage analysis of personality in bipolar disorder. *J. Affect. Disord.* 151 (2): 748–755.

Gudbjartsson, D.F., Jonasson, K., Frigge, M.L., and Kong, A. (2000). Allegro, a new computer program for multipoint linkage analysis. *Nat. Genet.* 25: 12–13.

Haines, J.L., Ter-Minassian, M., Bazyk, A. et al. (1996). A complete genomic screen for multiple sclerosis underscores a role for the major histocompatability complex. *Nat. Genet.* 13: 469–471.

Hanis, C.L., Boerwinkle, E., Chakraborty, R. et al. (1996). A genome-wide search for human non-insulin-dependent (type 2) diabetes genes reveals a major susceptibility locus on chromosome 2. *Nat. Genet.* 13: 161–166.

Haseman, J.K. and Elston, R.C. (1972). The investigation of linkage between a quantitative trait and a marker locus. *Behav. Genet.* 2: 3–19.

Hauser, E.R. and Boehnke, M. (1998). Genetic linkage analysis of complex genetic traits by using affected sibling pairs. *Biometrics* 54: 1238–1246.

Hauser, E.R., Boehnke, M., Guo, S.W., and Risch, N. (1996). Affected-sib-pair interval mapping and exclusion for complex genetic traits: sampling considerations. *Genet. Epidemiol.* 13: 117–137.

Haviland, M.B., Lussier-Cacan, S., Davignon, J., and Sing, C.F. (1995). Impact of apolipoprotein E genotype variation on means, variances, and correlations of plasma lipid, lipoprotein, and apolipoprotein traits in octogenarians. *Am. J. Med. Genet.* 58: 315–331.

Hodge, S.E. (1984). The information contained in multiple sibling pairs. *Genet. Epidemiol.* 1: 109–122.

Hodge, S.E. and Elston, R.C. (1994). Lods, wrods, and mods: the interpretation of LOD scores calculated under different models. *Genet. Epidemiol.* 11: 329–342.

Hodge, S.E., Abreu, P.C., and Greenberg, D.A. (1997). Magnitude of type I error when single-locus linkage analysis is maximized over models: a simulation study. *Am. J. Hum. Genet.* 60: 217–227.

Holmans, P. (1993). Asymptotic properties of affected-sib-pair linkage analysis. *Am. J. Hum. Genet.* 52: 362–374.

Horikawa, Y., Oda, N., Cox, N.J. et al. (2000). Genetic variation in the gene encoding calpain-10 is associated with type 2 diabetes mellitus. *Nat. Genet.* 26: 163–175.

Jeffreys, A.J., Wilson, V., and Thein, S.L. (1985). Individual-specific 'fingerprints' of human DNA. *Nature* 316: 76–79.

Kan, Y.W. and Dozy, A.M. (1978). Antenatal diagnosis of sickle-cell anaemia by DNA analysis of amniotic-fluid cells. *Lancet* 2 (8096): 910–912.

Keats, B.J.B., Morton, N.E., Rao, D.C., and Williams, W.R. (1979). *A Source Book for Linkage in Man.* Baltimore, MD: The Johns Hopkins University Press.

Khoury, M.J., Beaty, T.H., and Cohen, B.H. (1993). *Fundamentals of Genetic Epidemiology.* New York, NY: Oxford University Press.

Kissebah, A.H., Sonnenberg, G.E., Myklebust, J. et al. (2000). Quantitative trait loci on chromosomes 3 and 17 influence phenotypes of the metabolic syndrome. *Proc. Natl. Acad. Sci. U. S. A.* 97: 14478–14483.

Knowles, J.A., Vieland, V.J., and Gilliam, T.C. (1992). Perils of gene mapping with microsatellite markers. *Am. J. Hum. Genet.* 51: 905–909.

Kong, A. and Cox, N.J. (1997). Allele-sharing models: LOD scores and accurate linkage tests. *Am. J. Hum. Genet.* 61: 1179–1188.

Kruglyak, L. and Lander, E.S. (1995a). A nonparametric approach for mapping quantitative trait loci. *Genetics* 139: 1421–1428.

Kruglyak, L. and Lander, E.S. (1995b). Complete multipoint sib-pair analysis of qualitative and quantitative traits. *Am. J. Hum. Genet.* 57: 439–454.

Kruglyak, L., Daly, M.J., Reeve-Daly, M.P., and Lander, E.S. (1996). Parametric and nonparametric linkage analysis: a unified multipoint approach. *Am. J. Hum. Genet.* 58: 1347–1363.

Lander, E.S. and Green, P. (1987). Construction of multilocus genetic linkage maps in humans. *Proc. Natl. Acad. Sci. U. S. A.* 84 (8): 2363–2367.

Lange, K., Papp, J.C., Sinsheimer, J.S. et al. (2013). Mendel: the Swiss army knife of genetic analysis programs. *Bioinformatics* 29: 1568–1570.

Lathrop, G.M., Lalouel, J.M., Julier, C., and Ott, J. (1984). Strategies for multilocus linkage analysis in humans. *Proc. Natl. Acad. Sci. U. S. A.* 81: 3443–3446.

Li, W., Fann, C.S.J., and Ott, J. (1998). Low-order polynomial trends of female-to-male map distance ratios along human chromosomes. *Hum. Hered.* 48: 266–270.

Li, M., Lou, X.Y., and Lu, Q. (2014a). On epistasis: a methodological review for detecting gene-gene interactions underlying various types of phenotypic traits. *Recent Pat. Biotechnol.* 6 (3): 230–236.

Li, S., Mukherjee, B., Taylor, J.M. et al. (2014b). The role of environmental heterogeneity in meta-analysis of gene-environment interactions with quantitative traits. *Genet. Epidemiol.* 38 (5): 416–429.

Lynch, M. and Walsh, B. (1998). *Genetics and Analysis of Quantitative Traits*. Sunderland, MA: Sinauer Associates, Inc.

Markianos, K., Daly, M.J., and Kruglyak, L. (2001). Efficient multipoint linkage analysis through reduction of inheritance space. *Am. J. Hum. Genet.* 68: 963–977.

Matise, T.C. and Gitlin, J.A. (1999). MAP-O-MAT: marker-based linkage mapping on the world wide web. *Am. J. Hum. Genet.* 65: A435.

Matise, T.C., Perlin, M., and Chakravarti, A. (1994). Automated construction of genetic linkage maps using an expert system (MultiMap): a human genome linkage map. *Nat. Genet.* 6: 384–390.

Matise, T.C., Chen, F., Chen, W. et al. (2007). A second-generation combined linkage physical map of the human genome. *Genome Res.* 17 (12): 1783–1786.

McPeek, M.S. and Sun, L. (2000). Statistical tests for detection of misspecified relationships by use of genome-screen data. *Am. J. Hum. Genet.* 66: 1076–1094.

Mitchell, B.D., Kammerer, C.M., Blangero, J. et al. (1996). Genetic and environmental contributions to cardiovascular risk factors in Mexican Americans. The San Antonio family heart study. *Circulation* 94: 2159–2170.

Mitchell, B.D., Ghosh, S., Schneider, J.L. et al. (1997). Power of variance component linkage analysis to detect epistasis. *Genet. Epidemiol.* 14: 1017–1022.

Moore, J.H. (2003). Quantitative traits. In: *Genetics* (ed. R. Robinson), 1–3. New York, NY: Macmillan Science Library.

Moore, J.H., Smolkin, M., Lamb, J.M. et al. (2002a). The relationship between plasma t-PA and PAI-1 levels is dependent on epistatic effects of the ACE I/D and PAI-1 4G/5G polymorphisms. *Clin. Genet.* 62: 53–59.

Moore, J.H., Lamb, J.M., Brown, N.J., and Vaughan, D.E. (2002b). A comparison of combinatorial partitioning and linear regression for the detection of epistatic effects of the ACE I/D and PAI-1 4G/5G polymorphisms on plasma PAI-1 levels. *Clin. Genet.* 62: 74–79.

Morgan, T.H. (1911). Chromosomes and associative inheritance. *Science* 34 (880): 636–638.

Morton, N.E. (1955). Sequential tests for the detection of linkage. *Am. J. Hum. Genet.* 7: 277–318.

Morton, N.E. (1956). The detection and estimation of linkage between the genes for elliptocytosis and the Rh blood type. *Am. J. Hum. Genet.* 8: 80–96.

Mukhopadhyay, N., Buxbaum, S.G., and Weeks, D.E. (2004). Comparative study of multipoint methods for genotype error detection. *Hum. Hered.* 58 (3–4): 175–189.

Nachman, M.W. (2002). Variation in recombination rate across the genome: evidence and implications. *Curr. Opin. Genet. Dev.* 12: 657–663.

Nakagawa, K. and Ishizaki, T. (2000). Therapeutic relevance of pharmacogenetic factors in cardiovascular medicine. *Pharmacol. Ther.* 86: 1–28.

Nelson, M.R., Kardia, S.L., Ferrell, R.E., and Sing, C.F. (1999). Influence of apolipoprotein E genotype variation on the means, variances, and correlations of plasma lipids and apolipoproteins in children. *Ann. Hum. Genet.* 63: 311–328.

Nelson, M.R., Kardia, S.L., Ferrell, R.E., and Sing, C.F. (2001). A combinatorial partitioning method to identify multilocus genotypic partitions that predict quantitative trait variation. *Genome Res.* 11: 458–470.

Noble, J.A. (2015). Immunogenetics of type 1 diabetes: a comprehensive review. *J. Autoimmun.* 64: 101–112.

O'Connell, J.R. (2001). Rapid multipoint linkage analysis via inheritance vectors in the Elston- Stewart algorithm. *Hum. Hered.* 51: 226–240.

O'Connell, J.R. and Weeks, D.E. (1995). The VITESSE algorithm for rapid exact multilocus linkage analysis via genotype set-recoding and fuzzy inheritance. *Nat. Genet.* 11: 402–408.

O'Connell, J.R. and Weeks, D.E. (1998). PedCheck: a program for identification of genotype incompatibilities in linkage analysis. *Am. J. Hum. Genet.* 63: 259–266.

Olson, J.M. (1995). Multipoint linkage analysis using sibpairs: an interval mapping approach for dichotomous outcomes. *Am. J. Hum. Genet.* 56: 788–798.

Ott, J. (1974). Estimation of the recombination fraction in human pedigrees: efficient computation of the likelihood for human linkage studies. *Am. J. Hum. Genet.* 26: 588–597.

Ott, J. (1986). Linkage probability and its approximate confidence interval under possible heterogeneity. *Genet. Epidemiol. Suppl.* 1: 251–257.

Ott, J. (1989). Computer simulation methods in human linkage analysis. *Proc. Natl. Acad. Sci. U. S. A.* 86: 4175–4178.

Ott, J. (1992). Strategies for characterizing highly polymorphic markers in human gene mapping. *Am. J. Hum. Genet.* 51: 283–290.

Ott, J. (1995). Linkage analysis with biological markers. *Hum. Hered.* 45: 169–174.

Ott, J. (1999). *Analysis of Human Genetic Linkage*, 3e. Baltimore, MD: The Johns Hopkins University Press.

Pattin, K.A. and Moore, J.H. (2008). Exploiting the proteome to improve the genome-wide genetic analysis of epistasis in common human diseases. *Hum. Genet.* 124 (1): 19–29.

Penrose, L.S. (1935). The detection of autosomal linkage in data which consists of pairs of brothers and sisters of unspecified parentage. *Ann. Eugenics* 6: 133–138.

Penrose, L.S. (1953). The general purpose sib-pair linkage test. *Ann. Eugenics* 18: 120–124.

Pericak-Vance, M.A., Bebout, J.L., Gaskell, P.C. et al. (1991). Linkage studies in familial Alzheimer's disease: evidence for chromosome 19 linkage. *Am. J. Hum. Genet.* 48: 1034–1050.

Pigliucci, M. (2001). *Phenotypic Plasticity: Beyond Nature and Nurture*. Baltimore, MD: The Johns Hopkins University Press.
Ploughman, L.M. and Boehnke, M. (1989). Estimating the power of a proposed linkage study for a complex genetic trait. *Am. J. Hum. Genet.* 44 (4): 543–551.
Rainwater, D.L., Almasy, L., Blangero, J. et al. (1999). A genome search identifes major quantitative trait loci on human chromosomes 3 and 4 that influence cholesterol concentrations in small LDL particles. *Arterioscler. Thromb. Vasc. Biol.* 19: 777–783.
Reilly, S.L., Ferrell, R.E., Kottke, B.A. et al. (1991). The gender-specific apolipoprotein E genotype influence on the distribution of lipids and apolipoproteins in the population of Rochester, MN. I. Pleiotropic effects on means and variances. *Am. J. Hum. Genet.* 49: 1155–1166.
Rice, J.P. (1993). Phenotype definition for genetic studies. *Eur. Arch. Psychiatry Clin. Neurosci.* 243: 158–163.
Rimmler, J.B., Haynes, C.S., McDowell, J.G. et al. (1999). DataTracker: comprehensive software for data quality control protocols in complex disease studies. *Am. J. Hum. Genet.* 65 (Suppl): A442.
Risch, N. (1988). A new statistical test for linkage heterogeneity. *Am. J. Hum. Genet.* 42: 353–364.
Risch, N. (1990a). Linkage strategies for genetically complex traits I. Multilocus models. *Am. J. Hum. Genet.* 46: 222–228.
Risch, N. (1990b). Linkage strategies for genetically complex traits II. The power of affected relative pairs. *Am. J. Hum. Genet.* 46: 229–241.
Risch, N. (1990c). Linkage strategies for genetically complex traits III. The effect of marker polymorphism on analysis of affected pairs. *Am. J. Hum. Genet.* 46: 242–253.
Risch, N. and Giuffra, L. (1992). Model misspecification and multipoint linkage analysis. *Hum. Hered.* 42: 77–92.
Robertson, A. (1973). Linkage between marker loci and those affecting a quantitative trait. *Behav. Genet.* 3: 389–391.
Saccone, N.L., Kwon, J.M., Corbett, J. et al. (2000). A genome screen of maximum number of drinks as an alcoholism phenotype. *Am. J. Med. Genet.* 96: 632–637.
Sachidanandam, R., Weissman, D., Schmidt, S.C. et al. (2001). A map of human genome sequence variation containing 1.42 million single nucleotide polymorphisms. *Nature* 409 (6822): 928–933.
Saunders, I.W., Brohede, J., and Hannan, G.N. (2007). Estimating genotyping error rates from Mendelian errors in SNP array genotypes and their impact on inference. *Genomics* 90 (3): 291–296.
Sawcer, S., Jones, H.B., Feakes, R. et al. (1996). A genome screen in multiple sclerosis reveals susceptibility loci on chromosome 6p21 and 17q22. *Nat. Genet.* 13: 464–476.
Schaffer, A.A., Gupta, S.K., Shriram, K., and Cottingham, R.W. (1994). Avoiding recomputation in linkage analysis. *Hum. Hered.* 44: 225–237.
Schlichting, C.D. and Pigliucci, M. (1998). *Phenotypic Evolution: A Reaction Norm Perspective*. Sunderland, Massachusetts: Sinauer Associates Inc.
Schmidt, S., Shao, Y., Hauser, E.R. et al. (2001). Life after the screen: making sense of many p-values. *Genet. Epidemiol.* 21: S546–S551.
Schork, N.J., Boehnke, M., Terwilliger, J.D., and Ott, J. (1993). Two trait locus linkage analysis: a powerful strategy for mapping complex genetic traits. *Am. J. Hum. Genet.* 53: 1127–1136.
Sengul, H., Weeks, D.E., and Feingold, E. (2001). A survey of affected-sibship statistics for nonparametric linkage analysis. *Am. J. Hum. Genet.* 69: 179–190.
Sham, P.C., Purcell, S., Cherny, S.S., and Abecasis, G.R. (2002). Powerful regression-based quantitative-trait linkage analysis of general pedigrees. *Am. J. Hum. Genet.* 71: 238–253.
Shete, S. and Amos, C.I. (2002). Testing for genetic linkage in families by a variance-components approach in the presence of genomic imprinting. *Am. J. Hum. Genet.* 70: 751–757.

Silberstein, M., Weissbrod, O., Otten, L. et al. (2013). A system for exact and approximate genetic linkage analysis of SNP data in large pedigrees. *Bioinformatics* 29 (2): 197–205.

Smith, C.A.B. (1963). Testing for heterogeneity of recombination fractions values in human genetics. *Ann. Hum. Genet.* 27: 175–182.

Sobel, E. and Lange, K. (1996). Descent graphs in pedigree analysis: applications to haplotyping, location scores and marker-sharing statistics. *Am. J. Hum. Genet.* 58: 1323–1337.

Sobel, E., Sengul, H., and Weeks, D.E. (2001). Multipoint estimation of identity-by-descent probabilities at arbitrary positions among marker loci on general pedigrees. *Hum. Hered.* 52: 121–131.

Speer, M.C. 1, Vance, J.M., Grubber, J.M. et al. (1999). Identification of a new autosomal dominant limb-girdle muscular dystrophy locus on chromosome 7. *Am. J. Hum. Genet.* 64 (2): 556–562. https://doi.org/10.1086/302252.

Spence, M.A., Sparkes, R.S., Heckenlively, J.R. et al. (1977). Probable genetic linkage between autosomal dominant retinitis pigmentosa (RP) and amylase (AMY2): evidence of an RP locus on chromosome 1. *Am. J. Hum. Genet.* 29 (4): 397–404.

Spielman, R.S., Baker, L., and Zmijewski, C.M. (1980). Gene dosage and suceptibility to insulin-dependent diabetes. *Ann. Hum. Genet.* 44: 135–150.

St. George-Hyslop, P.H., Haines, J.L., Farrer, L.A. et al. (1990). Genetic linkage studies suggest that Alzheimer's disease is not a single homogenous disorder. *Nature* 347: 194–197.

Sturtevant, A.H. (1921). Genetic studies on Drosophila simulans. II. Sex-linked Group of Genes. *Genetics* 6 (1): 43–64.

Suarez, B.K., Rice, J.P., and Reich, T. (1978). The generalized sib pair IBD distribution: its use in the detection of linkage. *Ann. Hum. Genet.* 42: 87–94.

Templeton, A.R. (2000). Epistasis and complex traits. In: *Epistasis and the Evolutionary Process* (eds. M. Wade, I.I.I.B. Brodie and J. Wolf), 41–57. New York, NY: Oxford University Press.

Terwilliger, J.D. and Ott, J. (1994). *Handbook of Human Genetic Linkage*. Baltimore, MD: The Johns Hopkins University Press.

Tiwari, H.K. and Elston, R.C. (1998). Restrictions on components of variance for epistatic models. *Theor. Popul. Biol.* 54: 161–174.

Tsuang, M.T., Faraone, S.V., and Lyons, M.J. (1993). Dentification of the phenotype in psychiatric genetics. *Eur. Arch. Psychiatry Clin. Neurosci.* 243: 131–142.

Weber, J.L. and May, P.E. (1989). Abundant class of human DNA polymorphisms which can be typed using the polymerase chain reaction. *Am. J. Hum. Genet.* 44: 388–396.

Weeks, D.E. and Lange, K. (1992). A multilocus extension of the affected-pedigree-member method of linkage analysis. *Am. J. Hum. Genet.* 50: 859–868.

Weeks, D.E., Ott, J., and Lathrop, G.M. (1990a). SLINK: a general simulation program for linkage analysis. *Am. J. Hum. Genet.* 47: A204.

Weeks, D.E., Lehner, T., Squires-Wheeler, E. et al. (1990b). Measuring the inflation of the LOD score due to its maximization over model parameter values in human linkage analysis. *Genet. Epidemiol.* 7: 237–243.

Weeks, D.E., Sobel, E., O'Connell, J.R., and Lange, K. (1995). Computer programs for multilocus haplotyping of general pedigrees. *Am. J. Hum. Genet.* 56: 1506–1507.

Weiss, K.M. (1993). *Genetic Variation and Human Disease: Principles and Evolutionary Approaches*. New York, NY: Cambridge University Press.

Whittemore, A.S. and Halpern, J. (1994). A class of tests for linkage using affected pedigree members. *Biometrics* 50: 118–127.

Williams, J.T., Begleiter, H., Porjesz, B. et al. (1999). Joint multipoint linkage analysis of multivariate qualitative and quantitative traits. II. Alcoholism and event-related potentials. *Am. J. Hum. Genet.* 65: 1148–1160.

Williamson, J.A. and Amos, C.I. (1990). On the asymptotic behavior of the estimate of the recombination fraction under the null hypothesis of no linkage when the model is misspecified. *Genet. Epidemiol.* 7: 309–318.

Wright, S. (1932). The roles of mutation, inbreeding, crossbreeding and selection in evolution. *Proc. 6th Intl. Congr. Genet.* 1: 356–366.

Wyman, A.R. and White, R. (1980). A highly polymorphic locus in human DNA. *PNAS* 77: 6754–6758.

Xie, X. and Ott, J. (1990). Determining the effect of a change in affection status on the Lod score. *Am. J. Hum. Genet.* 47: A205.

Xu, J. (2000). Combined analysis of hereditary prostate cancer linkage to 1q24-25: results from 772 hereditary prostate cancer families from the international consortium for prostate cancer genetics. *Am. J. Hum. Genet.* 66: 945–957.

Xu, X., Weiss, S., Xu, X., and Wei, L.J. (2000). A unified Haseman-Elston method for testing linkage with quantitative traits. *Am. J. Hum. Genet.* 67: 1025–1028.

Yang, B.Z., Han, S., Kranzler, H.R. et al. (2012). Autosomal linkage scan for loci predisposing to comorbid dependence on multiple substances. *Am. J. Med. Genet. B Neuropsychiatr. Genet.* 159B (4): 361–369.

7

Data Management

Stephen D. Turner[1] *and William S. Bush*[2]

[1] *Signature Science, LLC, Charlottesville, VA, USA*
[2] *Department of Population and Quantitative Health Sciences, Case Western Reserve University, Cleveland, OH, USA*

Large-scale high-throughput genotyping has become commonplace in studies of the genetic underpinnings of complex disease. Genome-wide association studies (GWAS), where up to several million single-nucleotide polymorphisms (SNPs) are assayed in thousands of samples, are routinely used to investigate how relatively common genetic variation affects complex disease (Hindorff et al. 2009; Manolio 2010). Information on global patterns of genome-wide genetic variation in multiple populations is now available with the completion of the 1000 Genomes Project (Durbin et al. 2010; Abecasis et al. 2012; Consortium 2015). This in combination with the availability of next-generation DNA sequencing technologies that are decreasing in cost faster than Moore's law (Metzker 2010; Wetterstrand 2011) is opening the door to genetic analysis on the whole-genome scale to more researchers than ever. In addition to the large amount of data being collected on genetic exposures, research groups are now gathering data on many environmental exposures and clinical outcomes. For example, environment-wide association studies (EWAS) are used to investigate the effect of hundreds of environmental exposures on health outcomes (Patel et al. 2010; Hall et al. 2014; McGinnis et al. 2016), and phenome-wide association studies are used to explore genetic associations for hundreds of human diseases (Denny et al. 2010; Pendergrass et al. 2011; Pendergrass et al. 2013; Verma and Ritchie 2017; Verma et al. 2018). Furthermore, mining hospital electronic health records to automatically abstract phenotypic characteristics and detailed records of medical treatment is an exciting opportunity in complex disease genetic studies (Kohane 2011; Kho et al. 2012; Gottesman et al. 2013; Wei and Denny 2015; Wei et al. 2016). This approach is being effectively employed by a consortium whose mission is to enable genetic research by linking electronic health records with DNA from hospital biobanks (Kho et al. 2011; McCarty et al. 2011; Gottesman et al. 2013), and another private company replicated over 180 genetic associations to 50 medical phenotypes using self-reported medical data from web-based surveys in a re-contactable cohort (Tung et al. 2011).

The ready availability of large-scale clinical, demographic, environmental, and genomic data comes with new challenges in data management. Successful genetic analysis of complex disease requires sophisticated information management strategies, advanced analytics, and integration of computational solutions with existing biological knowledge. Designing and implementing thoughtful experiments requires that information be efficiently stored and readily accessible for downstream analysis. Functionality and usability requirements mandate advance planning for

Genetic Analysis of Complex Diseases, Third Edition. Edited by William K. Scott and Marylyn D. Ritchie.

hardware, software, and data retrieval protocols. Security and access control also become important when human data or other sensitive information are collected in a study. In this chapter, we will highlight several challenges and discuss data management strategies and solutions for each stage of a large genetic analysis project. Links and descriptions to all the software tools and resources discussed in this chapter are provided in Table 7.1. It is important to note that several commercial systems are available to facilitate routine access to multiple datasets, including rich features such as account-level restrictions, custom import and export tools, and data-sharing utilities. For the purposes of this chapter, we present tools for storage, access, and manipulation of a single dataset rather than more complex designs.

Developing a Data Organization Strategy

A Brief Overview of Data Normalization

Normalization is a fundamental principle in database design used to avoid unnecessary redundancy of data, which decreases overall data volume and reduces the number of data points that must change during data updates. The concept of normalization relies on splitting data into more fundamental groupings that contain unique items (called tuples in the vernacular of databases). Consider a collection of individuals and their corresponding street addresses in the United States. Multiple individuals could live in a single household, and multiple households belong to a single community. If we store the name, address, and community information for each and every person in the United States, we have duplicated many pieces of data, each of which consume a tiny bit of disk space, which in aggregate may require more space than is available or feasible. If however, we split the data into three separate groups or tables corresponding to individuals, households, and communities, and we eliminate much of the redundancy without losing any information. For this data splitting to work, each of these tables must contain perfectly unique tuples, meaning no two rows of the table can have the same values. The households table might contain a street number, street name; similarly, the community table might contain a town or city name, a state, and a zip code. But these tables need to be linked together to determine the community that corresponds to a given household. This linking is accomplished by defining a key – a value that links a row from one table to a unique entry from another table. For example, the community table may have thousands of rows in it, but each row has a unique numeric zip code; in this case, the zip code serves as the primary key for the community table. To assign a community to each entry in the household table, we would include the zip code in each row; the zip code serves as a foreign key identifying the city and state for a household.

In the case of the community table, the zip code (a useful piece of information already in our dataset) is an ideal key. For many other tables, this may not be the case. Each individual in our dataset has a corresponding household, but household addresses do not contain uniquely identifying information like a zip code. In this case, we must create a unique identifier for each row – a simple number will do. If we create a unique household ID as the primary key for the household table, we can add this information to the individuals table as a foreign key. Now, if we want to retrieve a full address for an individual, we can by linking the foreign key of the individuals table to the primary key of the household table and linking the foreign key of the household table to the primary key of the community table. Through this process, we can completely restore all of the original information, yet we have dramatically reduced data redundancy and disk usage compared to the original data structure. This idea of data normalization is a principle employed by

Table 7.1 Software and web resources.

Tool/resource name	Description	URL
MySQL	Popular, free, open-source, very well-documented relational database management system	http://http://www.mysql.com
Workbench	Graphical interface for designing, developing, administering, and querying MySQL databases	http://http://www.mysql.com/products/workbench
ODBC driver	MySQL ODBC driver	http://dev.mysql.com/downloads/connector/odbc
PLINK	Free, open-source genome-wide association study data management and analysis toolset	http://zzz.bwh.harvard.edu/plink
SAMtools	Set of utilities for manipulating alignments in the SAM format, including sorting, merging, indexing, and generating alignments.	http://www.htslib.org
VCFtools	Software tools for working with complex genetic variation data in the form of VCF files.	https://vcftools.github.io/index.html
R	Free, open-source environment for statistical computing	http://www.r-project.org
RODBC	ODBC interface for R	http://cran.r-project.org/web/packages/RODBC/index.html
Bioconductor	Suite of packages for the R environment focusing on management and analysis of next-generation sequencing data, microarrays, high-throughput assays, and annotation	http://www.bioconductor.org
plyr	R package for implementing a split-apply-merge strategy to data management and analysis	http://plyr.had.co.nz
Galaxy	Web-based workflow management platform with built-in tools for managing and analyzing large-scale genomics experiments	http://usegalaxy.org
Comparison of RDBMSs	Comprehensive comparison of over 40 relational database management systems (features, limitations, available data types, licenses, access control, and supported operating systems)	http://en.wikipedia.org/wiki/Comparison_of_relational_database_management_systems
Amazon Web Services	On-demand, scalable, and flexible Infrastructure-as-a-Service cloud computing provider. Many of the resources mentioned in this chapter (Galaxy, parallel R applications, etc.) can benefit from parallelization. AWS also offers essentially unlimited disk storage with redundant backup.	http://aws.amazon.com

relational database systems, which are by far the most ubiquitous way to store, update, and retrieve information in the world today.

The idea of information uniqueness exploited by the use of keys is a common part of information we use every day much like a zip code. While the United States Postal Service can use a map to identify every city and township in the country, geneticists and bioinformaticians are not quite as lucky because their maps continue to change and be revised and refined. Despite this difficulty, there are established primary keys in human genetics. Reference sequence (RS) numbers uniquely identify SNPs based on the flanking sequences used in genotyping assays. Resources like Ensembl and Entrez-Gene assign unique gene IDs to the transcribed and/or translated DNA sequences identified by their respective annotation pipelines, and gene symbols are now standardized as unique identifiers by the HUGO Gene Nomenclature Committee.

Normalization of genomic data can be achieved much like the zip code example exploiting existing key systems. What might be considered a "complete" genetic dataset for downstream statistical analysis might be split up into several database tables. For example, a dataset with information about exonic SNPs used in a pathway enrichment analysis may have the following four fields for each SNP: rs-number, gene symbol, full gene name, and alternative gene aliases. If 1000 SNPs are in a single gene, the full gene name and aliases fields will be unnecessarily duplicated 1000 times. For thousands of genes, this can result in gigabytes of redundant data duplication. In a normalized database, there would be two tables. The first table has two columns: rs-number and gene symbol. SNP rs-numbers are unique, and rs-number serves as primary key for this table. The gene symbol becomes a foreign key to link a second table containing three columns with information about genes: gene symbol (primary key), full gene name, and alternative aliases. Furthermore, if annotation information about a gene is updated (for example, if "TGTP" was named as a new alias for cholesteryl ester transfer protein, CETP), only one entry in database Table B needs to be updated, rather than updating multiple cells in the complete table for each row containing a CETP SNP. Finally, further consider that rs IDs may change with new builds of the genome. For this reason, it is also important to consider keeping a table linking rs-number to the chromosome number, base pair, and genome build version, and making sure to keep a record of which build was used for each analysis.

Database Management System (DBMS) and Structured Query Language (SQL)

Storing normalized data in a series of text files or spreadsheet documents would quickly become overwhelming, and retrieving, updating, or deleting data from such a series of files would be extremely slow and cumbersome. As such, software tools were developed to enable and greatly optimize the storage of normalized information. These tools are called database management systems (DBMSs) and typically operate as a continuously running application or server that can handle many hundreds of simultaneous data requests. Popular systems include Oracle, MySQL, PostgreSQL, and SQL Server. Each of these systems has distinct benefits and drawbacks (including the need to purchase licenses), but all perform the same basic functions – allowing users to create tables, primary and foreign keys, update and delete information, and perform queries to retrieve data. Remarkably, all these systems use the same basic language for manipulating data (with a few exceptions here or there) making structured query language (SQL), the most consistent way to retrieve information.

Full details of the SQL language can be found elsewhere (including a number of online resources). The basic retrieval command is the select statement, written "SELECT [fieldname]

FROM [tablename];", requesting rows from a single table. Given the ideas of data normalization, it is often useful to link tables together using an operation called a join. This is illustrated in Figure 7.1, where Table A (using gene_symbol as the foreign key) is related to Table B (using gene_symbol as the primary key). A simple SQL join similar to this – "SELECT * FROM tableA INNER JOIN tableB on tableA.gene_symbol=tableB.gene_symbol;" – could be used to recreate the complete table from the two normalized database tables.

Partitioning Data by Type

A diverse array of data types is collected in any genetic study, but all this information can be broadly classified into two ways: sequence-level data and sample-level data. Sequence-level data includes any information about genetic variants assayed in the study, including SNP genotypes, quality control metrics such as call rate or imputation accuracy, and analysis results (here we will not cover methodology for calling SNPs from next-generation sequencing data, as this has been reviewed elsewhere (Nielsen et al. 2011)). Sample-level data includes all phenotype, clinical, and covariate data, informed consent status, follow-up data from longitudinal designs, demographic descriptors, familial relationship information, per-sample quality control metrics, and any annotation about laboratory batch processing such as plating information, DNA extraction conditions, and laboratory personnel and equipment used to process each sample.

"Complete" table

rs_id	gene_symbol	gene_name	aliases
rs4068	ADIPOQ	ADIPOQ adiponectin, C1Q and collagen domain containing	ACDC; ADPN; APM1; APM-1; GBP28; ACRP30; ADIPQTL1
rs1501298	ADIPOQ	ADIPOQ adiponectin, C1Q and collagen domain containing	ACDC; ADPN; APM1; APM-1; GBP28; ACRP30; ADIPQTL1
rs4686440	ADIPOQ	ADIPOQ adiponectin, C1Q and collagen domain containing	ACDC; ADPN; APM1; APM-1; GBP28; ACRP30; ADIPQTL1
rs891141	CETP	CETP cholesteryl esler transfer protein, plasma	ACDC; ADPN; APM1; APM-1; GBP28; ACRP30; ADIPQTL1
rs35878799	CETP	CETP cholesteryl ester transfer protein, plasma	ACDC; ADPN; APM1; APM-1; GBP28; ACRP30; ADIPQTL1
rs34716057	CETP	CETP cholesteryl ester transfer protein, plasma	ACDC; ADPN; APM1; APM-1; GBP28; ACRP30; ADIPQTL1

Database table A

rs_id (primary key)	gene_symbol (foreign key)
rs4068	ADIPOQ
rs1501298	ADIPOQ
rs4686440	ADIPOQ
rs891141	CETP
rs35878799	CETP
rs34716057	CETP

Database table B

gene_symbol (primary key)	gene_name	aliases
ADIPOQ	ADIPOQ adiponectin, C1Q and collagen domain containing	ACDC; ADPN; APM1; APM-1; GBP28; ACRP30; ADIPQTL1
CETP	CETP cholesteryl ester transfer protein, plasma	BPIFF; HDLCQ10

Figure 7.1 Data normalization example for dataset with information mapping SNPs to genes. The "complete" table is not normalized; data are duplicated and integrity may be compromised if gene information is updated. Normalizing these data will yield two smaller tables. Table A has two columns: rs-number and gene symbol. SNP rs-numbers are unique and serve as the *primary key* for indexing Table A. The gene symbol (called a *foreign key* in Table A) is found only once in Table B, which contains three columns with information about genes: gene symbol (unique primary key in Table B), full gene name, and alternative aliases. The two tables are joined by gene symbol to recreate the "complete" annotation table.

Sequence-Level Data

A flexible and robust data management system must be able to keep track of the genetic variation information about each sample in a dataset. Genotypes stored in relational DBMSs (discussed later) are often dumped to flat text files for downstream analysis. Regardless of the underlying study design (such as family-based or population-based), the most commonly used format for storing and analyzing genetic data is the linkage, or pedigree file format (pedfile). This file contains one individual per row, where the first six columns are identifying information (family ID, individual ID, father ID, mother ID, sex, and phenotype), and the remaining columns are genotypes (two columns per genotype; one for each allele). The genotype column pairs correspond to an ordered set of SNPs present in an associated map file. Additional phenotypes can also be stored in separate files consisting of family ID, individual ID, then extra columns representing additional phenotypes. There are several variations on pedfile format, including transposed (long) formats (tped) and compressed (binary) formats. Descriptions of these file formats can be found on the PLINK homepage (Table 7.1). PLINK is a freely available, open-source, cross-platform application for QC and analysis of GWAS data (Purcell et al. 2007). More discussion of this can be found in Chapter 9.

It is critically important to assess data quality and integrity. A common mishap in management of genetic data is inconsistent strand orientation (i.e. forward or reverse complement). Choice of strand orientation might depend on the strand orientation of other data used in a combined analysis or of a reference set used for imputation. The goal is to ensure uniformity in genotype call reporting that is critical in downstream analyses, reporting, and annotation. Consistent strand orientation can be assessed by computing summary statistics such as allele frequencies and comparing these results to a similar population from the 1000 Genomes Project (Durbin et al. 2010). In addition, it is often useful to genotype at least a few samples in duplicate. Technical replicates are unnecessary when using modern genotyping technologies but having duplicate genotyping on a few individuals allows for assessment of reproducibility and can help identify inconsistencies when merging data from different sources. A thorough discussion of quality control for GWAS can be found elsewhere (Turner et al. 2011; Zuvich et al. 2011; Verma et al. 2014).

After an analysis is complete, results should be parsed and stored in a relational database indexed by SNP name or position. This allows results from association or linkage analyses to be easily merged (i.e. "INNER JOIN" in SQL) with quality assessment statistics such as genotype call rate or allele frequency. SNPs can be joined to other database tables containing positional information and further annotated with gene information or pathway involvement.

Sample-Level Data

Sample-level data should also be stored in a database indexed by sample. Sample-level data include phenotypes, clinical covariates, environmental exposures, familial relationship information, and quality assessment statistics. Familial relationship information is easily tracked using the pedigree file format discussed above. Each family is given a unique identifier, and each individual within each family is given an identifier that is unique within that family. Each individual is thus uniquely identified by both a family ID and individual ID. Each individual is also annotated with the ID of their mother and father (both within the same family ID). Further documentation of this format for tracking familial relationships can be found in the PLINK documentation (Table 7.1).

Several procedures allow for sample data integrity assessment. One of the easiest ways to discover potential sample handling issues that result in mix-ups is by checking the reported sex of each individual against that predicted by the genetic data. This can be done by examining X chromosome heterozygosity rates to determine sex empirically and comparing that inferred sex to those indicated in the sample annotation database. Checking X chromosome heterozygosity may also reveal sex chromosome anomalies such as Turner syndrome (females having karyotype XO), Klinefelter syndrome (males having karyotype XXY), mosaic individuals (XX/XO, XX/XXY), or females with large stretches of loss-of-heterozygosity on the X chromosome who are otherwise phenotypically normal (Turner et al. 2011).

Sample and pedigree data integrity can also be assessed by reconciling genomic data with self-reported relationships between individuals (if available). Using dense genotyping data obtained in GWAS, it is easy to compute the proportion of loci across the genome where two individuals share zero, one, or two alleles identical by descent (IBD). Individuals sharing two alleles IBD at every locus are either monozygotic twins, or the pair is actually a single sample processed twice. Individuals sharing zero alleles IBD at every locus are unrelated. Individuals sharing one allele IBD at every locus are parent–child pairs. On average, siblings share zero, one, and two alleles IBD at 25, 50, and 25% of the genome, respectively. Computing pairwise kinship estimates in this manner facilitates the assessment of sample-level data integrity (Turner et al. 2011).

Database Implementation

Hardware and Software Requirements

Hardware and software requirements should be considered before implementing and using a database to manage information in a genetic analysis project. Many relational DBMSs like MySQL, SQLite, or PostgreSQL will run on almost any platform from large Linux clusters to desktop PCs or even OS X laptops. An obvious hardware consideration is storage – the computer hosting the database should have enough disk space to hold all the data (tens to hundreds of gigabytes for GWAS; terabytes for next-generation sequencing data) and should be fast enough to respond to the user's database queries. Hard drives inevitably fail, so redundant (preferably offsite) backup is essential to prevent catastrophic loss of expensive data. For individuals with little institutional IT support, there are numerous commercial entities (e.g. Amazon Web Services, see Table 7.1) that offer distributed redundant storage on commodity hard disks based on a per-usage fee.

Implementation and Performance Tuning

In addition to selecting the proper hardware, a suitable relational DBMS must be chosen. MySQL is a free, open-source relational DBMS that is commonly used in bioinformatics (including Ensembl and UCSC). A comprehensive comparison of relational DBMSs, including features, limitations, access control, available data types, licenses, and supported operating systems can be found online (Table 7.1).

Many relational database operations (queries) do not perform well after the database is initially designed and data uploaded. The database may need adjustments to improve response time. This most often requires indexing fields that are being searched. In the previous example (Figure 7.1), Tables A and B are being joined based on gene_symbol (foreign key in Table A, primary key in Table B). While joining six rows from one table to two rows in another is trivial, joining millions of SNP association statistics to millions of SNP-gene annotations, and joining that to millions of rows of quality assessment statistics is not. Performance will suffer unless the SNP identifier (e.g. rs-number) is indexed. Indexing dramatically increases speed of data retrieval at the cost of increased storage space (an index is data written to disk that serves as a pointer to the location of records in the table). This increased speed is worth the increase in storage cost depending on how you query the data. Index fields are those that are constantly queried or joined. Not all columns need an index, especially columns that are rarely, if ever, involved in a JOIN, WHERE, or ORDER BY clause. Finally, using faster disks, redesigning the database schema, or rewriting the query are the only ways to improve the performance of a query.

Interacting with the Database Directly

Relational DBMSs usually come with a tool to facilitate interacting with the data. MySQL ships with the MySQL command line tool which can be used to connect to databases, load data, query the database, and save results. There are other front-ends or graphical user interfaces that allow the user to visually inspect, load data, query tables, save results, and debug queries. MySQL workbench (Table 7.1) is a suite of graphical tools for querying and administering a MySQL database. Other third-party tools allow users to query and administer databases that are implemented in disparate relational DBMSs.

ODBC (Open Database Connectivity) is a common interface between applications that retrieve or store data and the DBMS. Nearly, all DBMSs (including MySQL) provide ODBC connectors for most operating systems, making it possible for a wide variety of external applications to seamlessly interact with a MySQL database across Windows, Mac OS, Linux, and other platforms. This becomes extremely useful for downstream analysis where the database can be accessed directly by statistical analysis software, such as R, Stata, or SAS. For example, rather than querying a database, saving the returned data in a text file, and uploading that data into an R session for statistical analysis, one can query the database directly from the R session by issuing a query through the RODBC package (Ripley and Lapsley 1999) to the MySQL ODBC driver (available http://dev.mysql.com/downloads/connector/odbc). Data is returned as a data frame in the R session without the need for creating intermediate text files. This ensures analysis is always being performed on the most up-to-date version of the data stored in the database, saves disk space by eliminating unnecessary data duplication, and reduces disk I/O time because files do not have to be written to and read from the disk. We direct the reader to an excellent tutorial on setting up an ODBC driver and querying a MySQL database using R (Spector 2008). Using ODBC, databases are also easily queried from other software packages including STATA, SAS, and Microsoft Excel.

In addition to data retrieval, information such as summary statistics or analysis results can be directly written to new or existing database tables using ODBC. Because ODBC provides uniform access to a variety of applications and operating systems, DBMSs can be integrated into existing analyst workflows, providing better data organization and less user file management and duplication.

In addition to accessing a database from statistical analysis software, ODBC provides an interface for querying a database to any ODBC-compliant external application. Most general-purpose programming languages provide a method of querying a database either directly or through an ODBC driver. Many downstream applications for genetic analysis are written in Perl. Furthermore, Python has an entire library of open-source scientific tools (Jones et al. 2001). Directly accessing data through an ODBC connection in a general-purpose programming language provides the same benefits as mentioned above for statistical computing software.

Security

The internet has forever changed the data access and security requirements for every genetic study. Security requirements are study-specific and are often dictated by funding agencies that support the study. Providing a detailed security plan is outside the scope of this book but remains an important decision in the data management planning phase. What follows is a brief overview of some of the issues that should be considered when designing and implementing a database management plan for sensitive clinical and genetic data.

For data that is protected by privacy laws such as HIPAA, FERPA, etc. data security cannot be taken lightly. Database servers may be housed in a secured data center preferably with multifactor access control systems that are regularly audited. All systems should be password protected and role permissions should be applied individually to control access. Sensitive data should never be stored on mobile devices. Systems administrators should regularly install security patches and update the software and operating system. Institutional access should be behind a firewall. External access if available should be even more scrutinized, perhaps by requiring authentication via a virtual private network (VPN). Users should be HIPAA trained and should have a basic understanding of Linux permissions and security protocols. Different relational DBMSs allow for different levels of access control and security administration, and security should be considered when choosing one that is appropriate.

Other Tools for Data Management and Manipulation

Until now, we have discussed data management purely from the perspective of relational databases. The relational model is the most widely used method in the world for storing and accessing data, and relational DBMSs allow users to manage and query databases using SQL. However, there are several other platforms and tools for data management and analysis that are of particular interest to statistical geneticists examining the genetic basis of complex disease.

R

As mentioned earlier, *R* is a widely used, free, open-source environment for statistical analysis. *R* can connect to a relational database using an ODBC interface to query databases and return the results in a data frame. *R* also has many data management tools that allow complex, and in some cases, distributed and parallel data analysis. The apply family of functions (apply, sapply, lapply, tapply, mapply) are built-in routines that allow the analyst to execute a particular function over subsets of data and return the aggregated results. The function may be a particular analytical procedure or a data manipulation routine. The plyr package (Table 7.1) allows for the application of a split-apply-combine strategy to data analysis, where a large problem is broken up into manageable pieces, operations are carried out on each piece independently, and results are combined and returned (Wickham 2011). Additionally, the Bioconductor project (Table 7.1) has many packages that allow for management, analysis, and annotation of high-throughput genomic projects such as microarrays and next-generation sequencing (Gentleman et al. 2004). Finally, the Comprehensive R Archive Network (CRAN) has an entire page (a "task view") devoted to high-performance computing, which discusses packages and functionality related to connecting to databases and managing large, out-of-memory data (http://cran.r-project.org/web/views/HighPerformanceComputing.html).

PLINK

PLINK (Table 7.1) is a freely available toolset for manipulating and analyzing GWAS data. PLINK allows subsetting GWAS data by samples, subsetting by SNPs, subsetting according to allele frequency, grouping analyses based on familial or other specified cluster membership, and many other tasks that are common in genetic data analysis but tedious and error-prone using standard SQL approaches. PLINK takes as input pedigree (linkage) format files discussed earlier. Often these, pedigree format files can be constructed "on the fly" with an SQL query to the database where genetic and phenotypic information is stored.

SAMtools

This review has focused mostly on GWAS data. However, the ability to interrogate the entirety of variation in a genome is becoming more feasible as the cost of sequencing continues to plummet (Metzker 2010). One of the first steps in a next-generation sequencing analysis pipeline is aligning a large set of short nucleotide sequence reads to a reference genome. The SAM (Sequence Alignment/Map) format is a versatile file format for storing large nucleotide sequence alignments (Li et al. 2009). The BAM file is the compressed, optionally indexed, binary representation of the SAM file. While this is not intended to be a review of data management and analysis strategies for next-generation sequencing, it is worth mentioning here that BAM is the most widely used and near-standardized format for storing nucleotide sequence alignments from next-generation sequencing studies. SAMtools (Table 7.1) is a powerful set of utilities for manipulating SAM files (and the corresponding compact binary representation, BAM files), such as viewing, sorting, indexing, merging, and generating alignment reports in a per-position format (for downstream variant calling analyses). VCFtools (Table 7.1) offers further functionality for viewing, sorting, merging, filtering, joining, and other operations on variant data in Variant Call Format (VCF) files. Bioinformatics pipelines for storing, manipulating, and analyzing next-generation sequencing data are still being actively developed. As the technology and associated bioinformatics matures, the SAM/BAM and other formats (e.g. VCF for variant calls, GFF/GTF for feature annotation) for next-generation sequencing data management and analysis will become more standardized, and more tools will become available to handle and manipulate this data. More information on this is in Chapter 10.

Workflow Management and Cloud Computing

Rapid advances in genotyping and sequencing technology have created a challenge for bioinformaticians and information technology specialists. Many of the solutions discussed above are not feasible for researchers without strong institutional IT support. Furthermore, reproducibility is critical for the broad acceptance of biomedical research, and many of the tools discussed already do not provide a simple mechanism to share data, software versions, parameter settings, and data manipulation steps. Recently, several "workflow" management tools have been developed that allow researchers to construct pipelines, where the output of one tool (e.g. a short-read aligner) can be used as input to another tool (e.g. a genome assembly tool). Among others (Hull et al. 2006; Reich et al. 2006), perhaps the most widely used tool for workflow management in genomics is Galaxy (Goecks et al. 2010). Galaxy (Table 7.1) abstracts tools for managing and analyzing genomic data behind a consistent and easy-to-use web interface and facilitates transparent and reproducible computational genetic analyses. The Galaxy developers have also made available a virtual machine image that can be run on a number of cloud computing providers, such as Amazon Web Services. Cloud computing (Infrastructure-as-a-Service, IaaS) offers a flexible and scalable approach to large-scale genetic analysis to researchers without immediate access to computing clusters or enterprise-grade disk space with redundant backup. In cloud computing, virtualized resources are provided as a service over the internet, where compute cycles and disk space can be procured and released on demand. Building a large-scale computing infrastructure requires an up-front expenditure of hundreds of thousands of dollars to cover computing hardware, housing, cooling equipment, electricity, and systems administration staff. In contrast, cloud computing allows the researcher to access computing resources on demand, paying only for computing cycles or disk space used. Several other examples of how "cloud" computing can be used in data management and analysis in genetic data analysis can be found elsewhere (Dudley and Butte 2010; Dudley et al. 2010; Stein 2010; Fusaro et al. 2011)

Conclusion

Here we have presented an overview of information management for genetic studies of complex disease. We give an overview of data management planning, including the fundamental data types collected in a genetic study and how those data types can be represented in a relational database. We discussed database implementation, focusing on computing requirements, performance adjustments, security, and how to interact with the database directly. Finally, we presented a set of tools and resources tangential to the relational model for managing, manipulating, and analyzing genetic data. Some of the main challenges in the genetic analysis of complex disease are the complexity of human genotype–phenotype relationships, the enormous quantity of data generated by modern high-throughput technologies, and the scattered nature of existing data and knowledge. Advanced data management planning, strategic implementation, and knowledge of existing tools and resources are critical to the success of any genetic analysis research program.

References

Abecasis, G.R., Auton, A., Brooks, L.D. et al. (2012). An integrated map of genetic variation from 1,092 human genomes. *Nature* 491 (7422): 56–65.

Consortium, The 1000 Genomes Project (2015). A global reference for human genetic variation. *Nature* 526 (7571): 68–74. https://doi.org/10.1038/nature15393.

Denny, J.C., Ritchie, M.D., Basford, M.A. et al. (2010). PheWAS: demonstrating the feasibility of a phenome-wide scan to discover gene-disease associations. *Bioinformatics (Oxford, England)* 26 (9): 1205–1210.

Dudley, J.T. and Butte, A.J. (2010). In silico research in the Era of cloud computing. *Nat. Biotechnol.* 28 (11): 1181–1185.

Dudley, J.T., Pouliot, Y., Chen, R. et al. (2010). Translational bioinformatics in the cloud: an affordable alternative. *Genome Medicine* 2 (8): 51.

Durbin, R.M., Altshuler, D.L., Abecasis, G.R. et al. (2010). A map of human genome variation from population-scale sequencing. *Nature* 467 (7319): 1061–1073. https://doi.org/10.1038/nature09534.

Fusaro, V.A., Patil, P., Gafni, E. et al. (2011). Biomedical cloud computing with Amazon Web services." Edited by Fran Lewitter. *PLoS Comput. Biol.* 7 (8): e1002147.

Gentleman, R.C., Carey, V.J., Bates, D.M. et al. (2004). Bioconductor: open software development for computational biology and bioinformatics. *Genome Biol.* 5 (10): R80.

Goecks, J., Nekrutenko, A., and Taylor, J. (2010). Galaxy: a comprehensive approach for supporting accessible, reproducible, and transparent computational research in the life sciences. *Genome Biol.* 11 (8): R86.

Gottesman, O., Kuivaniemi, H., Tromp, G. et al. (2013). The electronic medical records and genomics (EMERGE) network: past, present, and future. *Genet. Med.* 15 (10): 761–771.

Hall, M.A., Dudek, S.M., Goodloe, R. et al. (2014). Environment-Wide Association Study (EWAS) for Type 2 Diabetes in the Marshfield Personalized Medicine Research Project Biobank. *Pacific Symposium on Biocomputing. Pacific Symposium on Biocomputing*, January, 200–211.

Hindorff, L.A., Sethupathy, P., Junkins, H.A. et al. (2009). Potential etiologic and functional implications of genome-wide association loci for human diseases and traits. *Proc. Natl. Acad. Sci. U. S. A.* 106 (23): 9362–9367. https://doi.org/10.1073/pnas.0903103106.

Hull, D., Wolstencroft, K., Stevens, R. et al. (2006). Taverna: a tool for building and running workflows of services. *Nucleic Acids Res.* 34 (Web Server issue): W729–W732.

Jones, E., Olophant, T., Peterson, P. et al (2001). SciPy: Open Source Scientific Tools for Python. Http://Www.Scipy.Org.

Kho, A.N., Pacheco, J.A., Peissig, P.L. et al. (2011). Electronic medical Records for Genetic Research: results of the EMERGE consortium. *Sci. Transl. Med.* 3 (79): 79re1.

Kho, A.N., Geoffrey Hayes, M., Rasmussen-Torvik, L. et al. (2012). Use of diverse electronic medical record systems to identify genetic risk for Type 2 diabetes within a genome-wide association study. *J. Am. Med. Inform. Assoc.* 19 (2): 212–218. https://doi.org/10.1136/amiajnl-2011-000439.

Kohane, I.S. (2011). Using electronic health records to drive discovery in disease genomics. *Nat. Rev. Genet.* 12 (6): 417–428. https://doi.org/10.1038/nrg2999.

Li, H., Handsaker, B., Wysoker, A. et al. (2009). The sequence alignment/map format and SAMtools. *Bioinformatics (Oxford, England)* 25 (16): 2078–2079. https://doi.org/10.1093/bioinformatics/btp352.

Manolio, T.A. (2010). Genomewide association studies and assessment of the risk of disease. *N. Engl. J. Med.* 363 (2): 166–176.

McCarty, C.A., Chisholm, R.L., Chute, C.G. et al. (2011). The EMERGE network: a consortium of biorepositories linked to electronic medical records data for conducting genomic studies. *BMC Med. Genet.* 4 (January): 13.

McGinnis, D.P., Brownstein, J.S., and Patel, C.J. (2016). Environment-wide association study of blood pressure in the National Health and nutrition examination survey (1999-2012). *Sci. Rep.* 6: 30373. https://doi.org/10.1038/srep30373.

Metzker, M.L. (2010). Sequencing technologies – the next generation. *Nat. Rev. Genet.* 11 (1): 31–46. https://doi.org/10.1038/nrg2626.

Nielsen, R., Paul, J.S., Albrechtsen, A., and Song, Y.S. (2011). Genotype and SNP calling from next-generation sequencing data. *Nat. Rev. Genet.* 12 (6): 443–451. https://doi.org/10.1038/nrg2986.

Patel, C.J., Bhattacharya, J., and Butte, A.J. (2010). An environment-wide association study (EWAS) on Type 2 diabetes mellitus. *PLoS One* 5 (5): e10746.

Pendergrass, S.A., Brown-Gentry, K., Dudek, S.M. et al. (2011). The use of phenome-wide association studies (PheWAS) for exploration of novel genotype-phenotype relationships and pleiotropy discovery. *Genet. Epidemiol.* 35 (5): 410–422.

Pendergrass, S.A., Brown-Gentry, K., Dudek, S. et al. (2013). Phenome-wide association study (PheWAS) for detection of pleiotropy within the population architecture using genomics and epidemiology (PAGE) network. *PLoS Genet.* 9 (1): e1003087.

Purcell, S., Neale, B., Todd-Brown, K. et al. (2007). PLINK: a tool set for whole-genome association and population-based linkage analyses. *Am. J. Hum. Genet.* 81 (3): 559–575. https://doi.org/10.1086/519795.

Reich, M., Liefeld, T., Gould, J. et al. (2006). GenePattern 2.0. *Nat. Genet.* 38 (5): 500–501.

Ripley, B. and Lapsley, M. (1999). "RODBC: ODBC Database Access. Http://Cran.r-Project.Org/Package=RODBC."

Spector, P. (2008). *Data Manipulation with R*. Springer Verlag.

Stein, L.D. (2010). The case for cloud computing in genome informatics. *Genome Biol.* 11 (5): 207.

Tung, J.Y., Do, C.B., Hinds, D.A. et al. (2011). Efficient replication of over 180 genetic associations with self-reported medical data." Edited by Pieter H. Reitsma. *PLoS One* 6 (8): e23473.

Turner, S.D., Armstrong, L.L., Bradford, Y. et al. (2011). Quality control procedures for genome-wide association studies. *Curr. Protoc. Hum. Genet.* 1 (Unit 19): 1–18. https://doi.org/10.1002/0471142905.hg0119s68.

Verma, A. and Ritchie, M.D. (2017). Current scope and challenges in phenome-wide association studies. *Curr. Epidemiol. Rep.* 4 (4): 321–329. https://doi.org/10.1007/s40471-017-0127-7.

Verma, S.S., de Andrade, M., Tromp, G. et al. (2014). Imputation and quality control steps for combining multiple genome-wide datasets. *Appl. Genet. Epidemiol.* 5: 370. https://doi.org/10.3389/fgene.2014.00370.

Verma, A., Lucas, A., Verma, S.S. et al. (2018). PheWAS and beyond: the landscape of associations with medical diagnoses and clinical measures across 38,662 individuals from Geisinger. *Am. J. Hum. Genet.* 102 (4): 592–608. https://doi.org/10.1016/j.ajhg.2018.02.017.

Wei, W.-Q. and Denny, J.C. (2015). Extracting research-quality phenotypes from electronic health records to support precision medicine. *Genome Medicine* 7 (1): 41. https://doi.org/10.1186/s13073-015-0166-y.

Wei, W.-Q., Teixeira, P.L., Mo, H. et al. (2016). Combining billing codes, clinical notes, and medications from electronic health records provides superior phenotyping performance. *J. Am. Med. Inform. Assoc.* 23 (e1): e20–e27. https://doi.org/10.1093/jamia/ocv130.

Wetterstrand, K.S. 2011. "DNA Sequencing Costs: Data from the NHGRI Large-Scale Genome Sequencing Program." http://genome.gov/sequencingcosts.

Wickham, H. (2011). The split-apply-combine strategy for data analysis. *J. Stat. Softw.* 40 (1): 1–29.

Zuvich, R.L., Armstrong, L.L., Bielinski, S.J. et al. (2011). Pitfalls of merging GWAS data: lessons learned in the EMERGE network and quality control procedures to maintain high data quality. *Genet. Epidemiol.* 35 (8): 887–898. https://doi.org/10.1002/gepi.20639.

8

Linkage Disequilibrium and Association Analysis

Eden R. Martin[1] *and Ren-Hua Chung*[2]

[1] *Dr. John T. Macdonald Foundation Department of Human Genetics, University of Miami Miller School of Medicine, Miami, FL, USA*
[2] *Institute of Population Health Sciences, Division of Biostatistics and Bioinformatics, National Health Research Institutes (Taiwan), Hsinchu, Taiwan*

Introduction

There are two primary approaches for mapping genes that either cause or increase susceptibility to human disease. The first is the linkage analysis approach, either a parametric logarithm of the odds (LOD) score analysis when the genetic model is known or model-independent affected relative pair analysis when the genetic model is unknown. The linkage approach has been discussed in detail in Chapter 6 of this book. The second approach is to use association analysis. Association studies look for a significantly increased or decreased frequency of a marker allele, genotype, or haplotype with a disease trait than would be expected by chance if there was no association between marker(s) and phenotype. Association analysis is a useful and often necessary tool in identifying disease gene loci, particularly in genetically complex diseases.

Association between genotype and phenotype can be explained either by direct biological action of the polymorphism (e.g. the *APOE-4* allele in Alzheimer disease) or by allelic association between the marker and a susceptibility gene. In this chapter we consider *allelic association* to be association occurring between alleles at different loci. Allelic association can occur between linked or unlinked loci. The term *linkage disequilibrium* often is used to refer to allelic associations between linked loci. Associations between alleles at unlinked loci decay rapidly, so allelic associations are typically only found between closely linked loci. For this reason, association studies can play a critical role in the analysis of genetically complex traits, in the evaluation of candidate gene loci, as well as in the fine-mapping of a region once linkage studies have indicated a region of interest for follow-up analysis. Certainly, as the Human Genome Project and other initiatives identified and characterized each gene in the human genome as well as millions of single-nucleotide polymorphisms (SNPs), this approach became widely used for the identification of complex disease genes.

Linkage Disequilibrium

Classically, linkage disequilibrium refers to allelic associations that occur between alleles at different loci, whether or not they are linked (e.g. Lewontin 1988; Weir 1996). This is sometimes referred to as gametic disequilibrium, since typically linkage disequilibrium is used for associations

Genetic Analysis of Complex Diseases, Third Edition. Edited by William K. Scott and Marylyn D. Ritchie.

between alleles from the same gamete. In the human genetics literature, however, linkage disequilibrium has been used to mean specifically allelic associations that occur between linked loci (Crow and Kimura 1970; Spielman and Ewens 1996), which is the definition that we use here.

Measures of Allelic Association

There are many different measures of linkage disequilibrium. All measures deviate from independent assortment of alleles at different loci. Table 8.1 gives the formulas for several common measures. Each depends on allele and haplotype frequencies (i.e. the haplotype frequency is the frequency at which two alleles occur on the same gamete) but in different ways. The disequilibrium coefficient directly measures the deviation of the haplotype frequency from the product of the allele frequencies. The disequilibrium coefficient is 0 if there is no allelic association, and positive or negative depending on whether the alleles tend to occur together more or less frequently than expected if they were independent. The disequilibrium coefficient can be difficult to interpret because its maximum value depends on allele frequencies. Thus, alternative measures that are standardized to lie between −1 and 1 are often used.

Two standardized measures that are commonly used are the correlation coefficient and Lewontin's D'. In a random mating population, the count of the number of times that a particular allele at a locus is observed in a sample of unrelated individuals is distributed as a binomial random variable. Therefore, the disequilibrium coefficient, D (Table 8.1), is proportional to the covariance of the allele counts at two loci, and the formula for the squared correlation coefficient, r^2, is given in Table 8.1. An advantage of this measure is that it is a standard statistical measure with well-known properties. Lewontin's D' (Table 8.1) is simply the disequilibrium coefficient standardized by its maximum value. The measure is easily interpretable as the proportion of the maximum possible level of association between two loci, given the allele frequencies. As with the disequilibrium coefficient, r^2 and D' are 0 if there is no allelic association (complete linkage equilibrium); r^2 and D' are 1 if there is allelic association (complete linkage disequilibrium). The correlation coefficient r and Lewontin's D' fall between −1 and 1, thus the maximum values are not dependent on allele frequencies; however, it is important to recognize that this does not mean that the measures themselves are independent of allele frequencies. Though these measures are often referred to as measures of linkage disequilibrium, they do not explicitly measure linkage. They can be used to measure allelic associations between loci, linked or unlinked. Several other measures of allelic association have been proposed, but these are not covered here.

When haplotype frequencies can be estimated, either through direct observation or through additional family data, we can use estimates of these measures of association to conduct statistical tests for significant allelic associations between different loci. Test statistics can be

Table 8.1 Measures of allelic association for alleles *A* and *B* at different loci.

Disequilibrium coefficient	$D = \Pr(AB) - \Pr(A)\Pr(B)$
Correlation coefficient	$r^2 = D/\{\Pr(A)[1-\Pr(A)]\Pr(B)[1-\Pr(B)]\}$
Le won tin's D'	$D' = \begin{cases} D/\min\{\Pr(A)[1-\Pr(B)],\ \Pr(B)[1-\Pr(A)]\} & \text{if } D < 0 \\ D/\min\{\Pr(A)[1-\Pr(B)],\ \Pr(B)[1-\Pr(A)]\} & \text{if } D < 0 \end{cases}$

Pr (AB) = frequency of AB haplotype: Pr (A) = frequency of A allele; Pr (B) = frequency of B allele.

constructed to test the null hypothesis of no allelic association (e.g. see (Weir 1996). In human genetic studies, it is typically the genotypes at the different loci that are measured in the lab, and therefore haplotypes are not directly observed. If one can assume that Hardy–Weinberg equilibrium (HWE) exists, then the Expectation–Maximization (E-M) algorithm (Dempster et al. 1977) can be used to estimate haplotype frequencies (e.g. see (Weir 1996)). Alternatively, tests can be based on a composite measure of linkage disequilibrium, which does not require the assumption of HWE (Weir and Cockerham 1989). To avoid making large-sample approximations, exact tests can be used to test for significant disequilibria between alleles at two loci (Zaykin et al. 1995).

Causes of Allelic Association

There are several forces that can generate allelic associations between loci. These include: (i) mutation, (ii) migration, (iii) selection, and (iv) genetic drift. After an initial event generates an allelic association, the association begins to decay. The rate of decay is related to the rate of recombination. In general, the more tightly linked two loci are, the slower the decay of association. To illustrate this, consider this hypothetical example in which association is generated by mutation. A new mutation occurs in a gene that results in a disease-causing phenotype. This allele occurs with a particular ancestral haplotype. At the time of the initial mutation, alleles at every marker in the genome are associated with the mutant allele. The chromosome with the mutation is then transmitted to the offspring of the original individual in whom the mutation occurred. Transmission over several generations gives the opportunity for recombination to occur and thus for the rearrangement of the alleles at the marker loci, which breaks up allelic associations. Alleles at marker loci that are farther away from the disease mutation will exchange faster than markers that are closer to the disease mutation. In general, the closer the marker is to the disease gene, the longer the marker allele/disease association will persist.

Consider the disequilibrium coefficient (*D*). For a large, random-mating population, in the absence of forces such as mutation, migration, and selection, the rate of decay of allelic association is dependent on the distance between loci and the number of generations since the allelic association arose. Specifically,

$$D_t = D_0(1-\theta)^t$$

where t is the current generation number, D_t is the current amount of association, D_0 is the association at generation 0, and θ is the recombination fraction between loci. Allelic association decays quite rapidly for unlinked loci ($\theta = 0.5$); however, there is little decay θ that is close to zero (Figure 8.1). The slowness of allelic association decay for tightly linked loci makes this a useful mapping tool. However, for tightly linked loci, forces such as recurrent mutation may play a more important role in the decay of allelic association than recombination.

The general rule of thumb is that the stronger the allelic association, the closer the marker is to the disease locus. This is not always the case, however, for several reasons. Mutation rates at the marker locus affect the level of association by increasing the chance that the associated marker allele will change and so diminish the apparent association. New mutations in the disease gene or selection in favor of a particular phenotype can either increase or decrease the level of allelic association. Additionally, the frequencies of the marker alleles have an impact on the amount of allelic association that can exist. For example, if the disease susceptibility allele is positively associated with a marker allele whose general population allele frequency is 0.50, then the greatest the

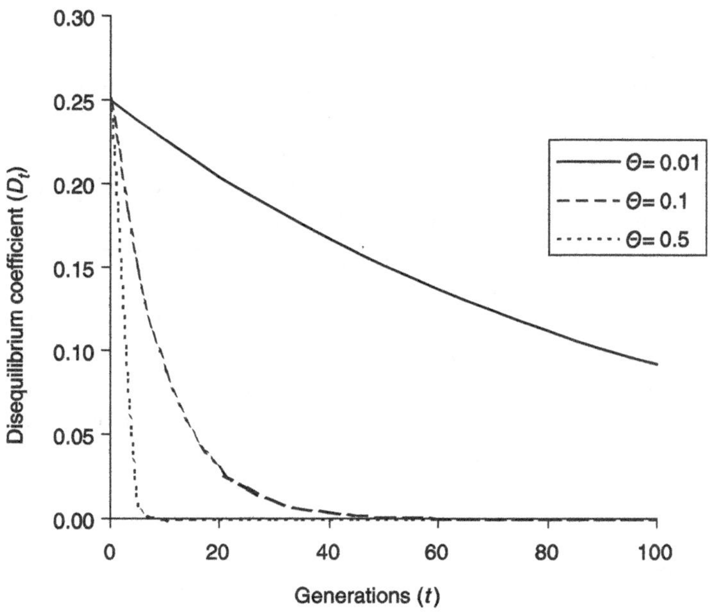

Figure 8.1 Decay of allelic association for recombination fractions (Θ) of 0.01, 0.1, and 0.5; D_0 = 0.25.

level of association can be seen when the allele has frequency 1 in the affected individuals, which is a 100% increase in the allele frequency. If, however, the positively associated marker allele has frequency 0.8 in the population, then again, the greatest level of association is observed when the frequency in affected individuals is 1, but this is only a 25% increase. Generally, the potential level of association is the greatest when the susceptibility allele is positively associated with a low-frequency marker allele, and the chance for substantial association decreases as the positively associated marker allele increases in frequency. It is also important to keep in mind that allelic association is population-specific. The level of allelic association between alleles at two loci can differ between populations. It depends on population-specific allele frequencies and depends critically on population dynamics.

Several studies have examined the relationship between linkage disequilibrium and physical distance. The identification of a multitude of SNPs and other genetic markers in recent years has enabled detailed studies of linkage disequilibrium in many genomic regions and in different populations. The general result that linkage disequilibrium decays in relation to distance is supported by empirical studies; however, the relationship is often not simple for very small distances (see for example, (Terwilliger and Weiss 1998; Daly et al. 2001; Reich et al. 2001)). Markers very close to one another may show very different levels of linkage disequilibrium with other markers in the region, for example, an allele at one marker may be strongly associated with an allele at another locus, while alleles at an immediately adjacent marker may show no association with alleles at the locus (e.g. (Martin et al. 2000a); (Reich et al. 2001). Not only does linkage disequilibrium vary between markers within a genomic region, but also it is highly variable between regions of the genome and between populations. Some theoretical and empirical studies have suggested that significant linkage disequilibrium will only be found over distances of only a few kilobases (kb) (Kruglyak 1999; Dunning et al. 2000), while others have found linkage disequilibrium spanning distances of 100 kb or larger (Collins et al. 1998; Abecasis et al. 2001). The variability observed highlights the important point that linkage disequilibrium depends on more than physical distance. There is great variability depending on population history, regional mutation rate, and recombination.

Studies examining association between SNPs have demonstrated that linkage disequilibrium may be organized in distinct regions of strong linkage disequilibrium, with little association occurring between SNPs in different regions (Daly et al. 2001; Jeffreys et al. 2001). These results suggest that the genome consists of conserved “haplotype blocks” in which little recombination has occurred over time separated by hotspots of recombination. Within these haplotype blocks there is low haplotype diversity, with only a few distinct haplotypes being found. An important implication is that only a subset of markers may be needed to capture much of the variation within a haplotype block, thus if these haplotype blocks can be defined across the genome, this may improve the efficiency of gene-mapping studies discussed below. Figure 8.2 shows an example of haplotype blocks in a 100 kb region on chromosome 1. The LD measures r^2 shown in the Figure were calculated based on the Utah residents with ancestry from northern and western Europe (CEPH) samples from the HapMap project, which will be introduced in the later section.

Mapping Genes Using Linkage Disequilibrium

In the late twentieth century, the primary use of association analysis was for fine-mapping in candidate regions and testing of candidate genes for biological relevance. Some candidate gene studies attempt to identify direct biological involvement by testing specific functional polymorphisms or mutations. Other studies try to exploit linkage disequilibrium that may exist between marker polymorphisms and disease susceptibility alleles in a gene or small region.

Early statistical methods of mapping genes using linkage disequilibrium attempted to invoke a simple relationship between linkage disequilibrium and the rate of recombination to estimate

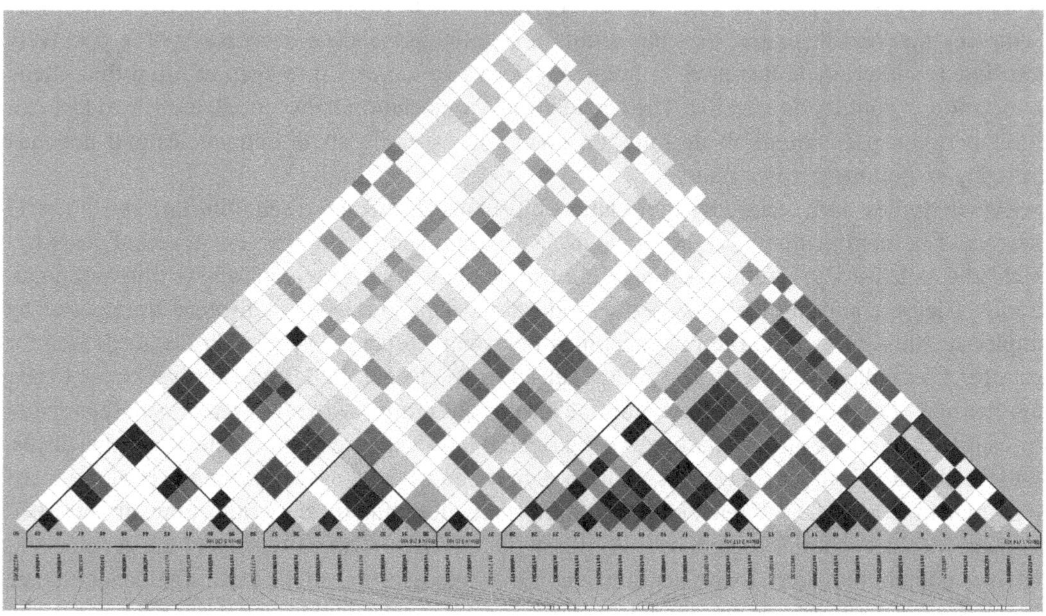

Figure 8.2 An example of haplotype blocks in a 100 kb region on chromosome 1. Darker color for a square represents higher level of LD between two SNPs.

the location of a disease locus relative to a genetic marker (Hastbacka et al. 1992; Hill and Weir 1994; Kaplan et al. 1995; Terwilliger 1995). These methods achieved some notable successes in mapping genes for Mendelian disorders such as cystic fibrosis (Kerem et al. 1989), diastrophic dysplasia (Hastbacka et al. 1992; Hastbacka et al. 1994), and have been applied to help localize genes for several other diseases including Friedreich ataxia (Fujita et al. 1990), myotonic dystrophy (Harley et al. 1991), and Huntington disease (Huntington's Disease Collaborative Research Group 1993). Such methods for linkage disequilibrium mapping can be useful for estimating the location of disease genes when there is a single disease mutation with a strong effect leading to the majority of disease cases in the population. These methods focus on the situation in which chromosomes carrying the disease mutation can be identified, as would be the case for a rare, fully penetrant, recessive disease. Furthermore, estimates of distance depend on accurate knowledge of the history of the disease gene in the population. For these reasons, many early successes with linkage disequilibrium mapping were achieved in isolated populations (e.g. the Finnish population). Isolated populations arising from a small founder group are more homogeneous in disease origin than general populations, since there are likely to be only a few founding individuals who carried the disease mutation. Though even in isolated populations, these methods are restricted to simple Mendelian disorders and have limited utility for disease localization for complex disease genes.

Because linkage disequilibrium rarely extends more than 1 centi-Morgan (cM) in general populations and is often found only over much smaller distances, association between the disease and marker alleles signals a significant decrease in the minimum candidate region making it useful for fine-mapping. However, this great strength is also its great drawback. Because the effect is so localized, it will be very hard to find against the background of the entire human genome without screening a very dense array of markers. The advent of panels of millions of SNPs that can be rapidly and cost-effectively genotyped has led to the ability to conduct genome-wide association studies (GWAS) for complex diseases as discussed later and in Chapter 9. However, in genetically complex diseases, detection of association is still complicated by genetic heterogeneity and/or gene–gene interaction as well as the signal-to-noise problem faced in GWAS. Successful identification of genes for complex diseases requires large, well-characterized samples. In the following sections we discuss several methods for testing for association that are used to identify complex disease genes in general populations.

Tests of Association

For complex diseases, one typically does not know which chromosomes carry the susceptibility allele. Therefore, we cannot directly assess allelic association between marker and disease alleles. Instead, we look for associations between marker alleles and disease phenotype. The phenotype might be a simple dichotomous trait such as affected or unaffected with disease, or it might be a quantitative trait such as blood pressure. Initially, we present tests for dichotomous traits.

Tests for association fall into three broad categories, depending on what types of samples and traits are used. Case–control tests use unrelated individuals who are affected (cases) and unaffected (controls). Case–control tests compare allele or genotype frequencies in the cases to the frequency in a set of matched controls. Family-based tests of association use affected individuals and their relatives. Allele frequencies in affected individuals are compared to family-based controls, typically parental "controls" or unaffected siblings. Association tests with quantitative traits use continuous measures, such as weight or height, and test the correlation between genotypes and the trait values in the sample. Unrelated individuals or family samples can be used for the association tests with quantitative traits.

Case-Control Tests

Test Statistics

In its simplest form, the case–control test is a standard test for the differences between allele and genotype frequencies in two samples. For balanced data, in which the same number of cases and controls are genotyped, an appropriate test for allele frequency differences is given below:

$$T_{cc} = \sum_{i=1}^{m} \frac{(a_i - u_i)^2}{(a_i + u_i)}$$

where m is the number of marker alleles, a_i and u_i are the number of times the ith allele is found in the cases and controls, respectively. The statistic can be compared to a Chi-square with m-1 df for large samples.

It is often difficult to obtain balanced case–control data in any real data set, and there is no need to restrict a study to balanced numbers. A similar case–control statistic can be constructed for unbalanced data, but it takes a slightly more complicated form (e.g. see (Schlesselman 1982)). Essentially, the test compares the observed allele frequencies in the cases and control to those expected under the null hypothesis that the frequencies are equal in cases and controls.

An example of the successful use of a case–control study is the identification of the *APOE-4* allele as the susceptibility gene in late-onset familial and sporadic Alzheimer's disease. Table 8.2 presents data on a large study of over 500 Alzheimer's disease patients and age and ethnically matched controls. Using standard Chi-square analysis, a significant association was found ($p < 0.001$). As indicated in Table 8.2, there appears to be an increase in the *APOE-4* allele, and a concomitant decrease in the *APOE-3* allele, in Alzheimer's patients.

Table 8.2 Case–control association studies: *APOE-4* allele and AD.

Allele	Cases	Controls	Total
a. Observed counts			
APOE4	240	60	300
Not *APOE-4*	360	340	700
Total	600	400	1000
b. Expected counts			
APOE4	180	120	300
Not *APOE-4*	420	280	700
Total	600	400	1000

χ^2 = (observed − expected)^2/expected = [(240 − 180)^2/180] + [(60 − 120)^2/120] + [(360 − 420)^2/420] + [(340 − 280)^2/280] = 71.5; $p < 0.0001$.

Measures of Disease Association and Impact

Case–control studies can also provide estimates of important measures of association and impact. Tests can be based on these measures through either standard hypothesis testing or the construction of confidence intervals (CIs). There are several measures that are often applied in epidemiologic research to quantify the magnitude of the association between the risk factor and the disease (see Table 8.3). The two most commonly used measures of association are the relative risk (RR) and the odds ratio (OR). Measures of impact quantify the importance or impact of risk factors in the occurrence of the disease relative to other risk factors for that disease. These measures include the attributable risk (AR) and population AR.

The RR is the ratio of the incidence of disease among the "exposed" group, which can be defined as individuals with the susceptibility alleles, compared to the incidence of disease in the "nonexposed" group, or individuals without the susceptibility alleles. Table 8.4 is a 2×2 table representing the joint distribution of the presence or absence of the genotype and disease in the study population. If the disease represents the occurrence of "newly" diagnosed cases of the disease over a defined period of time, the incidence of disease among the exposed (i.e. those with a mutation in a susceptibility gene or family history of disease) is I_e, and the incidence of disease in the unexposed (i.e. absence of a genetic mutation or a family history of disease) is I_0.

Incidence rates and RRs are derived from cohort study data (prospective studies), where the exposure is defined "before" disease occurs. Study subjects who are disease-free at baseline are followed over time and observed for subsequent disease onset. However, it is often the case in genetic studies that cohort data are not available or are impractical to obtain, especially for late-onset and rare diseases. Case–control data (retrospective studies), where disease status is established first and exposure subsequently determined, are often utilized to estimate the RR between a study factor and disease. If the data in Table 8.4 were obtained from a case–control study, the marginal totals for the exposed $(a+b)$ and nonexposed $(c+d)$ groups of the population would not be known,

Table 8.3 Summary of epidemiological measures.

Measure	Definition
Relative risk (RR)	Measure of association between genetic exposure and disease, the ratio of the incidence of the disease in the "exposed" and "unexposed" groups using prospective/cohort data: $\mathrm{RR} = \frac{I_e}{I_0} = \frac{a(a=b)}{c(c+d)}$ (see Table 8.4)
Odds ratio (OR)	Measure of association between genetic exposure and disease, the ratio of the odds of exposure (genetic susceptibility) among affected subjects or cases to the odds or exposure (genetic susceptibility) among unaffected individuals or controls calculated using data from case–control studies: $\mathrm{OR} = \frac{ad}{bc}$ (see Table 8.4)
Attributable risk (AR)	Measure of impact or the excess risk of disease among genetically susceptible individuals compared to those who are not. The genetic AR is dependent on the proportion of disease due to the susceptibility gene and the magnitude of the relative risk among gene carriers and noncarriers: $\mathrm{AR} = \frac{P_{\mathrm{Aa}}\left[1-2q\left(1-P_{\mathrm{aa}}\right)\right]}{P_{\mathrm{Aa}}\left(1-2q\right)}$ (Claus et al. 1996)

Table 8.4 2 × 2 Contingency table for case–control analysis.

	Disease		
Genetic test	**Yes**	**No**	**Total**
Positive	a	b	$a+b$
Negative	c	d	$c+d$
Total	$a+c$	$b+d$	$a+b+c+d$

and the incidence rate could not be calculated directly. In this scenario, an OR may be calculated to estimate the RR. The OR is defined as the ratio of the odds of exposure among the cases, the subjects diagnosed with disease, to that among the non-cases or controls. Based on the 2×2 table in Table 8.4, the OR is calculated by applying the following formula:

$$\mathrm{OR} = (a/c)/(b/d) = (a*d)/(b*c)$$

CIs, which specify the range of the true values for the OR, can be derived by estimating the variance and are calculated from the following equation:

$$\mathrm{CI} = (\mathrm{OR})\exp z\left[\pm\left(\mathrm{var}\left(\ln \mathrm{OR}\right)\right)\right]^{1/2}$$

where var represents the variance of the function, and z is the appropriate standard normal deviate, e.g. for a 95% confidence limit or a significance level of 0.05, $z = 1.96$. The CI for the OR can be approximated using a Taylor series expansion as follows (Hennekens and Buring 1987):

$$\mathrm{CI} = (ad/bc)\exp\left[\pm z\left(1/a + 1/b + 1/c + 1/d\right)^{1/2}\right]$$

If the CI does not include 1.0 (the null value for a measure of association), the association is statistically significant at the specified level.

In most case–control studies, in the absence of bias and provided that the disease is not too common, the OR provides a valid estimate of the RR (Hennekens and Buring 1987). The validity of the RR estimate and the inferences based on this estimate depend on the absence of measurement error, the representativeness of the study population (absence of selection bias), and the absence of an imbalance of other factors related to the disease and the exposure in the exposed and nonexposed groups (confounding bias).

The AR represents the excess risk of disease in those exposed compared with those nonexposed to the factor of interest. The AR depends not only on the magnitude of the RR, but on the prevalence of the exposure (the proportion of individuals in a population that are exposed at a specific instant in time such as the proportion carrying a susceptibility allele) in the study population. The AR reflects the "impact" of the exposure on the occurrence of the disease or how much disease would be "prevented" if that factor was somehow eliminated. The AR can be expressed as a percentage and is calculated (Rothman 1986) by incorporating the RR (or OR) using the following formula:

$$\mathrm{AR\%} = \left[(\mathrm{RR} - 1)/\mathrm{RR}\right] \times 100$$

The following equation, for the population attributable risk percent (PAR%), is used to estimate the excess rate of disease in the total study population of exposed and nonexposed individuals that is attributable to the exposure:

$$\text{PAR\%} = \left[(\text{P}_e)(\text{RR} - 1)/(\text{P}_e)(\text{RR} - 1) + 1\right] \times 100$$

where P_e is the prevalence of exposure in the population, which can be either estimated in the control group or obtained from another source (Hennekens and Buring 1987). The AR may also be expressed as the "risk difference" or $I_e - I_0$.

Assessing Confounding Bias

Confounding occurs when an observed association is due to the mixing of effects between the exposure (e.g. presence of the susceptibility allele), the disease, and a third factor (or confounder) that is associated with the exposure and independently affects the risk of developing the disease (Hennekens and Buring 1987).

Confounding bias in epidemiologic studies can be addressed through various statistical analytic strategies to control for such bias. For case–control studies, stratification and logistic regression are the most common methods for assessing an association between the disease and a risk factor while controlling for confounding factors. Survival analyses are often used to analyze data in prospective cohort studies.

Stratified analyses can be useful to control for confounding of either genes or environmental factors. Using this approach, the genetic susceptibility–disease association can be estimated within various levels of the confounding variable. Of particular concern in genetic epidemiology studies is potential confounding by ethnic background of the study participants; such confounding forms the motivation for family-based tests of association discussed in the following section.

An example of confounding by ethnicity is seen in a study of Native Americans and type 2 diabetes mellitus in which the strong negative association between the Gm haplotype Gm3;5,13,14 from the Gm system of human immunoglobin was found to be confounded by the effect of European admixture (Knowler et al. 1988). The crude OR for the association between Gm3;5,13,14 and diabetes in a sample of 4920 Native Americans was 0.21 (95% CI = 0.14 – 0.32) and was statistically significant, suggesting that the absence of this haplotype is a risk factor in this disease. However, Gm3;5,13,14 is also a marker for European admixture, and therefore the degree of Native American heritage is a potential confounder for this association. Stratum-specific ORs, according to the degree of Native American heritage, either none, or half, or full, revealed that there was no significant relationship between Gm3;5,13,14 and diabetes mellitus. The stratum-specific ORs and 95% CI for full, half, and no Native American heritage were 0.63 (95% CI = 0.34 – 1.18), 0.69 (95% CI = 0.24 – 1.96), and 0.75 (95% CI = 0.11 – 5.32), respectively. Further adjustments for confounding by age resulted in an even weaker association between the marker and disease within these strata. Therefore, the relationship between Gm3;5,13,14 and diabetes mellitus was explained by the inverse relationship between the genetic marker and European admixture (Knowler et al. 1988).

The example above illustrates the importance of ensuring that case and control populations are matched with respect to ethnicity/ancestry as well as other factors, such as age, which may be associated with affection status or marker–allele frequency. Similarly, we must ensure that cases and controls are chosen from a population that is homogeneous with respect to confounding factors. Specifically, spurious associations can result because of population stratification (i.e. the existence of non-intermating subpopulations in what is assumed to be a homogeneous

Table 8.5 Example of population stratification[a].

Population A			Population B			Population C (mixed)		
	A (0.80)	*a* (0.20)		*A* (0.20)	*a* (0.80)		*A* (0.50)	*a* (0.50)
B (0.80)	0.64	0.16	*B* (0.20)	0.04	0.16	*B* (0.50)	0.25	0.25
b (0.20)	0.16	0.04	*b* (0.80)	0.16	0.64	*b* (0.50)	0.15	0.25

[a] In this example, populations A and B have very different allele frequencies for the disease gene *A* and the unlinked marker *B*. If the populations mix evenly, then the overall allele frequencies are as seen in population C. If comparisons of the haplotype frequencies are made from population C to either population A or B, the results will be significant even if no linkage exists. For example, if the affected individuals are drawn from population C, they will have an *A/B* haplotype frequency of 0.25, assuming no association with the disease. If the controls are drawn from population B, the *A/B* frequency will be 0.04. The erroneous conclusion is that there is an association of the disease with the *B* allele (0.25 vs. 0.04), since they occur together more often than in population B. In most datasets, it is not possible to determine whether the sample was drawn from one, two, or more populations.

population). The existence of population stratification can lead to significant associations even for markers that are unassociated with disease within subpopulations (Table 8.5).

Family-Based Tests of Association

Family-based studies ensure that cases and controls are appropriately matched by using family-based controls. The case–control tests described above are not appropriate in family data since cases and family-based controls are not necessarily independent. Thus, several family-based association tests have been developed.

Early examples of family-based tests for association used family triads (trios) (i.e. an affected individual and both parents). In a family triad, there is a transmitted pair of alleles and a non-transmitted pair of alleles. The case genotype is made up of the transmitted pair of alleles and the non-transmitted pair of alleles makes up the family-based control genotype (Figure 8.3). These tests include the haplotype relative risk test (HRR) (Falk and Rubinstein 1987), the haplotype-based haplotype relative risk test (HHRR) (Terwilliger and Ott 1992), the affected family-based controls (AFBAC) method (Thomson 1995), and the transmission/disequilibrium test (TDT) (Spielman et al. 1993). The HRR, HHRR, and AFBAC approaches were developed to detect association while controlling for possible bias due to population stratification. The TDT approach has the advantage of being valid as a test of linkage as well as association, thus assures that associations detected occur between linked loci. All tests have little power unless linkage and association coexist. The statistical differences between these methods are subtle and are not described here. Reviews of these methods can be found elsewhere (Spielman and Ewens 1996).

The Transmission/Disequilibrium Test

The TDT is one of the most widely used of the family-based association tests. The TDT was originally proposed as a test of linkage in the presence of association (Spielman et al. 1993), meaning that it was intended to verify that significant associations between marker alleles and disease were due to linkage and not some other force such as population admixture or stratification. The TDT can be a more powerful test to detect linkage than the affected sib pair linkage tests, especially when the genetic effect is small (Spielman et al. 1993; Risch and Merikangas 1996), as is often the case with genetically complex traits. The disadvantage of this approach is that the TDT has little

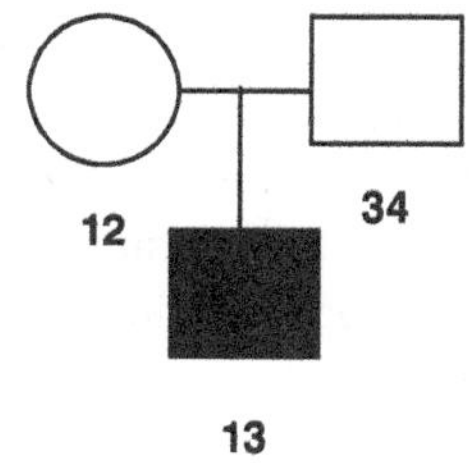

Figure 8.3 Transmitted and non-transmitted alleles in a family triad.

Transmitted pair of alleles 13

Nontransmitted pair of alleles 24

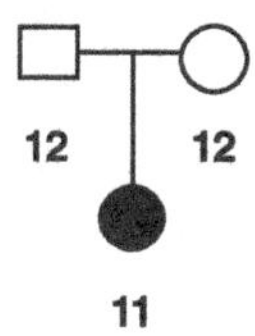

Transmitted	Not Transmitted	
	Allele 1	Allele 2
Allele 1	0	2
Allele 2	0	0

Figure 8.4 Example of scoring a TDT family.

power to detect linkage if allelic association is not present. For a marker with two alleles (allele 1 and allele 2), the TDT examines the number of transmissions of allele 1 or allele 2 from a heterozygous parent to an affected offspring. An example of the TDT is given in Figure 8.4. As a test of linkage, the counts needed in Figure 8.4 can come from simplex, multiplex, or even multigenerational family data. The statistical significance can be tested using standard Chi-square distribution with 1 degree of freedom for large samples. The data used in constructing the counts comes only from heterozygous parents. The TDT statistic is as follows:

$$\text{TDT } (b-c)^2/(b+c)$$

where b is the number of times a parent with the genotype 1/2 transmits allele 1 to an affected offspring, and c is the number of times a parent with the 1/2 genotype transmits allele 2 to an affected offspring.

The test statistic tests for deviations from the expected equal transmission rate into the two categories from the heterozygous parents. Homozygous parents are not used in the TDT. This is different from the HRR, HHRR, and AFBAC methods, which use both heterozygous and homozygous parents. A significant result indicates that the marker is linked to the disease locus. However, the test has little power to detect linkage unless there is also allelic association. In fact, if the data set is composed of only independent family triads, then the TDT can find linkage only in the presence of association. If there were only linkage and no allelic association, then across families we would expect no difference between b and c, since the allele in coupling with the disease allele in each family is random, preventing the detection of linkage. For larger families, the TDT has some power to detect linkage even without allelic association, though this power is expected to be minimal without significant association.

The TDT can serve as a test of allelic association in addition to linkage (i.e. linkage disequilibrium) if only a single family triad from any family is used in the test. This use of the TDT to test for linkage disequilibrium is critical for narrowing a broad region of interest identified by linkage analysis. Analysis of linkage disequilibrium could potentially identify the markers that are closest to the actual disease susceptibility locus. The TDT is not valid, however, as a test for allelic association if more than a single triad within a family is used since it treats triads independently, which is not the case if there is linkage. An extension of the TDT, the Tsp test, allows for multiple affected siblings while maintaining validity as a test of allelic association (and linkage) (Martin et al. 1997). The test properly adjusts for correlations that may occur between transmissions to affected siblings.

The TDT was originally developed to look at biallelic marker systems or situations of alleles that could be readily collapsed because of prior information regarding a known association. With the wide use of multiallelic markers in genetic studies, it is necessary to be able to assess association with multiple alleles. There are two approaches that are commonly used for markers with more than two alleles. One is to test for association at each allele versus the others. This provides a test for each marker allele separately, thus multiple testing should be considered when assessing significance of any of these tests. A Bonferroni correction for multiple tests could be applied; the appropriate adjusted significance level for a marker with m alleles is $\alpha/(m-1)$, where α is the desired experiment-wise significance level, e.g. $\alpha = 0.05$ (Schaid and Rowland 1998). Alternatively, simulations can be used to obtain appropriate critical values for the statistics (Spielman and Ewens 1996). The other approach that is used with multiallelic markers is to use a global statistic that provides an overall assessment of significance. Several global statistics have been proposed. These include the symmetry statistic, the marginal statistic of Bickeboller and Clerget-Darpoux (Bickeboller and Clerget-Darpoux 1995), the likelihood ratio statistic (Extended-TDT) of Sham and Curtis (Sham and Curtis 1995), and the marginal statistic with only heterozygous parents (T_{mhet}) of Spielman and Ewens (Spielman and Ewens 1996). For a comparison of these tests, see Kaplan et al. (1997). T_{mhet} was found to perform well in general. Table 8.6 gives the definition of T_{mhet}.

The TDT approach was a useful approach in identifying the relationship of the insulin gene in insulin-dependent diabetes mellitus (IDDM) (Spielman et al. 1993) (Table 8.7). The TDT method does not require researchers to go to the expense and effort of recruiting families with multiple affected individuals, indeed, the clinical status of the parents does not have to be known. In some cases, however, these advantages may be outweighed by practical problems. In the diabetes situation, both parents of the affected family members were available for study, making this approach ideally suited to IDDM. However, in the Alzheimer's/*APOE-4* example, since Alzheimer's disease is a late-onset disease, parental DNA is almost always unavailable, and thus the traditional TDT approach was impossible.

Tests Using Unaffected Sibling Controls

An approach that circumvents this difficulty is to use unaffected siblings as controls rather than relying on parental controls. Several test statistics have been proposed (Curtis 1997; Boehnke and Langefeld 1998; Spielman and Ewens 1998). These sib-based tests compare marker allele frequencies in affected and unaffected siblings. The tests require only a simple affected/unaffected sibling pair (discordant sibling pair – DSP), although power may be increased if additional siblings are available. These sib-based association tests retain the advantages of the TDT in that they provide tests for linkage in the presence of association and are immune to the effects of sampling bias. If only a single DSP from each family is used, then the tests provide tests for association in addition to linkage. For biallelic markers, if there is a single DSP in each family, the tests of Curtis, the DAT, and the Sib-TDT are equivalent. The tests differ in their treatment of multiple alleles and larger sibships. Monks et al. (1998) provide a comparison of these methods.

When larger sibships are available, the Sibship Disequilibrium Test (SDT) (Horvath and Laird 1998) can be used be used. The SDT is a sign test based on the sign of the difference in the proportion of times the allele occurs in affected siblings and the proportion of times the allele

Table 8.6 Multiallelic TDT: T_{mhet}[a].

	Transmitted allele			
Nontransmitted allele	**1**	**2**	**3**	**Total**
1	n_{11}	n_{12}	n_{13}	$n_{1.}$
2	n_{21}	n_{22}	n_{23}	$n_{2.}$
3	n_{31}	n_{32}	n_{33}	$n_{3.}$
Total	$n_{.1}$	$n_{.2}$	$n_{.3}$	$n_{..}$

[a] $T_{\text{mhet}} = \frac{(m-1)}{m}\sum_{i=1}^{m}\frac{(n_{i.}-n_{.i})2}{(n_{i.}+n_{.i}-2n_{ii})}$, where m is the number of alleles.

Table 8.7 Transmission disequilibrium test and diabetes[a].

	Nontransmitted	
Transmitted	***A1***	***A2***
A1	NI	78 (62)
A2	46 (62)	NI

[a] In this example, the number of *A*1 alleles transmitted from a heterozygous parent to the affected (IDDM) child is 78, while the number of *A*2 alleles transmitted is 46. If no association exists, the expectation is that each allele would be transmitted 62 times. This result is highly significant. NI, not informative; expected number in parentheses. *Source*: Spielman et al. (1993).

occurs in unaffected siblings within a sibship. This difference is equally likely to be positive or negative if there is either no allelic association or no linkage between the marker and disease locus, regardless of the size of the sibship. The SDT provides a valid test for linkage and association in sibships of arbitrary size, but it does not take full advantage of the information since it only considers the sign of the difference. An alternative is to use the Weighted-SDT (WSDT) (Martin et al. 1999), which is based on the magnitude as well as the sign of the difference.

A concern with using unaffected siblings is that one can never really be sure that they will remain unaffected, and since they are siblings of affected individuals, they are at increased risk. This will bias the test toward the null hypothesis, decreasing the power to detect an association. One strategy to help decrease the loss of power is to select unaffected siblings who are older than the age of disease onset in their affected siblings. Alternatively, methods that make inference about missing parental genotypes where possible without using the phenotype of the unaffected siblings can be used. These methods include the Reconstruction-Combined-TDT (RC-TDT) (Knapp 1999), which compares affected siblings to unaffected siblings only when parents' genotypes cannot be reconstructed from the sibship data and likelihood-based methods such as those discussed below.

Tests Using Extended Pedigrees

In some disease studies, extended pedigrees containing multiple nuclear families or sibships may be sampled. The tests discussed above are valid as tests of linkage but not valid tests of association when extended pedigree data are used. The difficulty is that contributions of multiple nuclear families or sibships to the statistic are not independent when there is linkage, even if the null hypothesis of no allelic association is true. Thus, in extended pedigrees tests such as the TDT and Sib-TDT that treat nuclear families or sibships, respectively, as independent are not valid tests of association at linked markers; i.e. they will not have the correct significance level.

For fine-mapping and candidate gene studies, it is desirable to establish linkage disequilibrium; therefore, tests have been developed to test for both linkage and association in extended pedigrees. Several tests for linkage disequilibrium that are appropriate for pedigrees of general structure have been proposed (Martin et al. 2000b; Abecasis et al. 2000; Rabinowitz and Laird 2000; Martin et al. 2001). The tests properly account for correlations within pedigrees that arise from linkage between disease and marker loci. The test statistics all take a similar form:

$$T = \frac{\left(\sum_{i=1}^{N} X_i\right)^2}{\sum_{i=1}^{N} X_i^2}$$

Where X_i is a measure of association for the ith pedigree, $i = 1, \ldots, N$, with mean 0 under the null hypothesis. The tests differ in the way that association is measured within pedigrees. For example, the pedigree disequilibrium test (PDT) sums the association statistics from all family triads and all DSPs within a pedigree to provide the overall measure of association for the pedigree (Martin et al. 2000b; Martin et al. 2001).

Regression and Likelihood-Based Methods

An appealing property of the TDT and related methods is their simplicity. Alternative methods based on statistical modeling can be employed to expand the uses of the TDT. In family triads or DSPs, logistic regression can be used to test for linkage disequilibrium (Schaid 1996; Schaid 1999b; Waldman et al. 1999). Regression methods offer several advantages over traditional TDT-like methods. Estimates of the regression parameters provide estimates for allelic or genotypic RRs. Other genetic or environmental factors can be included in the model to test for gene–gene or gene–environment interactions (Schaid 1999a). In the regression framework, it is straightforward to change the model based on what is known about the mode of inheritance of the disease (Schaid 1996) to improve statistical power. Similarly, one can incorporate covariates into the model that may be important to consider when looking for disease genes. A practical difficulty with regression methods is that they are limited to simple pedigree structure (family triads or DSPs) if one is testing for linkage disequilibrium.

Log-linear models have also been used to construct association tests in family triads (Weinberg et al. 1998; Weinberg 1999). Standard TDT methods treat parental transmissions independently and do not distinguish between maternal and paternal transmissions. The log-linear approach models the probabilities of each possible family triad, with respect to genotype, thus allowing one to test for parent-specific effects. In addition to testing for dependence of affection status on an individual's genotype, log-linear models can be used to test for dependence of affection status on maternal (or paternal) genotype. This allows one to test for important prenatal or other maternal effects. Similarly, log-linear models allow one to test for parent-of-origin effects, which may result from genetic imprinting and may play a role in many complex diseases. Genetic imprinting can lead to differential transmission patterns for marker alleles at loci in linkage disequilibrium with the susceptibility locus.

The likelihood framework is most useful when there is missing parental data. The likelihood-based approach used in the APL test (Martin et al. 2003; Chung et al. 2006) uses an iterative procedure to test for linkage disequilibrium using transmission from parents to affected offspring even if there is missing genotype data in the parents. Unaffected siblings are used only to improve inference about missing parental genotypes, thus the test is not as sensitive to misclassification of unaffected siblings as methods that directly compare affected and unaffected siblings. The drawback with making inference about missing data is that the test assumes that the families come from a single, random-mating population, so the test is sensitive to deviations from Hardy–Weinberg equilibrium in the parental population that may arise from population stratification or other forces.

Association Tests with Quantitative Traits

Most diagnoses of complex diseases rely on a number of different clinical criteria. Often quantitative measurements are an important part of these criteria, and identifying genes influencing these quantitative traits could be important for controlling disease, assessing disease risk, and understanding underlying disease mechanisms. Testing for association between genetic markers and a quantitative trait can be used in fine-mapping the quantitative trait loci (QTL) in the same way that we use association tests to look for genes for dichotomous traits.

If a sample of unrelated individuals is available, randomly selected with respect to the quantitative trait, then it is straightforward to test for association between the trait measurement and marker alleles or genotypes. Methods based on general linear models are commonly employed (Searle 1971). In disease studies, individuals are selected for a genetic study because they have the disease of interest. Therefore, some trait measurements may be different from random members of the population at large. Though this may influence the statistical power of the test, it does not generally affect validity if data are properly normalized.

Association tests for quantitative traits in unrelated individuals suffer from the same potential biases as case–control studies if there is population stratification. Family-based tests can be used to avoid bias and provide an additional test of linkage. Several tests have been proposed for family triads, where the offspring in the triad may have a random or selected trait value (Allison 1997; Rabinowitz 1997). If multiple siblings are available, correlations between trait values that may result because of linkage must be taken into account (Monks and Kaplan 2000). Association tests for quantitative traits can also be used when there is missing parental data, provided multiple siblings are available (Allison et al. 1999), and have been extended to general pedigree structure (Abecasis et al. 2000; Rabinowitz and Laird 2000).

Analysis of Haplotype Data

Traditionally, association tests have been single-locus tests, testing each marker locus independently. However, there have been several tests proposed that test for association with marker haplotypes composed of alleles at multiple markers. These tests look for association between disease status and a particular combination(s) of alleles at different loci, i.e. marker haplotype(s). Intuitively, one might expect that marker haplotypes may provide more information than the markers individually. We often imagine that association arises due to the presence of a susceptibility allele that occurs initially on one, or few, ancestral haplotypes. As discussed earlier, the amount of association between the susceptibility allele and a particular marker allele or marker haplotype depends strongly on the frequency of the associated allele or haplotype. The strongest associations are found when the positively associated allele or haplotype has low frequency in the population.

A haplotype necessarily has frequency no larger than any of frequencies for the individual alleles in the haplotype. Thus, there is a potential for greater association when examining haplotypes.

A difficulty to be overcome in testing for haplotype associations is that haplotypes are not generally observed directly. Instead, we typically have information on individual's genotypes. In some cases, the haplotypes can be inferred with certainty from the genotypes, but when the individual is heterozygous at more than one locus, the haplotype assignment is ambiguous. The E-M algorithm (Dempster et al. 1977) is often used to accommodate the phase uncertainty. Methods have been proposed to test for haplotype association, allowing for ambiguous haplotypes, in unrelated cases and controls (Fallin and Schork 1999; Zhao et al. 2000). Genotype information from family members can often aid in haplotype inference. The likelihood-based APL test (Martin et al. 2003; Chung et al. 2006) provides a family-based approach to testing for haplotype association. The test has the same properties as the single-locus APL test (discussed above), but in addition to assumptions of random mating, the haplotype test assumes no recombination between the markers in the sample.

Another set of pedigree-based association tests are available in the FBAT-tools package (Horvath et al. 2001; Lange et al. 2004), which contains both the FBAT and PBAT programs. As with the PDT and APL, FBAT extends the TDT to multiple sibs and entire pedigrees. FBAT has also been extended to handle both dichotomous and quantitative traits. It can examine multiple loci simultaneously, thus handling haplotypic data as well. FBAT considers extended pedigrees as a series of nuclear families. PBAT conditions on the founder genotypes and thus can examine extended pedigrees without breaking them into subpedigrees. PBAT also provides some powerful tools for obtaining power and sample size estimates (Lange et al. 2002). Both FBAT and PBAT can incorporate covariate data, although the methods for incorporating covariates are different.

An additional difficulty with haplotype association tests is that there can be many haplotypes to consider with multiple markers and/or multiple alleles, and often there is no a priori knowledge of which haplotype might be associated. Haplotypes with low frequency may contribute small numbers of observations to the analysis, thus caution must be used in applying and interpreting the results of haplotype tests. Often haplotypes with small numbers can be pooled to provide valid asymptotic tests, but this pooling could reduce power if negatively and positively associated haplotypes are pooled. Typically, global as well as individual haplotype tests are produced. Multiple testing should be considered in interpreting many tests for individual haplotypes.

Genome-Wide Association Studies (GWAS)

Historically, the most common use of association analysis was to test specific candidate loci or markers in a relatively small region. With the advance of genotyping technology over the past 15 years, millions of markers can now be genotyped across the genome for a GWAS. A GWAS does not rely on previous knowledge of candidate regions. Instead, SNPs are densely genotyped across all chromosomes in the hope of finding markers in linkage disequilibrium with the disease loci. Common SNPs (e.g. minor allele frequency > 0.01 or > 0.05) are targeted for GWAS based on the hypothesis that common SNPs may account for the majority of the heritability of common diseases (the common disease, common variant hypothesis). GWAS has been widely performed for many complex traits. As of August 2021, the GWAS Catalog, a collaborative project between the EMBL-EBI and NHGRI, contains 5273 publications and 276696 associations (www.ebi.ac.uk/gwas).

However, there are several potential problems with the use of association analysis in genomic scanning. First, the problem of multiple comparisons arises in this situation where one is testing many markers individually or testing many haplotypes. That is, when so many statistical tests are performed, false-positive results are likely by chance alone unless the usual significance level of

0.05 or 0.01 is modified. Thus, use of a significance level that is much smaller than the nominal significance level is warranted at each marker or haplotype. The standard in the field is to use a Bonferroni correction of $p < 5 \times 10^{-8}$; though as the allele frequency threshold shifts to look at lower and lower allele frequencies, there is debate about that threshold being too liberal. That said, the usual Bonferroni correction approach (simply dividing the desired nominal significance level by the number of tests performed) may be too conservative because it assumes each test to be independent of the others, which is not the case, since many of the markers are in linkage disequilibrium with each other. Several additional approaches to adjusting for multiple tests are available such as the False Discovery Rate (FDR) (Storey 2002), permutation-based methods (Westfall and Young 1993; McIntyre et al. 2000), or methods that estimate the independent tests (Gao et al. 2008). As millions of association tests are performed in a GWAS, GWAS also pays a high penalty for the multiple test correction, which makes the power to detect the true positives low unless very large samples are used.

The second problem is that the association analysis approach rests completely on the assumption that some level of linkage disequilibrium exists between the markers screened and the disease susceptibility allele(s). While this may be true in many cases, susceptibility alleles arising from frequent mutation events (e.g. rare alleles), existing as extremely old mutations, or arising in regions with very high recombination rates, will have little or any detectable linkage disequilibrium with GWAS markers. Additionally, allelic heterogeneity, i.e. multiple disease mutations within a gene each influence disease susceptibility, can also make detecting association difficult. For more details about GWAS, please see Chapter 9.

Special Populations

Without a very dense map of markers, associations may be difficult to detect in general, outbred populations. One way in which one can improve the ability to use linkage disequilibrium to detect disease genes is by studying populations that may be enriched for linkage disequilibrium. So-called "special" populations, such as isolated founder populations, admixed, or inbred populations, can be particularly useful for gene identification through association analysis. This is possible because the linkage disequilibrium tends to extend over larger areas (possibly several centimorgans) of the chromosome than in general outbred populations. We have previously discussed the use of isolated populations in linkage disequilibrium mapping. Admixed populations, for which disease prevalence is known to differ between mixing populations, have also been used.

Special populations can be especially useful in mapping complex traits. The basic premise is that genetically isolated populations will have fewer distinct alleles or genes contributing to a disease trait, and therefore the effect of each remaining susceptibility gene will be easier to detect. The value of these populations in genetic-mapping studies has long been realized. Homozygosity mapping was described early on in the molecular revolution (Lander and Botstein 1987). These advances have been expanded to include the use of pooling strategies (Sheffield et al. 1994) and the exploitation of the phenomena of linkage disequilibrium together with the isolated inbred nature of these groups through the approach of "shared segment" mapping of a complex phenotype (Houwen et al. 1994; Durham and Feingold 1997). Thus, the great advantage of the special population is its power to detect linkage. However, this power comes at the potential cost of specificity. Only one or a few of the entire suite of susceptibility genes may be found, and the effect of this gene or genes may be limited to the special population being studied.

HapMap

In 2002, the International HapMap project was launched in a multicountry effort to identify the common patterns of sequence variation in the human genome (The International HapMap Consortium 2003). The initial study (the phase I HapMap) included samples from the Yoruba, Japanese, Chinese, and individuals from US residents with northern and western European ancestry. SNPs with minor allele frequencies greater than 0.05 were targeted for the initial study. A total of 1 007 329 SNPs that passed the quality control filters and are polymorphic across the samples were genotyped, based on the goal of genotyping at least one common SNP every 5 kb across the genome. In phase II of the HapMap project, additional 2.1 million SNPs were genotyped on the same individuals (The International HapMap Consortium 2007). The additional SNPs genotyped in phase II have lower minor allele frequencies than the SNPs in phase I because the selection of SNPs was not restricted to the 0.05 minor frequency threshold. In 2010, more samples including 1184 individuals from 11 populations were densely genotyped to provide further variation information (The International HapMap 3 Consortium 2010).

The HapMap project and resulting data have contributed significantly to the human genetic studies in many aspects. First, in 2002, when the project started, there were about 2.8 million SNPs publicly available; however, in some regions the SNP coverage was poor. The HapMap project discovered new SNPs based on random shotgun sequencing from whole-genome libraries generated from DNA samples of individuals and has characterized more than 3.1 million SNPs (The International HapMap Consortium 2007). The discovery of the dense set of SNPs allows for the design of commercial chips for genotyping. Second, the HapMap samples have been commonly used for identifying population stratification. Including HapMap samples with the study samples in algorithms for identifying population stratification such as EIGENSTRAT (Price et al. 2006) enables a clear visualization of samples clustered with different HapMap populations. Third, the HapMap project provides an LD map of the human genome and accurate haplotypes. Untyped markers in the study samples can hence be imputed based on the haplotype structure in the HapMap samples (Marchini et al. 2007). Such imputation can increase the statistical power to identify SNPs that are in LD with the disease loci, which may not have been typed in the original samples. As additional population-based data sets such as HapMap have been created, such as the 1000 Genomes discussed in the next section, the HapMap project has ended and the data are no longer publicly available. The HapMap project was retired on 16 June 2016 (https://www.ncbi.nlm.nih.gov/variation/news/NCBI_retiring_HapMap).

1000 Genomes Project

With the advancement of sequencing technology (i.e. the next-generation sequencing or NGS), cost of sequencing has been reduced significantly, which allows for genome-wide sequencing in a large sample. Using NGS, the 1000 Genomes Project aimed to provide a deeper characterization of human genetic variation compared to the HapMap project (The 1000 Genomes Project Consortium 2010; 1000 Genomes Project Consortium 2015). Samples from 26 different populations, which can be classified as African, Admixed American, East Asian, South Asian, and European populations, were selected for sequencing. The pilot study of the Project reported about 15 million SNPs, 1 million short insertions and deletions (indels), and 20 000 structural variations (The 1000 Genomes Project Consortium 2010). Phase 1 of the Project further reported 38 million SNPs, 1.4 million indels, and more than 14 000 large deletions based on 1092 individuals selected from 14 out of the 26 populations (The 1000 Genomes Project Consortium 2012). Phase 3 of the Project contained more than 79 million variants in 2535 individuals from the 26 populations.

The Project has had a tremendous impact on human genetics studies worldwide, as the Project collected a wide spectrum of human populations. The catalog of genetic variants, including rare variants that have been thought to contribute significantly to complex disease etiology, produced by the Project has been very informative to complex disease studies.

Summary

Allelic associations can arise due to forces such as mutation, selection, admixture, or genetic drift. The decay of association then depends on the degree of linkage between the loci. Allelic associations in the presence of linkage, i.e. linkage disequilibrium, can be exploited to aid in the mapping of genetically complex diseases. Both case–control and family-based methods can be used in either candidate gene or GWAS approaches. In this chapter we have focused on detecting associations for single loci; however, methods of association can also be used to detect complex interactions between different genes. Methods using association to detect gene–gene interaction are reviewed in Chapter 11, and these methods continue to be refined and advanced.

References

Abecasis, G.R., Cookson, W.O.C., and Cardon, L.R. (2000). Pedigree tests of transmission disequilibrium. *Eur. J. Hum. Genet.* 8: 545–551.

Abecasis, G.R., Noguchi, E., Heinzmann, A. et al. (2001). Extent and distribution of linkage disequilibrium in three genomic regions. *Am. J. Hum. Genet.* 68: 191–197.

Allison, D.B. (1997). Transmission-disequilibrium tests for quantitative traits. *Am. J. Hum. Genet.* 60: 676–690.

Allison, D.B., Heo, M., Kaplan, N., and Martin, E.R. (1999). Sibling-based tests of linkage and association for quantitative trials. *Am. J. Hum. Genet.* 64: 1754–1764.

Bickeboller, H. and Clerget-Darpoux, F. (1995). Statistical properties of the allelic and genotypic transmission/disequilibrium test for multiallelic markers. *Genet. Epidemiol.* 12: 865–870.

Boehnke, M. and Langefeld, C.D. (1998). Genetic association mapping based on discordant sib pairs: the discordant alleles test (DAT). *Am. J. Hum. Genet.* 62: 950–961.

Chung, R.H., Hauser, E.R., and Martin, E.R. (2006). The APL test: extension to general nuclear families and haplotypes and examination of its robustness. *Hum. Hered.* 61 (4): 189–199.

Claus EB, Schildkraut JM, Thompson WD, Risch NJ. The genetic attributable risk of breast and ovarian cancer. Cancer. 1996 Jun 1;77(11):2318-24. doi: 10.1002/(SICI)1097-0142(19960601)77:11<2318::AID CNCR21> 3.0.CO;2-Z. PMID: 8635102.

Collins, F.S., Brooks, L.D., and Chakravarti, A. (1998). A DNA polymorphism discovery resource for research on human genetic variation. *Genome Res.* 8: 1229–1231.

Crow, J.F. and Kimura, M. (1970). An Introduction to Population Genetics Theory. New York: Harper & Row.

Curtis, D. (1997). Use of siblings as controls in case-control association studies. *Ann. Hum. Genet.* 61: 319–333.

Daly, M.J., Rioux, J.D., Schaffner, S.F. et al. (2001). High-resolution haplotype structure in the human genome. *Nat. Genet.* 29: 229–232.

Dempster, A.P., Laird, N.M., and Rubin, D.B. (1977). Maximum likelihood from incomplete data via the EM algorithm (with discussion). *J. R. Stat. Soc. B* B39: 1–38.

Dunning, A.M., Durocher, F., Healey, C.S. et al. (2000). The extent of linkage disequilibrium in four populations with distinct demographic histories. *Am. J. Hum. Genet.* 67: 1544–1554.

Durham, L.K. and Feingold, E. (1997). Genome scanning for segments shared identical by descent among distant relatives in isolated populations. *Am. J. Hum. Genet.* 61: 830–842.

Falk, C.T. and Rubinstein, P. (1987). Haplotype relative risks: an easy reliable way to construct a proper control sample for risk calculations. *Ann. Hum. Genet.* 51: 227–233.

Fallin, D. and Schork, N. (1999). Haplotype analyses in case-control populations: factors which influence accuracy of haplotype estimation. *Am. J. Hum. Genet.* 65 (A83): 439.

Fujita, R., Hanauer, A., Sirugo, G. et al. (1990). Additional polymorphisms at marker loci D9S5 and D9S15 generate extended haplotypes in linkage disequilibrium with Friedreich ataxia. *Proc. Natl. Acad. Sci. U. S. A.* 87: 1796–1800.

Gao, X., Starmer, J., and Martin, E.R. (2008). A multiple testing correction method for genetic association studies using correlated single nucleotide polymorphisms. *Genet. Epidemiol.* 32: 361–369.

Harley, H.G., Brook, J.D., Floyd, J. et al. (1991). Detection of linkage disequilibrium between the myotonic dystrophy locus and a new polymorphic DNA marker. *Am. J. Hum. Genet.* 49: 68–75.

Hastbacka, J., De la Chapelle, A., Kaitila, I. et al. (1992). Linkage disequilibrium mapping in isolated founder populations: diastrophic dysplasia in Finland. *Nat. Genet.* 2: 204–211.

Hastbacka, J., De la Chapelle, A., Mahtani, M.M. et al. (1994). The diastrophic dysplasia gene encodes a novel sulfate transporter: positional cloning by fine-structure linkage disequilibrium mapping. *Cell* 78: 1073–1087.

Hennekens, C.H. and Buring, J.E. (1987). Measures of disese frequency and association. In: Epidemiology in Medicine, 1e (ed. S.L. Mayrent), 54–98. Little, Brown and Co.

Hill, W.G. and Weir, B.S. (1994). Maximum-likelihood estimation of gene location by linkage disequilibrium [published erratum appears in Am J Hum Genet 1994 Jul;55(1):217]. *Am. J. Hum. Genet.* 54: 705–714.

Horvath, S.M. and Laird, N.M. (1998). A discordant-sibship test for disequilibrium and linkage: no need for parental data. *Am. J. Hum. Genet.* 63: 1886–1897.

Horvath, S., Xu, X., and Laird, N.M. (2001). The family based association test method: strategies for studying general genotype – phenotype associations. *Eur. J. Hum. Genet.* 9: 301–306.

Houwen, R.H.J., Baharloo, S., Blankenship, K. et al. (1994). Genome screening by searching for shared segments: mapping a gene for benign recurrent intrahepatic cholestasis. *Nat. Genet.* 8: 380–386.

Huntington's Disease Collaborative Research Group (1993). A novel gene containing a trinucleotide repeat that is expanded and unstable on Huntington's disease chromosomes. *Cell* 72: 971–983.

Jeffreys, A.J., Kauppi, L., and Neumann, R. (2001). Intensely punctate meiotic recombination in the class II region of the major histocompatibility complex. *Nature* 29: 217–222.

Kaplan, N.L., Hill, W.G., and Weir, B.S. (1995). Likelihood methods for locating disease genes in nonequilibrium populations. *Am. J. Hum. Genet.* 56: 18–32.

Kaplan, N.L., Martin, E.R., and Weir, B.S. (1997). Power studies for the transmission/disequilibrium tests with multiple alleles. *Am. J. Hum. Genet.* 60: 691–702.

Kerem, B., Rommens, J.M., Buchanan, J.A. et al. (1989). Identification of the cystic fibrosis gene: genetic analysis. *Science* 245: 1073–1080.

Knapp, M. (1999). The transmission/disequilibrium test and parental-genotype reconstruction: the reconstruction-combined transmission/disequilibrium test. *Am. J. Hum. Genet.* 64: 861–870.

Knowler, W.C., William, R.C., Pettitt, D.J., and Steinberg, A. (1988). GM3;5,13,14 adn type 2 diabetes mellitus: an association in American Indians by genetic admixture. *Am. J. Hum. Genet.* 43: 520–526.

Kruglyak, L. (1999). Prospects for whole-genome linkage disequilibrium mapping of common disease genes. *Nat. Genet.* 22: 139–144.

Lander, E.S. and Botstein, D. (1987). Homozygosity mapping: a way to map human recessive traits with the DNA of inbred children. *Science* 236: 1567–1570.

Lange, C., Demeo, D.L., and Laird, N.M. (2002). Power and design considerations for a general class of family-based association tests: quantitative traits. *Am. J. Hum. Genet.* 71: 1330–1341.

Lange, C., DeMeo, D., Silverman, E.K. et al. (2004). PBAT: tools for family-based association studies. *Am. J. Hum. Genet.* 74: 367–369.

Lewontin, R.C. (1988). On measures of gametic disequilibrium. *Genetics* 120: 849–852.

Marchini, J., Howie, B., Myers, S. et al. (2007). A new multipoint method for genome-wide association studies via imputation of genotypes. *Nat. Genet.* 39: 906–913.

Martin, E.R., Kaplan, N.L., and Weir, B.S. (1997). Tests for linkage and association in nuclear families. *Am. J. Hum. Genet.* 61 (2): 439–448.

Martin, E.R., Monks, S.A., Warren, L.L., and Kaplan, N.L. (1999). A weighted sibship disequilibrium test for linkage and association in discordant sibships. *Am. J. Hum. Genet.* 65: A434.

Martin, E.R., Lai, E.H., Gilbert, J.R. et al. (2000a). SNPing away at complex disease: analysis of SNPs around APOE in Alzheimer disease. *Am. J. Hum. Genet.* 67: 383–394.

Martin, E.R., Monks, S.A., Warren, L.L., and Kaplan, N.L. (2000b). A test for linkage and association in general pedigrees: the pedigree disequilibrium test. *Am. J. Hum. Genet.* 67: 146–154.

Martin, E.R., Bass, M.P., and Kaplan, N.L. (2001). Correcting for a potential bias in the pedigree disequilibrium test. *Am. J. Hum. Genet.* 68: 1065–1067.

Martin, E.R., Bass, M.P., Hauser, E.R., and Kaplan, N.L. (2003). Accounting for linkage in family-based tests of association with missing parental genotypes. *Am. J. Hum. Genet.* 73 (5): 1016–1026.

McIntyre, L.M., Martin, E.R., Simonsen, K., and Kaplan, N.L. (2000). Circumventing the multiple testing problem: a multilocus transmission/disequilibrium test. *Genet. Epidemiol.* 19: 18–29.

Monks, S.A. and Kaplan, N.L. (2000). Removing the sampling restrictions from family-based tests of association for a quantitative trait locus. *Am. J. Hum. Genet.* 66: 576–592.

Monks, S.A., Kaplan, N.L., and Weir, B.S. (1998). A comparative study of sibship tests of linkage and /or association. *Am. J. Hum. Genet.* 63: 1507–1516.

Price, A.L., Patterson, N.J., Plenge, R.M. et al. (2006). Principal components analysis corrects for stratification in genome-wide association studies. *Nat. Genet.* 38: 904–909.

Rabinowitz, D. (1997). A transmission disequilibrium test for quantitative trait loci. *Hum. Hered.* 47: 342–350.

Rabinowitz, D. and Laird, N. (2000). A unified approach to adjusting association tests for population admixture with arbitrary pedigree structure and arbitrary missing marker information. *Hum. Hered.* 50: 211–223.

Reich, D.E., Cargill, M., Stacey, B. et al. (2001). Linkage disequilibrium in the human genome. *Nature* 411: 199–204.

Risch, N. and Merikangas, K. (1996). The future of genetic studies of complex human disorders. *Science* 273: 1516–1517.

Rothman, K.J. (1986). Modern Epidemiology. Boston/Toronto: Little, Brown and Company.

Schaid, D.J. (1996). General score tests for associations of genetic markers with disease using cases and their parents. *Genet. Epidemiol.* 13: 423–449.

Schaid, D.J. (1999a). Case-parents design for gene-environment interaction. *Genet. Epidemiol.* 16: 261–273.

Schaid, D.J. (1999b). Likelihoods and TDT for the case-parents design. *Genet. Epidemiol.* 16: 250–260.

Schaid, D.J. and Rowland, C. (1998). Use of parents, sibs, and unrelated controls for detection of associations between genetic markers and disease. *Am. J. Hum. Genet.* 63: 1492–1506.

Schlesselman, J.J. (1982). Case-Control Studies. Design, Conduct, Analysis. New York. Oxford: Oxford University Press.

Searle, S.R. (1971). Linear Models. New York: Wiley.

Sham, P.C. and Curtis, D. (1995). An extended transmission/disequilibrium test (TDT) for multi-allele marker loci. *Ann. Hum. Genet.* 59: 323–336.
Sheffield, V.C., Carmi, R., Kwitek-Black, A.E. et al. (1994). Identification of a Bardet-Biedl syndrome locus on chromosome 3 and evaluation of an efficient approach to homozygosity mapping. *Hum. Mol. Genet.* 3: 1331–1335.
Spielman, R.S. and Ewens, W.J. (1996). The TDT and other family-based tests for linkage disequilibrium and association [editorial]. *Am. J. Hum. Genet.* 59: 983–989.
Spielman, R.S. and Ewens, W.J. (1998). A sibship test for linkage in the presence of association: the sib transmission/disequilibrium test. *Am. J. Hum. Genet.* 62: 450–458.
Spielman, R.S., McGinnis, R.E., and Ewens, W.J. (1993). Transmission test for linkage disequilibrium: the insulin gene region and insulin-dependent diabetes mellitus (IDDM). *Am. J. Hum. Genet.* 52: 506–516.
Storey, J.D. (2002). A direct approach to false discovery rates. *J. R. Stat. Soc. Ser. B Methodol.* 64 (3): 479–498.
Terwilliger, J.D. (1995). A powerful likelihood method for the analysis of linkage disequilibrium between trait loci and one or more polymorphic marker loci. *Am. J. Hum. Genet.* 56: 777–787.
Terwilliger, J.D. and Ott, J. (1992). A haplotype-based 'haplotype relative risk' approach to detecting allelic associations. *Hum. Hered.* 42: 337–346.
Terwilliger, J.D. and Weiss, K.M. (1998). Linkage disequilibrium mapping of complex disease: fantasy or reality? *Curr. Opin. Biotechnol.* 9: 578–594.
The 1000 Genomes Project Consortium (2010). A map of human genome variation from population-scale sequencing. *Nature* 467: 1061–1073.
The 1000 Genomes Project Consortium (2012). An integrated map of genetic variation from 1,092 human genomes. *Nature* 491: 56–65.
The 1000 Genomes Project Consortium (2015). A global reference for human genetic variation. *Nature* 526: 68–74.
The International HapMap 3 Consortium (2010). Integrating common and rare genetic variation in diverse human populations. *Nature* 467: 52–58.
The International HapMap Consortium (2003). The international HapMap project. *Nature* 426: 789–796.
The International HapMap Consortium (2007). A second generation human haplotype map of over 3.1 million SNPs. *Nature* 449: 851–862.
Thomson, G. (1995). Mapping disease genes: family-based association studies. *Am. J. Hum. Genet.* 57: 487–498.
Waldman, I.D., Robinson, B.F., and Rowe, D.C. (1999). A logistic regression based extension of the TDT for continuous and categorical traits. *Ann. Hum. Genet.* 63 (Pt 4): 329–340.
Weinberg, C. (1999). Allowing for missing parents in genetic studies of case-parent triads. *Am. J. Hum. Genet.* 64: 1186–1193.
Weinberg, C.R., Wilcox, A.J., and Lie, R.T. (1998). A log-linear approach to case-parent-triad data: assessing effects of disease genes that act either directly or through maternal effects and that may be subject to parental imprinting. *Am. J. Hum. Genet.* 62: 969–978.
Weir, B.S. (1996). Genetic Data Analysis II: Methods for Discrete Population Genetic Data. Sunderland, Mass: Sinaur Associates.
Weir, B.S. and Cockerham, C.C. (1989). Complete characterization of disequilibrium at two loci. In: Mathematical Evolutionary Theory (ed. M.E. Feldman), 86–110. Princeton: Princeton University Press.
Westfall, P.H. and Young, S.S. (1993). Resampling-Based Multiple Testing: Examples and Methods for P-Value Adjustment, Wiley series of in probabilities and statistics. Wiley-Interscience.
Zaykin, D., Zhivotovsky, L., and Weir, B.S. (1995). Exact tests for association between alleles at arbitrary numbers of loci. *Genetica* 96: 169–178.
Zhao, J.H., Curtis, D., and Sham, P.C. (2000). Model-free analysis and permutation tests for allelic associations. *Hum. Hered.* 50: 133–139.

9

Genome-Wide Association Studies

Jacob L. McCauley[1], *Yogasudha Veturi*[2], *Shefali Setia Verma*[3], *and Marylyn D. Ritchie*[2]

[1] *Dr. John T. Macdonald Foundation Department of Human Genetics, University of Miami Miller School of Medicine, Miami, FL, USA*
[2] *Department of Genetics, Perelman School of Medicine at the University of Pennsylvania, Philadelphia, PA, USA*
[3] *Department of Pathology and Laboratory Medicine, Perelman School of Medicine at the University of Pennsylvania, Philadelphia, PA, USA*

Introduction

Genome-wide genotyping and analysis approaches offer the opportunity to examine patterns of variation across a large number of genomic positions for association with occurrence of disease or the distribution of traits of interest. Initial genome-wide approaches utilized genotyping panels consisting of hundreds of highly polymorphic microsatellites, or short tandem repeat (STR), markers across the genome. These panels were used to observe broad patterns of co-segregation among affected pedigree members or related individuals, followed by higher-density genotyping and other "fine-mapping" approaches in declared regions of interest (described in Chapter 6). Follow-up analyses of genetic variants underneath previously observed linkage peaks have been successful in identifying major genetic contributors in mono- and oligogenic disease and, to a lesser degree, in complex disease, including the contribution of *APOE* in late-onset Alzheimer's disease (AD) (Corder et al. 1993) and the role of *TCF7L2* in type 2 diabetes (Grant et al. 2006), both first identified through linkage approaches. While these approaches utilizing broad, but sparse, genotyping have proved useful in characterizing important heritable components of complex diseases and complex traits, they have not explained much of the heritability of these conditions, leading to the need for new tools to identify genetic contributors to disease. Biotechnology advances, facilitated by the draft genome of the Human Genome Project (Lander et al. 2001) and subsequent HapMap project (International HapMap Consortium 2005), have enabled the simultaneous genotyping of tens of thousands to several million single nucleotide polymorphisms (SNPs) densely distributed throughout the genome. This has given rise to a new class of studies that interrogate high-density panels of markers to determine whether any of the genotyped markers are associated with a disease or trait, using either related individuals or unrelated cases and controls. These studies, called genome-wide association studies (GWAS), have proven highly successful in characterizing a larger portion of the heritability in numerous complex diseases and traits (e.g. age-related macular degeneration [AMD], multiple sclerosis [MS], human height, and lipid levels to name a few) (Hindorff et al. 2009; MacArthur et al. 2017). While it is becoming increasingly clear that GWAS are not likely to

Genetic Analysis of Complex Diseases, Third Edition. Edited by William K. Scott and Marylyn D. Ritchie.

characterize all genetic effects on disease, GWAS continue to remain highly useful for elucidating common variants with modest or small effects on disease (Manolio et al. 2009). Modifications and extensions of GWAS in use and under development can further enhance the ability of these studies to capture genetic variants contributing to disease (Visscher et al. 2017).

Definition of GWAS

A GWAS is a genetic association study examining genotype data at thousands or millions of positions in the human genome for correlation/association with complex phenotypes (diseases or biological traits of interest). The simplest and often primary form of association testing examines the associations of individual SNPs with a phenotype of interest, although more advanced approaches exist for examining the associations of multiple SNPs simultaneously including haplotype association, multi-locus association testing (including gene–gene interaction testing and polygenic analysis [discussed in Chapter 11]), and pathway analysis (discussed in Chapter 10). These approaches are also frequently and widely employed in GWAS study designs, often as extensions of single SNP association analyses (Visscher et al. 2017).

Purpose of GWAS

The purpose of GWAS is to identify genomic variations that contribute to susceptibility of diseases of interest or the distribution of biological traits under examination. At their outset, these studies aspire to characterize patterns of association with common variants in the genome directly or indirectly by exploiting the physical phenomenon of linkage disequilibrium – the co-transmission of adjacent markers more frequently than expected by chance alone due to limited recombination between those markers. As discussed in Chapter 8, direct association refers to the genetic variant being associated with the trait of interest; such variants are often known a priori to have direct, significant effects on gene expression or protein structure or function. Indirect association refers to the detection of the positive association due to linkage disequilibrium with nearby functional variant(s). GWAS examine the entire genome, without regard to biological mechanism, putative gene/variant function, or preconceived hypotheses. GWAS are considered a high-throughput, hypothesis-generating analysis tool.

Design

GWAS efforts require rigorous planning and careful study design at their outset. Decisions must be made taking into consideration the parameters of your specific project, including both the size and structure of the available data set, the type of disease or trait being studied, and the fiscal resources available to conduct the project. All these elements can drive the project timeline, costs, and statistical power. The primary design elements for consideration include: genotyping technology or platform, data set sample design (family-based, case–control, cohort), and statistical analysis strategy. The following sections will discuss each of these design elements in more detail.

Technologies for High-Density Genotyping

Collecting genotype data for GWAS requires the use of one or more high-density SNP genotyping arrays, often colloquially referred to as "genotyping chips." These use hundreds of thousands (or millions) of SNP-specific oligonucleotide probes embedded on a substrate to hybridize with short, specific fragments of amplified genomic DNA. The probes are laid out on the substrate in assigned locations and are typically duplicated several times on each chip as a form of quality control (QC). This redundancy is necessary, as optimal hybridization conditions vary from probe to probe, and similarities in sequence between different targeted DNA fragments may lead to the hybridization of target DNA with mismatched probes. Genotypes are determined from the stimulation and measurement of light emission from one or two fluorophores used as part of the probe to differentiate between alleles (Rapley and Harbron 2004). Following the hybridization reaction, the fluorescence intensities of the two fluorophores for each probe are captured and recorded, and these are used to assign SNP genotypes. High intensities of one fluorophore or the other typically correspond to homozygosity for one allele or the other, whereas moderate intensity of both fluorophores typically indicates heterozygosity at the SNP. Genotype calling for each SNP is performed using sophisticated calling algorithms that look for clustering of fluorescence intensities across multiple individuals at the same SNP. See Figure 9.1 for an example of the output of a single genetic variant from a GWAS array with a high genotyping quality variant (left) and a low genotyping quality variant (right).

Several high-density SNP genotyping chips are commercially available from manufacturers such as Illumina or Applied Biosystems (formerly Affymetrix), and these chips can examine preselected panels of SNPs, customized panels of SNPs, or some mixture of both (see Chapter 10 for more details). With the increase in availability of whole exome and/or whole genome sequencing data, many studies are now deriving genotypes from the sequence data. In this process, a researcher selects a particular allele frequency threshold for which they wish to include

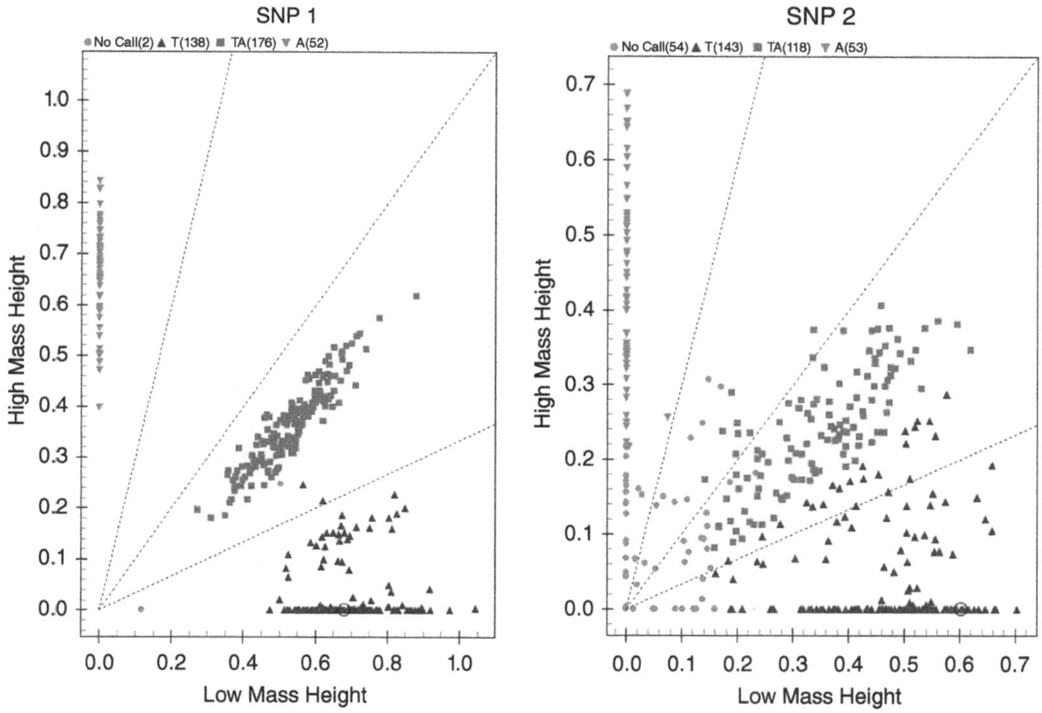

Figure 9.1 Example of raw genotype intensity clusters for a single SNP. Image on the left shows an example with high genotyping quality. The image on the right shows an example with low genotyping quality.

in their study only variants that exceed that frequency; these variants can then be subset out of the sequencing files and used for GWAS.

Discrete and Quantitative Trait Analysis

The type of disease or trait of interest being investigated will fundamentally influence many critical elements in the design of a GWAS to identify genetic associations for that disease or trait. While many studies may focus on discrete traits such as case–control status for complex disease (e.g. coronary heart disease, type 2 diabetes), many may focus on genetic factors influencing anthropometric traits that predict disease (e.g. body weight), common risk factors (e.g. HDL and LDL cholesterol levels) and even continuous traits that underlie case definitions (e.g. fasting plasma glucose levels).

In general, quantitative traits increase statistical power to detect genetic associations as they provide more information content than discrete traits; however, power may be reduced in the presence of measurement error or missing data. Conversely, power for some quantitative trait analyses may be reduced relative to discrete traits as certain traits may only be informative in subsets of individuals available for study; for instance, fasting plasma glucose levels in samples from case–control data sets of type 2 diabetes may only be valid if taken from individuals without diabetes or with newly diagnosed disease, as dysregulation of glucose levels is a feature of prolonged disease.

Detailed phenotypic characterization of disease may provide quantitative trait data for genetic analyses as some characteristics and pathologic features may be under genetic control. For instance, analyses of the *APOE* ε4 allele in AD have demonstrated that not only does each copy of *APOE* ε4 contribute to a considerable increase in AD risk (OR ~ 3–4), but each additional ε4 copy also leads to a reduction in age at onset by approximately eight years (Corder et al. 1994), suggesting an important temporal component to the risk effects of this variant.

It may be noted that quantitative traits can also be analyzed as discrete traits, often by categorization using threshold values or sampling extremes of the quantitative trait distribution. If the variance of a quantitative trait of interest is relatively small but a sufficiently large sample of characterized individuals is available, considerable additional power might be gained by sampling from extremes of the distribution (Chen and Li 2011). This approach applied to body mass index (BMI) measures in relatively small samples of individuals with very low BMI ($\leq 25\,\text{kg/m}^2$; N = 540) or very high ($\geq 35\,\text{kg/m}^2$; N = 520) was effective in identifying strong associations in *FTO* and *NRXN3* (Wang et al. 2011), previously only observed in studies of many thousands of cases and controls.

Examinations of qualitative traits also require a multitude of considerations for analyses. As in most epidemiologic studies, affection status (the most common qualitative trait examined) is subject to potential selection bias, as the sample of cases or controls may not be representative of the affected and unaffected parts of the population. Also, disease misclassification (the incorrect assignment of affection status) can be a problem. Furthermore, studying qualitative traits may require more scrupulous inclusion and exclusion criteria if the current understandings of the genetics of the disease suggest heterogeneity among cases, such as locus heterogeneity or even etiologic heterogeneity. In studies of AD, for example, people with age at onset of dementia younger than 60 years have often been found to be affected by highly penetrant mutations in any one of several genes: *APP*, *PSEN1*, *PSEN2*, *GRN*, and *MAPT*. Therefore, inclusion of cases with onset before age 60 in studies of sporadic or "late-onset" AD is often inadvisable without prior screening for mutations in these five genes, as younger cases may bias any associations observed.

Case–Control, Family-Based, and Cohort Study Designs

For surveying the wide variety of GWAS that have already been performed, several different common study designs (reviewed in Chapter 4) have already been implemented: case–control studies, family-based association studies, and prospective analyses of genome-wide association in existing cohort studies.

The most common designs implemented in GWAS thus far have been case–control designs. These are most common since samples are easier, less expensive, and less time-consuming to collect compared to sample collection in cohort studies and may be analyzed once genotype and phenotype data are collected and QC implemented. Compared to cohort designs, which typically sample participants independently of disease outcomes and follow unaffected individuals forward over time, the case–control design is more sensitive to potential sources of bias, such as population stratification (confounding by racial/ethnic/ancestry differences between comparison groups) and selection bias. While these sources of bias represent potential challenges to the use of this design, strategies can be implemented to reduce their potential effects; for instance, to limit selection bias in studies of more common diseases, investigators may randomly sample incident cases from the population. Regardless of the study design, dealing with the effects of population stratification on association testing may require multiple strategies, including genomic control (GC) approaches, stratified association, selecting matched controls, and the modeling of population substructure for covariate adjustment, described at length in Chapter 8. Early case–control designs for GWAS for a number of complex diseases such as AD (Avramopoulos 2009) were also limited in their ability to detect associations due to low statistical power from small sample sizes. Increases in the number of cases and controls in GWAS of complex disease, through both the enlargement of case–control data sets within groups and the development of inter-institutional collaborations to combine data sets, have been instrumental in improving the ability of case–control designs to detect strong genome-wide associations of more modest effect size. However, as discussed later in this chapter, meta-analysis or pooled analyses across data sets require careful harmonization of case–control definitions, sample selection criteria, and potential confounding variables with different distributions across participating sites. A successful example of such efforts is the Global Lipids Genetics Consortium, which assembled GWAS data from 188 577 individuals from 60 different studies (Willer et al. 2013). Through this large world-wide collaboration, 157 lipid loci have been identified (Willer et al. 2013). Similarly, in the GIANT consortium, they assembled GWAS results from 700 000 individuals in a meta-analysis to identify genetic loci associated with BMI. Over 940 SNPs associated with complex traits have been identified and replicated in these large multicenter efforts (Yengo et al. 2018).

Increased sampling of cases and controls is not always feasible. For less prevalent diseases and traits, sampling a larger number of controls relative to cases may have beneficial effects on statistical power to detect association, which is modestly increased if two or three times as many controls are sampled as there are cases. However, simulation studies reveal that marginal increases in power are considerably reduced above a control: case ratio of 3 : 1, providing reduced value without comparable marginal reductions in genotyping cost. Whenever feasible, additional recruitment of cases and controls in similar quantities (~1 : 1) provides larger increases in power relative to the recruitment of additional controls only. The use of existing genotype data on external controls that meet study inclusion criteria may also provide an additional inexpensive boost to power, but may contribute bias from genotyping batch effects, differences in genotyping platform and in population allele frequency.

Family-based association study designs (detailed in Chapter 8) represent another, albeit less common approach to GWAS. Rather than focusing on differences between the frequencies of allelic variants among cases and population-based controls, family-based studies analyze whether the frequency of particular allelic variants transmitted from a heterozygous parent to an affected offspring is greater than expected by chance alone. While the basic units of transmission examined require data only from parent–offspring pairs or trios, genotype data on siblings and first- and second-degree relatives in extended pedigrees can also be used to improve power to detect genetic associations. Family-based association study designs in GWAS provide several advantages. If a relatively rare, discrete (qualitative) trait is being examined, family-based studies with high burden of disease may provide an opportunity to enrich the number of cases being examined. The increased ability to identify genotyping errors at many loci by examining consistency with Mendelian inheritance patterns is an additional benefit inherent to family-based designs (Gordon and Ott 2001; Gordon et al. 2001, 2004). These family-based designs are more robust to population stratification as allelic transmissions from parents to children, not case and control genotypes, are being examined in most basic analysis approaches (see Chapter 8 for more details on how family-based tests control for population structure). Extended modeling approaches exploring genetic effects between-family as well as within-family may be more susceptible to the effects of population substructure. Weaknesses of family-based designs include the limited information value of trios or pedigrees with one or both parents being homozygous for variants examined; the reduction in the number of informative trios or pedigrees can also be a function of lower minor allele frequencies at SNPs of interest. However, there is some evidence that relatively modest number of families can prove valuable in identifying common variants potentially contributing to disease, as with *INSIG2* variation in childhood and adult obesity (Wang et al. 2011).

While most cohort studies were not specifically designed for GWAS, association analyses of genome-wide genotype data within existing longitudinal cohort studies provide a third approach for identifying genomic variants contributing to complex traits. Extensive data collection, large sample sizes, and considerable length of follow-up make large GWAS cohort designs prohibitively expensive and challenging; however, high-density genotyping and genome-wide association analyses in existing cohorts have been a useful approach to performing GWAS without the need for additional study recruitment. As most cohort studies already have extensive phenotypic characterization as well as tracking of incident disease, they allow for more varied statistical approaches beyond prevalent case–control analysis such as incident case–control analysis and case–cohort designs. Cohort study sampling, by not sampling on disease status, also tends to produce a study sample more representative of the population being investigated, reducing the potential for selection bias. Furthermore, detailed phenotype data allow for the study of not just disease traits of interest but also endophenotypes and related traits, some of which may be genetically driven predictors of disease. Possible drawbacks to genome-wide association analysis in cohort studies include power limitations for less prevalent diseases and conditions, where even with many noncases for comparison, the number of cases examined may be small, and poor characterization or measurement error in specific traits of interest that are being investigated. While large case–control studies in late-onset AD first identified common genomic variants affecting AD risk outside of *APOE* (Harold et al. 2009; Lambert et al. 2009), analysis of AD in multiple cohort studies first detected associations at *BIN1* (Seshadri et al. 2010), which have since replicated in larger follow-up studies (Hollingworth et al. 2011; Naj et al. 2011).

As discussed in Chapter 4, an emerging cohort design involves health-system-based biorepositories linked to electronic health records (EHR). This design for GWAS allows for a large cohort to be collected without preselection based on disease/phenotype. Instead, through EHR phenotype mining, different case–control studies can be derived from the cohort. The Electronic Medical Records and Genomics (eMERGE) network, funded by NHGRI, has been performing GWAS using this study design for several years (Gottesman et al. 2013). Please see more about the strengths and weaknesses of this cohort design in Chapter 4.

Statistical Power for Association and Correction for Testing Multiple Hypotheses

While several different kinds of statistical approaches may be used to test for association between genomic variants and disease, the large-scale testing of variants for association has the potential to generate a substantial type I error; that is, failure to account for the simultaneous testing of a million or more hypotheses may produce a large number of "false-positive" associations among which true associations may be difficult or impossible to discern. The most common approach for dealing with this is the Bonferroni correction of a 5% type I error rate ($\alpha = 0.05$) for the total number of statistical tests performed. For association testing of a million SNPs, this would be equivalent to $\alpha = 0.05/1000000 = 5.0 \times 10^{-8}$ (Dudbridge et al. 2008; Gao et al. 2008; Pe'er et al. 2008). Work using the HapMap project data to estimate genome-wide patterns of linkage disequilibrium suggested a testing burden of only a million independent statistical tests in analyses of data sets comprising persons of Northern European background, and approximately two million tests in analyses of African ancestry data sets (Pe'er et al. 2008), lending empirical support to the use of this threshold in GWAS. Though the threshold of $\alpha = 5.0 \times 10^{-8}$ has become a common standard for testing of genome-wide statistical significance, it is far from the only approach. Other approaches to controlling type I error in GWAS include use of the false discovery rate (Benjamini and Hochberg 1995; Storey and Tibshirani 2003), *P*-value weighting approaches (Genovese et al. 2006) (examined elsewhere in detail (Roeder and Wasserman 2009), and permutation tests to get an empirical null distribution of the test statistic (Seaman and Müller-Myhsok 2005). The features of these approaches (among others) are discussed in detail elsewhere (Sham and Purcell 2014). For instance, Bonferroni correction is conservative and may overcorrect in the presence of correlated tests, whereas permutation tests are very computationally intensive. The Bonferroni corrected threshold of $\alpha = 5.0 \times 10^{-8}$ has become the de facto standard for GWAS reporting, and application of alternative thresholds or corrections should be explained and well justified in publications.

Stringent thresholds for statistical significance imposed in GWAS consequently require considerable numbers of cases and controls for the detection of associations of large effect size, more than would be required for candidate gene associations of the same effect, to prevent type II (false-negative) errors due to low statistical power. Even larger sample sizes are needed for the detection of associations with modest and small effect size. Some complex diseases first examined for genome-wide association in data sets with many thousands of individuals were able to identify only a few association signals with $P < 5.0 \times 10^{-8}$. The Wellcome Trust Case–Control Consortium, for example, examined seven diseases with data on 2000 cases for each disease and a common set of 3000 controls, and in their analysis of type 2 diabetes, they identified only one signal with $P < 5.0 \times 10^{-8}$, that of a SNP in *TCF7L2* (OR = 1.36, $P = 5.7 \times 10^{-13}$) (Wellcome Trust Case–Control Consortium 2007). Several signals of marginal significance were also observed. In a subsequent meta-analysis of 10 128 individuals by the DIAGRAM Consortium, six new loci with genome-wide statistically significant associations were observed with double the sample size (Zeggini et al. 2008). More recently, the DIAGRAM+ Consortium extended its sample size to 34 412 cases and 59 925 controls, and in doing so, identified an additional 12 loci with associations of $P < 5.0 \times 10^{-8}$ (Voight et al. 2010). While *TCF7L2* demonstrated genome-wide statistical significance in the first study with an effect size of OR = 1.36, the DIAGRAM+ study was able to detect associations with ORs = 1.09–1.25. This pattern of identifying new associations with weak effects has been observed in GWAS across a broad spectrum of complex disease; it has revealed that genetic contributions of relatively common variants include few genetic contributors of strong effect on risk and many variants with modest or weak effects on risk. For example, the more recent meta-analysis of type II diabetes (T2D), which was conducted in 62 892 T2D cases and 596 424 controls, was able to identify 139 common variants at the genome-wide significance level of 5×10^{-8} (Xue et al. 2018).

For this reason, studies with adequate power (≥80%) to detect the multitude of these weak effects (OR < 1.3) in complex disease require samples of 10 000+ cases and controls for relatively common variants. For variants of rarer frequency (minor allele frequency [MAF] < 0.10), a study of 10 000 cases and 10 000 controls can be used to identify associations with large effect size (OR > 1.3) at 80% power, but for effect sizes similar to those observed for most common variants (OR = 1.09–1.25), larger samples may be necessary. In general, it has been well established that associations with variants of lower frequency and associations of weak or modest effects require larger sample sizes and thus are more difficult to identify. Several power calculation tools have been established for GWAS analysis. Please refer to Chapter 12 for more discussion on power calculations and specifically these tools to perform power calculations for GWAS.

Alternative approaches to avoiding type I error without requiring stringent thresholds for identifying association include replicating associations of subsets of markers that exceed more liberal, prespecified thresholds for significance (e.g. $\alpha = 10^{-5}$). Two-stage designs performing high-density genotyping in discovery samples and targeted genotyping of subsets of markers in replication samples provide the advantage of reduced genotyping costs within replication samples, and where appropriate case–control data are available, may permit replication sample sizes many times the size of the set of samples used for discovery. These two-stage designs have been used for many traits with mixed success. Another approach to improving power is to combine results across data sets using mega- or meta-analysis approaches, discussed later in this chapter.

Data Analysis

Quality Control on Genotyping Call Data

Due to the possibility of spurious associations, identifying and removing genotyping error require the implementation of QC measures on both sample and SNP levels. Several papers have described genotype QC protocols including Anderson et al. 2010. Turner et al. 2011, and Marees et al. 2018. Figure 9.2, which is republished with permission from Turner et al. 2011, shows one potential workflow for GWAS QC.

Genotype QC protocols typically address issues with individual samples before addressing SNP QC issues, as many problems with samples have the potential to exert effects on most or all of the SNPs genotyped on high-density arrays. Thus, performing QC on samples first and SNPs second has become common practice. Study-level problems include genotyping failures in sets of samples genotyped together, which may induce "batch effects." Typical strategies to proactively prevent the occurrence of these issues include the plating of duplicates within batches, randomization of samples to various batches, and the genotyping of control samples such as HapMap reference samples, for which genotype data from sequencing are available in order to assess the quality of genotyping through measures of genotype concordance. Sample-level problems that may introduce genotyping error include the contamination of samples, mislabeling or duplication of samples, and low DNA quality.

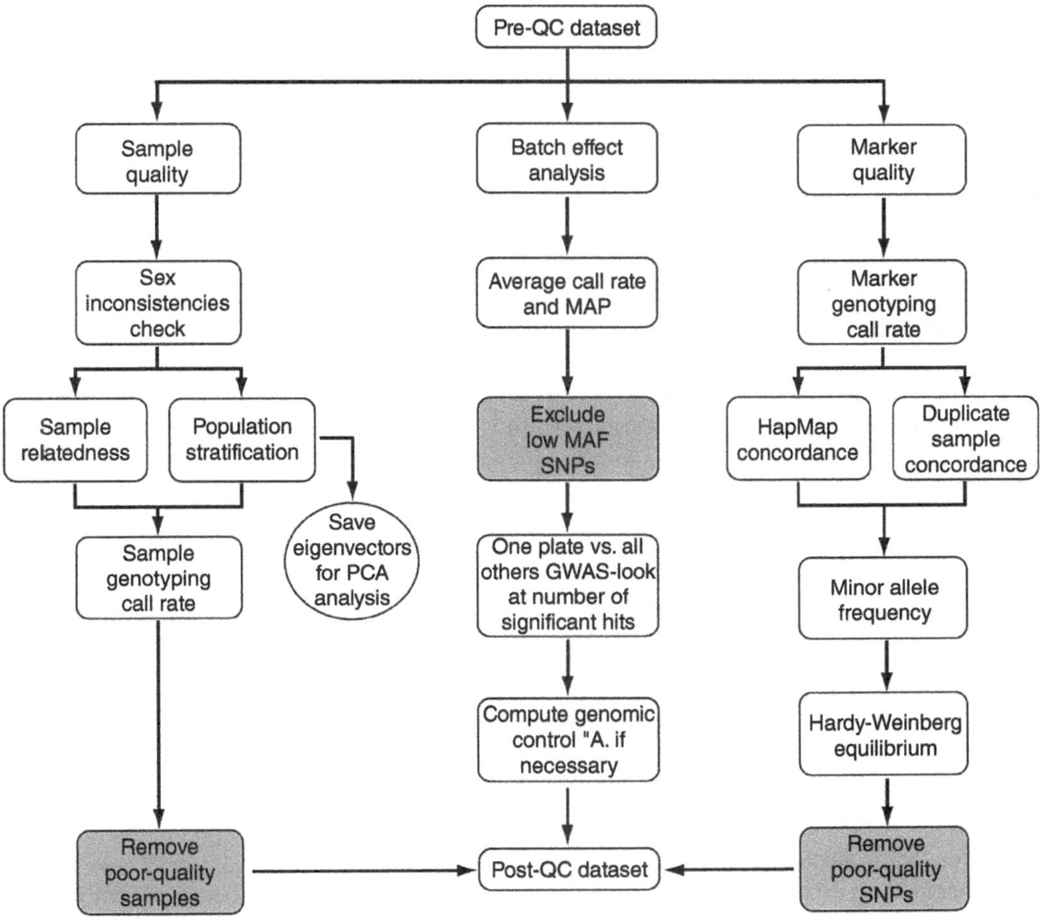

Figure 9.2 A flowchart overview of the entire GWAS QC process. Each topic is discussed in detail in the corresponding section in the text. Squares represent steps, ovals represent input or output data, and trapezoids represent filtering of data.

Initial Genotyping Quality Control

Post-genotyping QC strategies often first assess genotype call rates among paneled SNPs to ensure that any sample-level genotyping QC measures are implemented on accurate genotypes. Depending on the genotyping platform used, SNPs with missing data rates between 2 and 5% are excluded; stringency may be varied based on the past performance of the genotyping platform used and DNA quality of the samples examined. However, most recent GWAS arrays have extremely good analytic validity, and thus it is not common to see any SNPs missing more than 1% of the data be dropped from subsequent analyses. SNP missingness may also be a function of whether default clustering profiles for genotype calling are used or genotype calling algorithms are applied to fluorescence intensity data from the samples examined. One strategy for improved genotype calling involves censoring fluorescence intensity values of samples with the lowest overall SNP call rates, retraining genotype calling on samples with low missingness, and then recalling genotypes among the samples with the lowest overall SNP call rates. As genotype calling issues can elevate type I error, reasonable post-hoc QC suggests that cluster plots should be visually inspected for any SNPs found to have statistically significant associations with the phenotype of interest (NCI-NHGRI Working Group on Replication in Association Studies et al. 2007).

Sample-Level Quality Control

Post-genotyping examinations of sample-level genotype data may often give clear indications of the severity of these problems, should they exist. Sample duplication can be detected by estimating Pi-hat, the proportion of genotypes shared identical by state (IBS) in a sample of uncorrelated SNPs, through pairwise comparisons of all samples in a data set. A Pi-hat estimate for a pair of samples at or near 1 indicates that all or nearly all genotypes are shared IBS, which indicates that the samples are likely either identical twins (always 100% IBS) or duplicates of the same sample. Pi-hat estimates less than 1 may also be informative. Among pairs of related individuals, one would expect Pi-hat~0.5 for all first-degree relatives (parent–offspring and sibling–sibling), a Pi-hat~0.25 among second-degree relatives (avuncular pairs, grandparent–grandchild); a Pi-hat~0.125 among third-degree relatives (first cousin pairs), and smaller expected values of Pi-hat for more distant relationships. Although few consistent standards exist for excluding one sample from pairs of individuals based on estimates of Pi-hat<1, many publications have used Pi-hat >0.2 as a threshold for examination of potential related pairs and elimination of one or both of the members of the pair depending on (i) whether the relationship can be verified and (ii) one can be chosen to retain in the study based on criteria unrelated to the outcome of interest (e.g. choose based on % missing genotypes instead of keeping all cases) (Marees et al. 2018, International Journal of Methods in Psychiatric Research, https://doi.org/10.1002/mpr.1608). It should be noted that IBS differs from identity by descent (IBD) as it refers to two individuals sharing a common allele at the same position but not necessarily the same copy from a common ancestor. This nuance is of greater relevance when the genomic variant being examined may be a proxy for an actual causal variant being transmitted along the same haplotype due to linkage disequilibrium.

Another tool for post hoc sample-level genotyping QC is the estimation of the coefficient of inbreeding (*F*), which is the probability that two genes at a given locus are IBD (Bulmer). Whereas Pi-hat estimation can quantify interindividual relatedness, *F*, which is estimated from homozygous genotypes, can estimate relatedness levels study-wide as an indicator of widespread duplication or contamination issues. A low level of relatedness between individuals in a study would be reflected as a greater genome-wide level of heterozygosity than expected by chance alone, signified by a large negative value for *F*.

Sample-level genotyping QC should include several additional filters: checking the concordance of reported sex with genotype-imputed sex; if genotyping data for families is being used, checking for Mendelian inconsistencies; and finally, the exclusion of samples with pronounced differences in racial/ethnic background from analysis of population substructure. This last sample-level QC step is sufficiently complicated to be discussed in detail in a subsequent section.

Checking concordance between reported and genotype-imputed sex is done often by estimating *F*, the inbreeding coefficient, using X-chromosome genotypes outside of the pseudo-autosomal region (PAR). Due to the hemizygosity of all X-chromosome SNPs in males outside of the PAR, *F* should typically have a value of 0.8 or greater in males (indicating little or no heterozygosity), while *F* is expected to be less than 0.2 in females (indicating little or no excess homozygosity beyond what is expected). Where genotyped samples labeled male and female show deviation from expected *F* values, a more thorough investigation of genotyping quality may be necessary for those samples (including examination of the limited number of SNPs on the Y chromosome, expected to be present only in males), and those samples may need to be censored in analyses.

In family-based studies or data sets containing some related individuals, samples may be excluded if substantial numbers of SNPs demonstrate genotypes inconsistent with expectations under Mendelian inheritance patterns. Thresholds for excluding samples due to Mendelian inconsistency should be based on the error rate among the total number of SNPs examined, with consideration of complexity of family data available for the sample under consideration.

SNP-Level Quality Control

SNP-level genotyping QC includes several important filtration steps: testing for differential rates of genotype missingness by affection status; examining extreme deviation of genotype frequencies from expectation under Hardy–Weinberg Equilibrium (HWE); and filtering SNPs with Mendelian inconsistencies across multiple sets of related individuals in family-based datasets.

Using family data or data with some related samples (such as sibling controls), SNPs with Mendelian inconsistencies between large numbers of related individuals should be excluded from further analysis. In contrast to sample-level QC, exclusion of SNPs with evidence of Mendelian inconsistency should be based on the error rate among the number of related pairs examined.

Another QC filter for potential genotyping error involves the removal of SNPs with extreme deviation from HWE. Deviation from HWE may be attributable to a variety of sources, including population stratification, excess relatedness, and the effects of positive or negative selection. However, deviation from HWE may also reflect true association of an SNP with the disease or trait being examined, so the exclusion of SNPs with modest deviation from HWE may exclude real associations and elevate type II error. Thus, exclusion of SNPs based on HWE should be done carefully and thoughtfully. Although thresholds for SNP exclusion for deviation from HWE vary from study to study, a common threshold implemented includes the removal of SNPs from consideration with P for HWE $<10^{-6}$ (Wittke-Thompson et al. 2005). An alternative strategy for evaluating association at these SNPs may include flagging SNPs and reporting deviation from HWE alongside association results for these SNPs.

Depending on the sample sizes available for analysis, SNPs with low minor allele frequency (MAF) (1%–5%) may be excluded from analyses if there is inadequate statistical power (typically $1-\beta<0.80$) to detect associations at the effect sizes expected. Based on the patterns of associations observed in even the largest case–control GWAS of disease to date, the range of detectable effect sizes expected for previously unidentified loci is OR = 1.10–1.20 (Hindorff et al. 2009, 2011). For this reason, most GWAS (except the largest ones with >100 000 samples) are underpowered to detect associations with SNPs with MAF < 0.01. Also, stronger potential effects of genotyping error on low-frequency variants suggest that SNPs with MAF < 0.01 should be entirely excluded from analyses.

Finally, for analyses of case–control data sets, it is also important to consider whether missing genotype rates differ in cases and controls (i.e. data missing not at random), as failure to exclude SNPs with differential missingness between cases and controls may introduce systematic error into analyses. Determining differential missingness can be done using a Pearson's Chi-square test with one degree of freedom (Clayton et al. 2005).

Software Programs for Quality Control

GWAS require a powerful array of software and data management resources to be able to perform the extensive quality control required. Several software packages for GWAS integrate functions that can be used to perform some or all the QC steps. The most widely used of these is PLINK (Purcell et al. 2007), which is able to transform data into multiple formats and output styles; combine, standardize, and reformat data sets; and apply a wide variety of SNP-level QC measures (filtering on deviation from HWE, Mendelian inconsistencies, differential missingness between cases and controls, and high overall missingness) and sample-level QC measurements (reported-imputed gender discordance, Mendelian inconsistencies, excess relatedness). Other commonly used packages include PLATO (PLatform for the Analysis, Translation, and Organization of large-scale data) (Grady et al. 2010; Hall et al. 2017), a software package that incorporates a battery of similar SNP-level and sample-level QC functions as described above. Similar commercially available packages exist, such as SNP and Variation Suite from GoldenHelix (2017), which includes similar QC functionality and can also perform quality control on different types of data, such as gene expression data.

For characterizing relatedness and filtering and removal of related samples in independent case–control studies, several programs with varying approaches are available for use. GRR (Graphical Representation of Relationships) (Abecasis et al. 2001) is a Windows-based software program that can visually characterize samples based on allele sharing at a randomly sampled subset of SNPs from the data set, identifying potential sample duplication or unexpected relatedness among pairs of independent samples and detecting and correcting potential pedigree errors among pairs of family members in a study. Another widely used program for characterizing relationships, identifying pedigree errors, and observing cryptic relatedness in GWAS or linkage data is PREST and its successor, PREST-plus. PREST-plus (McPeek and Sun 2000) computes accurate IBD estimates using an optimized maximum likelihood (ML) estimator for IBD probabilities and can examine any number and combination of SNP or microsatellite marker data. Another program, KING (Manichaikul et al. 2010), implements a rapid algorithm for inferring relationships that allows the presence of unknown population substructure. KING's very efficient implementation allows for rapid comparisons of millions of individuals simultaneously, reducing time for estimation of relatedness compared to other programs by an order of 10. Even more recent programs include SNPrelate, SMARTREL in the Eigensoft package, and PLINK.

Population Structure

Population structure, also known as population stratification, is the presence of a systematic difference in allele frequencies between subsets of a population sample, which may result due to differences in ancestry between the subsets and unknown degrees of relatedness between study participants, often referred to as *cryptic* relatedness. When these differences correlate with case–control status or the distribution of a quantitative trait of interest, but are not otherwise independently associated with the phenotype of interest, this type of confounding can lead to the observation of spurious associations, which may be especially difficult to detect when performing millions of parallel tests of association. Accounting for population structure in GWAS can be done through many approaches, some of which can be combined to form a strong defense against false-positive associations. These approaches can be variously implemented in the design, analysis, and reporting phases of GWAS.

In the study design phase, it is important to ensure that populations from which controls are selected should be the same as the source of cases. After genotyping and prior to association analysis, population substructure among cases and controls can be characterized using several ways including structured association and principal components analysis. Structured association involves first clustering samples into groups distinct by ancestry and then performing association analyses within strata. A commonly used program for the first part is STRUCTURE (Pritchard et al. 2000a; Rosenberg et al. 2002), a freely available software package that uses genotype data at typically many thousands of loci to infer population substructure, identifying distinct subpopulation clusters within data sets and migrant or admixed individuals whose diverse ancestries may influence patterns of genetic association observed. Other genetic clustering algorithms include Bayesian Analysis of Population Structure (BAPS – Corander et al. 2004), TESS, which uses a Bayesian clustering algorithm for *spatial* population genetic studies (Chen et al. 2007; François and Durand 2010), STRUCTURE HARVESTER (Earl and vonHoldt 2012) and DISTRUCT (Rosenberg 2003) for visualizing membership coefficients, and CLUMPP (Jakobsson and Rosenberg 2007) for post-processing to avoid distinct outcomes in independent analyses of the same data. In the analysis phase of the GWAS, association may then be performed in applications such as STRAT (Pritchard et al. 2000b), which are capable of performing associations within strata and then combining results to get measures of association across the entire data set.

A second approach to evaluating and controlling for population structure is to group individuals into more homogeneous clusters using principal components analysis. One commonly used software package to characterize population substructure using principal components is EIGENSOFT/EIGENSTRAT (Price et al. 2006). EIGENSTRAT implements a principal-components-based approach using randomly sampled, independent SNPs to model ancestry along continuously distributed "axes of variation"; often, after the exclusion of outliers, these axes of variation can be used in covariate adjustment in regression modeling of genotype–phenotype associations in the analysis phase of the GWAS. Figure 9.3 shows an example of how a GWAS data set can be visualized on the first two principal components (PC1 on the x-axis and PC-2 on the y-axis). Similar approaches to

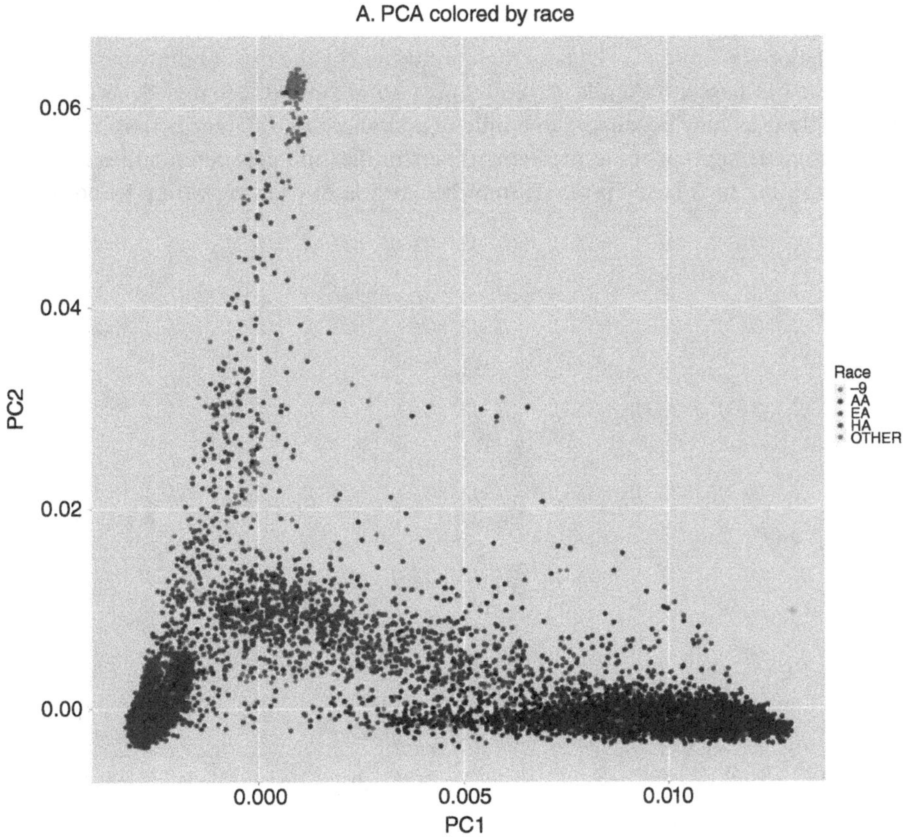

Figure 9.3 Example of a principal components analysis plot – visualizing the first two principal components from a GWAS dataset.

principal components analysis include multidimensional scaling (MDS; implemented in PLINK) and extensions of MDS (Miclaus et al. 2009; Lee et al. 2010).

The GC (Devlin and Roeder 1999; Roeder and Wasserman 2009) method allows for a study-wide assessment of the effects of population structure and other potential contributors of bias in the post-analysis reporting phase of a GWAS. This is done by computing λ_{GC}, the genomic inflation factor, which is defined as the ratio of the median Cochran–Armitage χ^2 (1 degree of freedom) association statistic (based on the additive model of allelic effects) across all examined SNPs to the median χ^2 expected under the null distribution. Extensions of this approach have been proposed for other models, such as the 2 df genotype model (Zheng et al. 2006, doi: 10.1086/50005). A $\lambda_{GC} \approx 1$ suggests little or no evidence of stratification, while a value of $\lambda_{GC} > 1$ may reflect the presence of population stratification, differential bias, or potential confounders such as cryptic relatedness or unaccounted effects of family structure. A $\lambda_{GC} < 1.10$ is considered an indicator of an acceptable level of departure from expectation, and a $\lambda_{GC} < 1.05$ is ideal in most studies. Reporting of λ_{GC} is often accompanied by the presentation of quantile–quantile (Q-Q plots), which observed – $\log_{10}$ p-values for each SNP association against the rank-ordered expected $-\log_{10}$ p-value under the null hypothesis of no association (see the example in Figure 9.4). Visual inspection of Q-Q plots may reveal departures from expected association that indicate the presence of systematic error rather than modest inflation of association statistics and are therefore highly recommended for detecting potential bias in GWAS association results. In Figure 9.4, the QQ plot on the left shows little deviation from expectation, and thus low λ_{GC}, whereas the QQ plot on the right shows a high λ_{GC} and considerable deviation from expectation.

The effects of population structure may be largely circumvented by the use of family-based study designs (described more in Chapter 8). While within-family tests are immune to the effects of stratification, between-family tests may be subject to ethnicity and ancestry differences that may exist between families examined. Approaches to protect against stratification between families recently proposed include combining rank-based p-values from between-family testing with p-values from

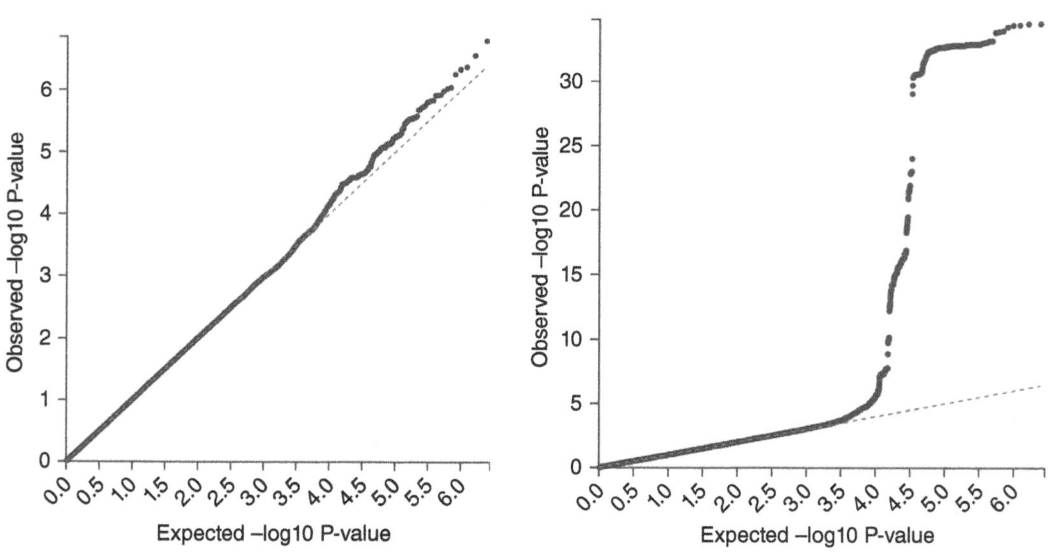

Figure 9.4 Example of two Q-Q plots for a GWAS dataset. The image on the left shows little deviation from expectation. The image on the right shows considerable deviation from expectation.

within-family testing to provide study-wide protection from population stratification (Won et al. 2009) and hybrid case–control family-based analysis designs (Lasky-Su et al. 2010).

A variety of more advanced approaches are also available, including mixed model approaches such as EMMAX (Kang et al. 2010), TASSEL (Zhang et al. 2010), and ROADTRIPS (Thornton and McPeek 2010), and are discussed in detail in reviews about the treatment of population structure in GWAS (Price et al. 2010). More contemporary approaches include variations on mixed models whereby the pedigree structure is regressed out and the residuals are used for association testing. This is available in packages such as GenABEL (Karssen et al. 2016), GENESIS (Gogarten et al. 2019) or ProbABEL (Aulchenko et al. 2010).

Imputation

Although most genome-wide SNP genotyping arrays collect data on markers at a high density throughout the genome (e.g. the Illumina HumanOmni2.5 has 2.45 million SNPs at a mean spacing of one SNP every 1.18 kb), even the densest chips examine only a subset of all common genetic variants. Associations at ungenotyped SNPs are thus only observed through indirect associations at proxy SNPs. Furthermore, GWAS analyses that combine data from sets of samples genotyped on multiple platforms may have little overlap in SNPs genotyped, rendering the added statistical power to detect associations from enhanced sample size moot. Genotype imputation is a common strategy to deal with these issues and can be performed using a variety of approaches. Genotype imputation uses missing data algorithms to predict genotypes at markers (SNPs usually) not directly genotyped by high-density genotyping assays. Imputation combines information on genome-wide patterns of linkage disequilibrium and data on phased haplotypes from large samples of individuals examined in genomic resequencing projects such as the International HapMap project (International Hap Map Consortium 2005), 1000 Genomes (1 kg) project (1000 Genomes Project Consortium et al. 2012), and the Haplotype Reference Consortium (HRC) (Loh et al. 2016).

Imputation approaches all follow a similar general protocol (Naj 2019). Using data from the samples under study, genotypes from SNPs directly assayed on GWAS genotyping arrays are first phased, predicting the most probable chromosomal haplotypes for each chromosome for each individual. These chromosomal haplotypes are compared against much more SNP-dense phased haplotypes from a reference panel of hundreds of sequenced individuals from HapMap or 1 kg or HRC. Based on the haplotypes observed in the reference panel, the probability of carrying one or the other alleles at non-genotyped SNPs is estimated for each non-genotyped SNP on all participants' chromosomal haplotypes, and these are in turn used to estimate genotype probabilities for each ungenotyped SNP. In this manner, genotype data from a million genotyped SNPs might be used to predict genotypes at as many as 37 million additional imputed SNPs (1000 Genomes Project Consortium et al. 2012). The resulting genotypic probabilities may then be tested for genetic association in a number of ways. Imputed genotypes may be called if the probability of one of the three genotypes at a position is sufficiently high. A common threshold used for calling a genotype is the observation of a probability ≥0.90 for one genotype. However, for many SNPs, this level of certainty may not be available. More recently, genome-wide analyses using genotypic probabilities instead of calling imputed SNPs have become more common, as this allows for the incorporation of uncertainty from imputation into analysis without the loss of power that missing data would cause (de Bakker et al. 2008). This is available in most contemporary GWAS software tools including SNPTest, EPACTS, ProbABEL, and PLINK.

Several publicly available software packages are available for imputation using varied approaches to predict genotypes at ungenotyped SNPs. Among the more commonly used applications are

IMPUTE (Marchini and Howie 2010) and IMPUTE2 (Howie et al. 2012). Version 1 of this program uses a hidden Markov model (HMM) of an individual's vector of genotypes for a set of SNPS within a predetermined window, conditioning on phased haplotypes from a reference panel, estimates of recombination in the region examined, and a parameter allowing for deviation of some genotypes from expectation, infers genotypes at loci missing from the individual but present from the reference haplotype panel. This approach is used to infer both systematically missing (i.e. ungenotyped SNPs) as well as randomly missing (i.e. genotyped SNPs with missing calls) genotypes. Version 2 (Howie et al. 2009) of IMPUTE extends the approach from version 1 by phasing genotyped SNPs to estimate haplotypes in the study sample, which allows for more rapid genotype imputation at non-genotyped SNPs conditioning on haplotypes from the reference panel. Phase uncertainty is dealt with by iterating haplotype estimation steps using a Markov chain Monte Carlo (MCMC) approach. An added benefit of haplotype estimation in study samples is the ability to use genotype data at unselected samples (such as controls in independent, unrelated data sets) genotyped on a different subset of markers in the region of interest as unphased, secondary reference panel for imputation. Another commonly used software application is MACH (Markov Chain Haplotyping) (Li et al. 2009, 2010), which uses HMMs similar to those of IMPUTE. MACH phases genotypes in each individual in the study sample to estimate haplotypes, iteratively improving the phasing based on the haplotype estimates of all other samples current to the iteration. Imputation of ungenotyped SNPs for an individual sample occurs through the inclusion of reference panel haplotypes to the set of haplotype estimates of all other samples, allowing for the marginal distribution of the unobserved genotypes to be estimated from the haplotypes sampled in each iteration. Other imputation software packages, though less frequently used than IMPUTE and MACH, provide quite powerful alternatives to the more common approaches described. BEAGLE (Browning 2006; Browning and Browning 2007, 2009) fits graphical Markov models to haplotype data by constructing a tree graph characterizing observed haplotypes marker by marker, adding SNPs in linear order down the chromosome. The resulting tree is pruned by comparing haplotype frequencies downstream and collapsing nodes when frequencies are similar. Probabilities of missing genotypes are estimated from the haplotype frequencies from the model fit at the last iteration of this approach. Imputation using fastPHASE (Scheet and Stephens 2006), as incorporated in the association-testing program BIMBAM (Servin and Stephens 2007; Guan and Stephens 2008), clusters common haplotypes and uses HMMs to estimate unobserved genotypes conditioning on haplotype clusters rather than individual reference haplotypes. More accurately, haplotype clusters are estimated over different windows of SNPs across a region, and genotype probabilities are based on the clusters within each window and switching between "states" (membership in haplotype clusters within each window). By varying the number of haplotype clusters allowed per window and the length of haplotypes examined, this method trades off between speed of imputation and accuracy of the imputed genotype.

While this is a brief overview of imputation approaches, aspect of these approaches (imputation accuracy, speed, and strengths and weaknesses) are explored elsewhere in great detail (Marchini and Howie 2010). A pipeline for performing imputation across multiple genotyping platforms and data sets was developed by the eMERGE genomics workgroup. Details of this pipeline are described in detail here (Verma et al. 2014). A web-based service for imputation that facilitates access to multiple reference panels is also available (Das et al. 2016) or the TOPMED Imputation Server (https://imputation.biodatacatalyst.nhlbi.nih.gov/#!).

Genetic Association Testing

Once the genotype data and samples have gone through QC, the data are ready for GWAS analysis. Depending on whether the disease or trait is binary or quantitative will drive the type of statistical analysis performed, as described above in the Discrete and Quantitative Trait Analysis section. Binary/discrete traits tend to be analyzed using logistic regression approaches, which allow for adjustment of important covariates such as age, sex, and principal components for controlling for population stratification. Quantitative traits tend to be analyzed using linear regression-based approaches, again to allow for the adjustment of covariates. The software packages to perform these analyses vary, based on user preferences. For example, the PLINK software package performs several different GWAS analyses (Purcell et al. 2007; Chang et al. 2015; Slifer 2018). This software package is open-source and thus users can make modifications or additions to the analysis methods. Likewise, PLATO performs the standard logistic and linear regression analyses, as well as firth regression and likelihood ratio tests (Hall et al. 2017). Many researchers use a variety of available packages or create their own tools in R for GWAS analysis. The statistical analyses used for GWAS in case–control or cohort samples have become mostly standardized in recent years. However, alternative statistical analysis approaches have begun to emerge as well. These will be described in more detail in the next section.

Meta-Analysis and "Mega-Analysis"

As briefly mentioned above, due to the small effect sizes and modest sample sizes, the statistical power to detect genetic associations in standard GWAS has been limited. Thus, many research groups have begun to collaborate and share either statistical association results (summary data) or individual level genotype data. The type of data shared determines whether a meta-analysis or mega-analysis is performed.

Meta-analysis is a set of methods to quantitatively combine the summary statistics on results from analysis of multiple distinct data sets (Lunetta 2013). Meta-analysis has been used in many fields for decades, but specifically in human genetics, once the GWAS became a common study design, meta-analysis quickly became an important tool to aggregate results across individual GWAS (Zeggini and Ioannidis 2009). Meta-analysis in GWAS has been used for the past several years in an attempt to increase the power to detect association signals by increasing the sample size through the aggregation of multiple GWAS data sets (Zeggini and Ioannidis 2009; Lunetta 2013).

Different data sets may have used different platforms and may have thus genotyped different variants. However, current approaches (Anderson et al. 2008) allow imputing genotypes at untyped variants using a reference such as HapMap, as described more above. Directly typed or imputed genotype information can currently be combined across data sets on up to several millions of common variants. Meta-analysis can be conducted in a sequential, cumulative manner; as more data sets become available, these may be included in the calculations resulting in more discoveries of previously unrecognized variants. Cumulative meta-analysis thus potentially represents a "replication ad infinitum" process.

Alternatively, in mega-analysis, the individual level data, either genotyped or imputed, is combined and the GWAS analysis is conducted on the combined data set. This mega-analysis approach has been employed by collaborative groups such as the eMERGE network for a number of different traits (Verma et al. 2014; Stanaway et al. 2019). Mega-analyses require careful attention to systematic differences in study design, case definition, and distribution of relevant covariates across studies. Harmonization of variable definitions, careful QC, and incorporation of confounding variables in analysis are essential to avoid bias in these studies. The advantage of a mega-analysis is that the sample size of the discovery data set is considerably larger; however, it then requires either a replication data set or other data sets to combine through meta-analysis to reduce the likelihood that a signal is a false-positive result. It is generally not sufficient to include only results from a mega-analysis. As with all GWAS, results in a single data set without replication or meta-analysis may result in a large number of false-positive SNPs (Kraft et al. 2009).

Whole-Genome Regression-Based GWAS

More recently, whole-genome regression-based methods (e.g. BOLT-LMM (Loh et al. 2015), fast-GWA (Jiang et al. 2019), and SAIGE (Zhou et al. 2018)) are gaining in popularity; these tools condition on associated SNPs across the genome to boost power of signal detection, while better accounting for population structure and relatedness. Thus, it may not be necessary to filter out related individuals while using these methods. Additionally, SAIGE can also better accommodate unbalancedness between numbers of cases and controls using saddlepoint approximation. Since these methods use generalized mixed models instead of single marker regressions to perform GWAS, they can be computationally very intensive. More recently, this drawback was addressed by REGENIE (Mbatchou et al. 2020), a tool that employs a new machine learning algorithm that can efficiently speed up whole-genome regressions by several orders of magnitude.

Conclusion

In summary, genome-wide association studies have become a predominant study design in the genetic analysis of human disease since their inception in 2005. As of 2021, the number of SNPs detected with association to common diseases and traits using GWAS is on the order of over 170000 variants from over 5200 publications. These data are curated and cataloged in the NHGRI-EBI GWAS Catalog (www.ebi.ac.uk/gwas). While our understanding of the heritability of most common diseases is still limited, the field has identified a large number of genes and variants important for common, complex disease (Hindorff et al. 2009; Manolio et al. 2009; Welter et al. 2014; MacArthur et al. 2017; Buniello et al. 2019).

In addition to the standard GWAS approaches described in this chapter, the SNP data generated by a genome-wide SNP array can be used for many other different types of analyses. These include polygenic risk scores, pathway analysis, gene set analysis, and transcriptome-wide association analysis, to name a few. Some of these alternative analysis approaches go beyond using SNP data and can accommodate additional "omic" information. These methods are discussed in Chapters 10 and 11.

References

1000 Genomes Project Consortium, Abecasis, G.R., Auton, A. et al. (2012). An integrated map of genetic variation from 1,092 human genomes. *Nature* 491: 56–65.

Abecasis, G.R., Cherny, S.S., Cookson, W.O., and Cardon, L.R. (2001). GRR: graphical representation of relationship errors. *Bioinformatics* 17: 742–743.

Anderson, C.A., Pettersson, F.H., Barrett, J.C. et al. (2008). Evaluating the effects of imputation on the power, coverage, and cost efficiency of genome-wide SNP platforms. *Am. J. Hum. Genet.* 83 (1): 112–119.

Anderson, C.A., Pettersson, F.H., Clarke, G.M. et al. (2010). Data quality control in genetic case-control association studies. *Nat. Protoc.* 5 (9): 1564–1573.

Aulchenko, Y.S., Struchalin, M.V., and van Duijn, C.M. (2010). ProbABEL package for genome-wide association analysis of imputed data. *BMC Bioinformatics* 11: 134.

Avramopoulos, D. (2009). Genetics of Alzheimer's disease: recent advances. *Genome medicine* 1: 34.

de Bakker, P.I., Ferreira, M.A., Jia, X. et al. (2008). Practical aspects of imputation-driven meta-analysis of genome-wide association studies. *Hum. Mol. Genet.* 17: R122–R128.

Benjamini, Y. and Hochberg, Y. (1995). Controlling the false discovery rate – a practical and powerful approach to multiple testing. *J. Roy. Stat. Soc. B Met.* 57: 289–300.

Browning, S.R. (2006). Multilocus association mapping using variable-length Markov chains. *Am. J. Hum. Genet.* 78: 903–913.

Browning, S.R. and Browning, B.L. (2007). Rapid and accurate haplotype phasing and missing-data inference for whole-genome association studies by use of localized haplotype clustering. *Am. J. Hum. Genet.* 81: 1084–1097.

Browning, B.L. and Browning, S.R. (2009). A unified approach to genotype imputation and haplotype-phase inference for large data sets of trios and unrelated individuals. *Am. J. Hum. Genet.* 84: 210–223.

Buniello, A., MacArthur, J.A.L., Cerezo, M. et al. (2019). The NHGRI-EBI GWAS catalog of published genome-wide association studies, targeted arrays and summary statistics. *Nucleic Acids Res.* 47 (Database issue): D1005–D1012.

Chang, C.C., Chow, C.C., Tellier, L.C. et al. (2015). Second-generation PLINK: rising to the challenge of larger and richer datasets. *Gigascience.* 4: 7.

Chen, H.Y. and Li, M. (2011). Improving power and robustness for detecting genetic association with extreme-value sampling design. *Genet. Epidemiol.* 35: 823–830.

Chen, C., Durand, E., Forbes, F., and François, O. (2007). Bayesian clustering algorithms ascertaining spatial population structure: a new computer program and a comparison study, PDF. *Mol. Ecol. Notes* 7: 747–756.

Clayton, D.G., Walker, N.M., Smyth, D.J. et al. (2005). Population structure, differential bias and genomic control in a large-scale, case-control association study. *Nat. Genet.* 37: 1243–1246.

Corander, J., Waldmann, P., Marttinen, P., and Sillanpää, M.J. (2004). BAPS 2: enhanced possibilities for the analysis of genetic population structure. *Bioinformatics* 20 (15): 2363–2369.

Corder, E.H., Saunders, A.M., Strittmatter, W.J. et al. (1993). Gene dose of apolipoprotein E type 4 allele and the risk of Alzheimer's disease in late onset families. *Science* 261: 921–923.

Corder, E.H., Saunders, A.M., Risch, N.J. et al. (1994). Protective effect of apolipoprotein E type 2 allele for late onset Alzheimer disease. *Nat. Genet.* 7: 180–184.

Cox, D.G. and Kraft, P. (2006). Quantification of the power of Hardy-Weinberg equilibrium testing to detect genotyping error. *Hum. Hered.* 61: 10–14.

Das, S., Forer, L., Schönherr, S. et al. (2016). Next-generation genotype imputation service and methods. *Nat. Genet.* 48 (10): 1284–1287.

Devlin, B. and Roeder, K. (1999). Genomic control for association studies. *Biometrics* 55: 997–1004.

Dudbridge, F., Gusnanto, A., Pe'er et al. (2008). Estimation of significance thresholds for genomewide association scans. *Genet. Epidemiol.* 32 (3): 227–234. https://doi.org/10.1002/gepi.20297.

Earl, D.A., vonHoldt, B.M. STRUCTURE HARVESTER: a website and program for visualizing STRUCTURE output and implementing the Evanno method. Conservation Genet Resour 4, 359–361 (2012).

François, O. and Durand, E. (2010). THE STATE OF THE ART - Spatially explicit Bayesian clustering models in population genetics PDF. *Mol. Ecol. Resour.* 10: 773–784.

Gao, X., Starmer, J., and Martin, E.R. (2008). A multiple testing correction method for genetic association studies using correlated single nucleotide polymorphisms. *Genet. Epidemiol.* 32 (4): 361–369.

Genovese, C.R., Roeder, K., and Wasserman, L. (2006). False discovery control with p-value weighting. *Biometrika* 93: 509–524.

Gogarten, S.M., Sofer, T., Chen, H. et al. (2019). Genetic association testing using the GENESIS R/Bioconductor package. *Bioinformatics* https://doi.org/10.1093/bioinformatics/btz567.

GoldenHelix (2017). *SNP & Variation Suite*. Bozeman, MT: Golden Helix, Inc.

Gordon, D. and Ott, J. (2001). Assessment and management of single nucleotide polymorphism genotype errors in genetic association analysis. *Pac. Symp. Biocomput.*: 18–29.

Gordon, D., Heath, S.C., Liu, X., and Ott, J. (2001). A transmission/disequilibrium test that allows for genotyping errors in the analysis of single-nucleotide polymorphism data. *Am. J. Hum. Genet.* 69: 371–380.

Gordon, D., Haynes, C., Johnnidis, C. et al. (2004). A transmission disequilibrium test for general pedigrees that is robust to the presence of random genotyping errors and any number of untyped parents. *Eur. J. Hum. Genet.* 12: 752–761.

Gottesman, O., Kuivaniemi, H., Tromp, G. et al. (2013). The electronic medical records and genomics (eMERGE) Network: past, present, and future. *Genet Med.* 15 (10): 761–771.

Grady, B.J., Torstenson, E., Dudek, S.M. et al. (2010). Finding unique filter sets in plato: a precursor to efficient interaction analysis in gwas data. *Pac. Symp. Biocomput.*: 315–326.

Grant, S.F., Thorleifsson, G., Reynisdottir, I. et al. (2006). Variant of transcription factor 7-like 2 (TCF7L2) gene confers risk of type 2 diabetes. *Nat. Genet.* 38: 320–323.

Guan, Y. and Stephens, M. (2008). Practical issues in imputation-based association mapping. *PLoS Genet.* 4: e1000279.

Hall, M.A., Wallace, J., Lucas, A. et al. (2017). PLATO software provides analytic framework for investigating complexity beyond genome-wide association studies. *Nat. Commun.* 8 (1): 1167.

Harold, D., Abraham, R., Hollingworth, P. et al. (2009). Genome-wide association study identifies variants at CLU and PICALM associated with Alzheimer's disease. *Nat. Genet.* 41: 1088–1093.

Hindorff, L.A., Sethupathy, P., Junkins, H.A. et al. (2009). Potential etiologic and functional implications of genome-wide association loci for human diseases and traits. *Proc. Natl. Acad. Sci. U. S. A.* 106 (23): 9362–9367. https://doi.org/10.1073/pnas.0903103106.

Hindorff, L.A., Gillanders, E.M., and Manolio, T.A. (2011). Genetic architecture of cancer and other complex diseases: lessons learned and future directions. *Carcinogenesis* 32 (7): 945–954. https://doi.org/10.1093/carcin/bgr056. Epub 2011 Mar 31.

Hindorff, L.A., MacArthur, J., Morales, J. etal. (2017). The new NHGRI-EBI Catalog of published genomewideassociation studies (GWAS Catalog). NucleicAcids Res. 4;45(D1):D896-D901. doi: 10.1093/nar/gkw1133. Epub 2016 Nov 29.PMID: 27899670; PMCID: PMC5210590.

Hollingworth, P., Harold, D., Sims, R. et al. (2011). Common variants at ABCA7, MS4A6A/MS4A4E, EPHA1, CD33 and CD2AP are associated with Alzheimer's disease. *Nat. Genet.* 43: 429–435.

Howie, B.N., Donnelly, P., and Marchini, J. (2009). A flexible and accurate genotype imputation method for the next generation of genome-wide association studies. *PLoS Genet.* 5: e1000529.

Howie, B., Fuchsberger, C., Stephens, M. et al. (2012). Fast and accurate genotype imputation in genome-wide association studies through pre-phasing. *Nat. Genet.* 44 (8): 955–959.

International HapMap Consortium (2005). A haplotype map of the human genome. *Nature* 437: 1299–1320.

Jakobsson, M. and Rosenberg, N.A. (2007). CLUMPP: a cluster matching and permutation program for dealing with label switching and multimodality in analysis of population structure. *Bioinformatics* 23 (14): 1801–1806. https://doi.org/10.1093/bioinformatics/btm233. Epub 2007 May 7.

Jiang, L., Zheng, Z., Qi, T. et al. (2019). A resource-efficient tool for mixed model association analysis of large-scale data. *Nat. Genet.* 51: 1749–1755.

Kang, H.M., Sul, J.H., Service SK et al. (2010). Variance component model to account for sample structure in genome-wide association studies. *Nat. Genet.* 42: 348–354.

Karssen, L.C., van Duijn, C.M., and Aulchenko, Y.S. (2016). The GenABEL project for statistical genomics. *F1000Res.* 5: 914.

Kraft, P., Zeggini, E., and Ioannidis, J.P. (2009). Replication in genome-wide association studies. *Stat. Sci.* 24 (4): 561–573.

Lambert, J.C., Heath, S., Even, G. et al. (2009). Genome-wide association study identifies variants at CLU and CR1 associated with Alzheimer's disease. *Nat. Genet.* 41: 1094–1099.

Lander, E.S., Linton, L.M., Birren, B. et al. (2001). Initial sequencing and analysis of the human genome. *Nature* 409: 860–921.

Lasky-Su, J., Won, S., Mick, E. et al. (2010). On genome-wide association studies for family-based designs: an integrative analysis approach combining ascertained family samples with unselected controls. *Am. J. Hum. Genet.* 86: 573–580.

Lee, A.B., Luca, D., Klei, L. et al. (2010). Discovering genetic ancestry using spectral graph theory. *Genet. Epidemiol.* 34: 51–59.

Li, Y., Willer, C., Sanna, S., and Abecasis, G. (2009). Genotype imputation. *Annu. Rev. Genomics Hum. Genet.* 10: 387–406.

Li, Y., Willer, C.J., Ding, J. et al. (2010). MaCH: using sequence and genotype data to estimate haplotypes and unobserved genotypes. *Genet. Epidemiol.* 34: 816–834.

Loh, P.-R., Tucker, G., Bulik-Sullivan, B.K. et al. (2015). Efficient Bayesian mixed-model analysis increases association power in large cohorts. *Nat. Genet.* 47 (3): 284–290. https://doi.org/10.1038/ng.3190. Epub 2015 Feb 2.

Loh, P.R., Danecek, P., Palamara, P.F. et al. (2016). Reference-based phasing using the haplotype reference consortium panel. *Nat. Genet.* 48 (11): 1443–1448.

Lunetta, K.L. (2013). Methods for meta-analysis of genetic data. *Curr. Protoc. Hum. Genet.* Chapter 1: Unit1.24.

MacArthur, J., Bowler, E., Cerezo, M. et al. (2017). The new NHGRI-EBI catalog of published genome-wide association studies (GWAS catalog). *Nucleic Acids Res.* 45 (D1): D896–D901.

Manichaikul, A., Mychaleckyj, J.C., Rich, S.S. et al. (2010). Robust relationship inference in genome-wide association studies. *Bioinformatics* 26: 2867–2873.

Manolio, T.A., Collins, F.S., Cox, N.J. et al. (2009). Finding the missing heritability of complex diseases. *Nature* 461 (7265): 747–753.

Marchini, J. and Howie, B. (2010). Genotype imputation for genome-wide association studies. *Nat. Rev. Genet.* 11: 499–511.

Marchini, J., Howie, B., Myers, S. et al. (2007). A new multipoint method for genome-wide association studies by imputation of genotypes. *Nat. Genet.* 39: 906–913.

Marees, A.T., de Kluiver, H., Stringer, S. et al. (2018). A tutorial on conducting genome-wide association studies: quality control and statistical analysis. *Int. J. Methods Psychiatr. Res.* 27 (2): e1608.

Mbatchou, L.B., Backman, J., Marcketta, A. et al. (2020). Computationally efficient whole genome regression for quantitative and binary traits. *bioRxiv* https://doi.org/10.1101/2020.06.19.162354.

McCarthy, M.I., Abecasis, G.R., Cardon, L.R. et al. (2008). Genome-wide association studies for complex traits: consensus, uncertainty and challenges. *Nat. Rev. Genet.* 9: 356–369.

McPeek, M.S. and Sun, L. (2000). Statistical tests for detection of misspecified relationships by use of genome-screen data. *Am. J. Hum. Genet.* 66: 1076–1094.

Miclaus, K., Wolfinger, R., and Czika, W. (2009). SNP selection and multidimensional scaling to quantify population structure. *Genet. Epidemiol.* 33: 488–496.

Naj, A.C. (2019). Genotype imputation in genome-wide association studies. *Curr. Protoc. Hum. Genet.* 102 (1): e84.

Naj, A.C., Jun, G., Beecham, G.W. et al. (2011). Common variants at MS4A4/MS4A6E, CD2AP, CD33 and EPHA1 are associated with late-onset Alzheimer's disease. *Nat. Genet.* 43: 436–441.

NCI-NHGRI Working Group on Replication in Association Studies, Chanock, S.J., Manolio, T. et al. (2007). Replicating genotype-phenotype associations. *Nature* 447: 655–660.

Pe'er, I., Yelensky, R., Altshuler, D., and Daly, M.J. (2008). Estimation of the multiple testing burden for genomewide association studies of nearly all common variants. *Genet. Epidemiol.* 32: 381–385.

Price, A.L., Patterson, N.J., Plenge, R.M. et al. (2006). Principal components analysis corrects for stratification in genome-wide association studies. *Nat. Genet.* 38: 904–909.

Price, A.L., Zaitlen, N.A., Reich, D., and Patterson, N. (2010). New approaches to population stratification in genome-wide association studies. *Nat. Rev. Genet.* 11: 459–463.

Pritchard, J.K., Stephens, M., and Donnelly, P. (2000a). Inference of population structure using multilocus genotype data. *Genetics* 155: 945–959.

Pritchard, J.K., Stephens, M., Rosenberg, N.A., and Donnelly, P. (2000b). Association mapping in structured populations. *Am. J. Hum. Genet.* 67: 170–181.

Purcell, S., Neale, B., Todd-Brown, K. et al. (2007). PLINK: a tool set for whole-genome association and population-based linkage analyses. *Am. J. Hum. Genet.* 81: 559–575.

Rapley, R. and Harbron, S. (2004). *Molecular Analysis and Genome Discovery*. Chichester, UK: John Wiley & Sons.

Roeder, K. and Wasserman, L. (2009). Genome-wide significance levels and weighted hypothesis testing. *Stat. Sci.* 24: 398–413.

Rosenberg, N.A. (2003). The shapes of neutral gene genealogies in two species: probabilities of monophyly, paraphyly, and polyphyly in a coalescent model. *Evolution* 57: 1465–1477.

Rosenberg, N.A., Pritchard, J.K., Weber, J.L. et al. (2002). Genetic structure of human populations. *Science* 298: 2381–2385.

Scheet, P. and Stephens, M. (2006). A fast and flexible statistical model for large-scale population genotype data: applications to inferring missing genotypes and haplotypic phase. *Am. J. Hum. Genet.* 78: 629–644.

Seaman, S.R. and Müller-Myhsok, B. (2005). Rapid simulation of P values for product methods and multiple-testing adjustment in association studies. *Am. J. Hum. Genet.* 76 (3): 399–408. https://doi.org/10.1086/428140. Epub 2005 Jan 11.

Servin, B. and Stephens, M. (2007). Imputation-based analysis of association studies: candidate regions and quantitative traits. *PLoS Genet.* 3: e114.

Seshadri, S., Fitzpatrick, A.L., Ikram, M.A. et al. (2010). Genome-wide analysis of genetic loci associated with Alzheimer disease. *JAMA* 303: 1832–1840.

Sham, P.C. and Purcell, S.M. (2014). Statistical power and significance testing in large-scale genetic studies. *Nat. Rev. Genet.* 15 (5): 335–346. https://doi.org/10.1038/nrg3706.

Skol, A.D., Scott, L.J., Abecasis, G.R., and Boehnke, M. (2006). Joint analysis is more efficient than replication-based analysis for two-stage genome-wide association studies. *Nat. Genet.* 38: 209–213.

Slifer, S.H. (2018). PLINK: key functions for data analysis. *Curr. Protoc. Hum. Genet.* 97 (1): e59.

Stanaway, I.B., Hall, T.O., Rosenthal, E.A. et al. (2019). The eMERGE genotype set of 83,717 subjects imputed to ~40 million variants genome wide and association with the herpes zoster medical record phenotype. *Genet. Epidemiol.* 43 (1): 63–81.

Storey, J.D. and Tibshirani, R. (2003). Statistical significance for genomewide studies. *Proc. Natl. Acad. Sci. U. S. A.* 100: 9440–9445.

Thornton, T. and McPeek, M.S. (2010). ROADTRIPS: case-control association testing with partially or completely unknown population and pedigree structure. *Am. J. Hum. Genet.* 86: 172–184.

Turner, S., Armstrong, L.L., Bradford, Y. et al. (2011). Quality Control procedures for genome wide association studies. *Curr. Protoc. Hum. Genet.* Chapter 1: Unit1.19.

Verma, S.S., de Andrade, M., Tromp, G. et al. (2014). Imputation and quality control steps for combining multiple genome-wide datasets. *Front. Genet.* 5: 370.

Visscher, P.M., Wray, N.R., Zhang, Q. et al. (2017). 10 years of GWAS discovery: biology, function, and translation. *AJHG* 101: 5–22.

Voight, B.F., Scott, L.J., Steinthorsdottir, V. et al. (2010). Twelve type 2 diabetes susceptibility loci identified through large-scale association analysis. *Nat. Genet.* 42: 579–589.

Wang, K., Li, W.D., Zhang, C.K. et al. (2011). A genome-wide association study on obesity and obesity-related traits. *PLoS One* 6: e18939.

Wellcome Trust Case Control Consortium (2007). Genome-wide association study of 14,000 cases of seven common diseases and 3,000 shared controls. *Nature* 447 (7145): 661–678. https://doi.org/10.1038/nature05911.

Welter, D., MacArthur, J., Morales, J. et al. (2014). The NHGRI GWAS catalog, a curated resource of SNP-trait associations. *Nucleic Acids Res.* 42 (Database issue): D1001–D1006.

Willer, C.J., Schmidt, E.M., Sengupta, S. et al. (2013). Discovery and refinement of loci associated with lipid levels. *Nat. Genet.* 45 (11): 1274–1283.

Wittke-Thompson, J.K., Pluzhnikov, A., and Cox, N.J. (2005). Rational inferences about departures from Hardy-Weinberg equilibrium. *Am. J. Hum. Genet.* 76: 967–986.

Won, S., Wilk, J.B., Mathias, R.A. et al. (2009). On the analysis of genome-wide association studies in family-based designs: a universal, robust analysis approach and an application to four genome-wide association studies. *PLoS Genet.* 5: e1000741.

Xue, A., Wu, Y., Zhu, Z. et al. (2018). Genome-wide association analyses identify 143 risk variants and putative regulatory mechanisms for type 2 diabetes. *Nat. Commun.* 9 (1): 2941.

Yengo, L., Sidorenko, J., Kemper, K.E. et al. (2018). Meta-analysis of genome-wide association studies for height and body mass index in ~700000 individuals of European ancestry. *Hum. Mol. Genet.* 27 (20): 3641–3649.

Zeggini, E. and Ioannidis, J.P. (2009). Meta-analysis in genome-wide association studies. *Pharmacogenomics* 10 (2): 191–201.

Zeggini, E., Scott, L.J., Saxena, R. et al. (2008). Meta-analysis of genome-wide association data and large-scale replication identifies additional susceptibility loci for type 2 diabetes. *Nat. Genet.* 40: 638–645.

Zhang, Z., Ersoz, E., Lai, C.Q. et al. (2010). Mixed linear model approach adapted for genome-wide association studies. *Nat. Genet.* 42: 355–360.

Zheng, G. 1, Freidlin, B., and Gastwirth, J.L. (2006). Robust genomic control for association studies. *Am. J. Hum. Genet.* 78 (2): 350–356. https://doi.org/10.1086/500054. Epub 2005 Dec 22.

Zhou, W., Nielsen, J.B., Fritsche, L.G. et al. (2018). Efficiently controlling for case-control imbalance and sample relatedness in large-scale genetic association studies. *Nat. Genet.* 50: 1335–1341.

10

Bioinformatics of Human Genetic Disease Studies

Dale J. Hedges

Center for Applied Bioinformatics, St. Jude Children's Research Hospital, Memphis, TN, USA

Introduction

Modern human genetics and genomics are thoroughly dependent upon information technology to process and interpret the ever-expanding volume of data being produced. The various applications of computational technologies to biological questions have become the basis of research programs in their own right. Genomic data processing and certain aspects of the downstream analysis and visualization of genomic data are frequently categorized under the umbrella term of "bioinformatics." In its broadest definition, bioinformatics has an enormous domain, spanning diverse areas ranging from biostatistics to the development and use of tools for the annotation and representation of genomic data. From another perspective, however, bioinformatics is a loosely knit collection of disciplines, all held together by a common theme of applying computation and information technology to address particular biological questions. With this notion in mind, the aim of this chapter is to provide the reader with a selection of those bioinformatics methods and tools relevant for studying the underlying genomic basis of human disease. Even under the more restrictive scope of human disease studies, the breadth of material remains large and precludes an at-length treatment of any one topic. The author hopes, however, that the outline provided here will serve as a starting point for readers by (i) acquainting them with the variety of considerations that must be made when dealing with genome-scale data sets and (ii) highlighting a representative selection of methods and tools that are at their disposal to aid their research.

This chapter is divided into three main sections. The first section addresses some common threads impacting genomics data processing in general, such as the manipulation of data formats and the increasingly important consideration of computational infrastructure requirements during the study design and budgeting process. The second section covers aspects of the acquisition, processing, and analysis of data associated with commonly used modern genomics platforms. The processing and analysis section is further divided between those approaches targeting array-based experimental methods and those targeting sequencing-based experimental methods. Although there is a fair amount of overlap among these approaches, particularly at the conceptual level, there are nevertheless enough differences to warrant separate treatment. While the use of DNA and gene expression arrays remains widespread, there has been a rapid transition toward high-throughput, massively parallel sequencing methods. In acknowledgement of this continuing trend, the array material is covered more briefly than previous editions of this text so that sequencing-based

Genetic Analysis of Complex Diseases, Third Edition. Edited by William K. Scott and Marylyn D. Ritchie.

methodologies can be more thoroughly covered. Partly as a consequence of the diversity inherent in the bioinformatics arena, the term "genomic analysis" can mean very different things to different people. In the present chapter, the focus is largely upon those aspects traditionally considered the domain of bioinformatics processing, as opposed to statistical genetics. Many aspects of genotype-based statistical inference that are applicable to genomic data, including association and linkage methodologies, are not covered here, as they are addressed at length elsewhere in this volume. The lines separating bioinformatics, biostatistics, statistical genetics, and data science in general are somewhat arbitrarily drawn. In the third and final section of the chapter, an overview of the more commonly used bioinformatics tools and resources is provided. These include useful databases, visualization tools, and software frameworks for data workflows relevant for human disease studies. It is ultimately the challenge of the bioinformatics component of a research project to bring these often disparate resources together to bear on the biological question at hand.

Common Threads Genome Analysis

A Brief Note on Study Design

No technology or analysis method, however sophisticated, can fully compensate for a fundamentally flawed study design. The reader is advised to consult Chapter 4 of this volume, *Study Design for Genetic Studies*, for a comprehensive overview of genetic study design issues. Here, a few issues commonly arising during analysis phase are noted, as they can most easily – or in some cases, only – be handled prior to the laboratory phase of the study. Most technologies described below, and the data associated with them, are subject to systematic biases or batch effects, often due to changes in laboratory reagents or even personnel that occur over time. As such, proper randomization of samples (most commonly with respect to cases and controls) is always warranted to reduce the chance that these systematic biases might influence biological inferences. In addition to batch effects, insufficient sample size is another problem that cannot be surmounted at the post-experimental level. Statistical assessment of power by qualified personnel should therefore be conducted prior to initiating a study. It is also worth emphasizing that data integrity begins with sample handling. This includes from the moment that tissue is extracted from the individual on through the laboratory processing. When handling samples at scale, as has become increasingly the norm for human genetic disease studies, it is necessary to take proactive measures to detect and correct sample handling mishaps. These measures can include common sense checks, such as making sure the sex listed on sample manifests remains consistent with the results obtained. More involved precautions can include an initial "fingerprinting" of samples with a low-cost genotyping panel so that downstream data generated can be cross-referenced with the initial fingerprint to rule out sample swaps. In high-volume situations, such as within core facilities, the implementation of a Laboratory Information Management Systems (LIMS) is becoming increasingly commonplace. Incorporating a LIMS can help detect and localize problems more rapidly before they can influence downstream analysis. This can even include diagnosing which factors (reagent or otherwise) might be associated with an observed batch effect.

Data Format Manipulation

Modern investigations of human genetic disease involve the use of multiple software tools during the course of data analyses. The resulting output of one program frequently needs to be fed as input to a second program in order for the analysis to proceed. One of the common frustrations faced by biologists is that the data formats output by these various programs can be quite diverse, while the requirements for input files for any given program are often quite stringent. Hence, some amount of format manipulation is frequently necessary for analysis to progress. The problem has been partially mitigated by the development of common "currency" formats (e.g. SAM/BAM in the sequencing arena [Li et al. 2009]) that serve as standards for input and output, but the diversity of formats nevertheless remains an obstacle for many researchers, and it is an annoyance that is unlikely to disappear any time in the near future. In some cases, necessary reformatting can be readily accomplished by manual manipulation in widely available office software, such as Microsoft Excel. This avenue is increasingly untenable, however, as the size of size of human genetic data sets continues to expand. When these tools are not adequate, the next step is to look for existing software that can make the specific format conversion required. Pre-written conversion utilities, as with most things acquired over the internet, can vary greatly in reliability, and it is incumbent upon the researcher to do adequate checking to ensure data integrity is maintained. Even with the variety of toolkits and scripts available online, there are nevertheless data manipulations required, such as complex filtering procedures that are not accommodated by existing tools. In these situations, it is highly recommended that anyone planning to perform a significant amount of bioinformatics work develop proficiency with a scripting language. Unlike fully typed programming languages, such as C and Java, scripting languages are more amenable to rapid development of customized solutions that may not warrant the time investment that development in a language like C or Java often require. Two widely popular languages allowing data transformation are PERL and Python. R is another popular option, primarily due to the wealth of available analysis libraries. All three languages have active user bases of bioinformaticians, and numerous pre-written modules for many common bioinformatics tasks can be found through the open source BioPerl, BioPython, and Bioconductor projects (Mangalam 2002; Stajich et al. 2002). Over the past decade, the Python language has gradually replaced Perl as the language of choice for practicing bioinformaticians. This has occurred for a variety reasons, not least of which has been the popularity of Python in the broader data science community. In addition, Python offers a gentler learning curve and is widely considered to encourage a more readable – and thus more maintainable – coding style. The plethora of freely available online resources for developing coding skills has now made the learning process more convenient than ever. While no preexisting software can ultimately match the flexibility available from command of a scripting language, there are now software frameworks available that address many of the most frequently encountered format conversion scenarios that would previously have required custom scripting. A large number of the most commonly required transformations are now encompassed in the Galaxy framework – a freely available workflow system that will be discussed later in the third section of this chapter (Lazarus et al. 2008; Blankenberg et al. 2010; Goecks et al. 2010). Cloud-based offerings, such as Illumina's base space, DNANexus, and Basepair Genomics, have provided services more tailored toward high-volume data processing. These provide attractive alternatives, particularly when the requisite infrastructure required for large-scale sequencing analyses, are not available on site. Some of these platforms (e.g. DNANexus) provide varying degrees of flexibility, allowing end users to design their own automation workflows, using either APIs or GUI-based interfaces.

Planning for Adequate Computational Resources

When designing a genomic study, it is increasingly important to consider the computer resources necessary to house, process, and ultimately archive the associated data. While not a significant problem for array data, the cost of data processing and storage has become an important budgetary factor for high-throughput sequencing data, as the size of data sets has expanded considerably and the hardware requirements to manipulate and analyze data have become more demanding. The following section describes some of the considerations that must be made when planning for projects that are anticipated to produce large volumes of data.

Storage

Deciding what data to keep and how it will be stored is an important consideration for any research project. Storage requirements will often balloon during active computational analysis before ultimately condensing into a final product that has a smaller storage footprint. A host of new compression tools are available to help reduce the footprint of data. Some, such as those implemented as part of the CRAM toolkit (Cochrane et al. 2013), allow for both lossy and lossless compression of the BAM format, which is an important result format that holds sequence, alignment, and associated metadata for experiments. While these compression tools can help control the final size of the data footprint, the researcher needs to make an estimate of total storage needs prior to beginning the study. This begins by determining what data will be retained (e.g. which file formats), what the footprint of that data will look like during processing and downstream analysis, and, finally, what the footprint of that data will look once it is finally completed and archived. With next-generation data in particular, the information typically exists within multiple data formats, and it is often replicated (albeit in different form) across various stages of analysis. There is usually a trade-off between maintaining the data in (i) the rawest possible form, so that no information is lost if new analyses need be conducted and (ii) results from later stage processing, which can allow one to avoid the need to repeat computationally expensive procedures. In the context of high-throughput genomic sequencing experiments, the files most commonly maintained include the initial fastq format files, which contain base calls and associated phred-scaled quality scores, the binary BAM format files, which house alignment and information, and the vcf files housing genotyping calls, which are the result. As raw sequence reads (i.e., base calls) can generally be fully recovered from BAM files (as long as unaligned sequences are also maintained during BAM generation), some groups choose to retain a compressed BAM file as opposed to retaining the fastq sequence. Such BAMs need not necessarily be aligned, and they can contain all the information necessary to recreate the original fastq files. Equivalent end point files to VCF and gene lists exist in the context of gene expression or other analyses, with the formats depending on the specific software used for analysis.

Regarding the storage hardware itself, there are typically different varieties of storage required, available at different cost points. A mistake that is often made is to expect the cost of storage to be commensurate with per terabyte costs available at local electronics stores, where two Terabytes of storage can presently be acquired for well under $100. This fails to take into account several important considerations, including reliable backup of the data and the personnel and physical resource costs of maintaining readily accessible (i.e. "live" or "online") data storage for an extended period of time. Finally, the type of storage often has a latency that will perform poorly during active analysis, which is quite demanding on input/output (i/o) capacity. Consumer hard drives are therefore more suitable for archival purposes than for use during active processing, particularly when the data will be processed with high-performance computing environment, where the drive may be simultaneously accessed from many CPUs in parallel. As a result of these considerations, procuring terabytes of quality storage in a local server facility may comprise a significant component of a research study's budget. Higher-speed storage, such as that used during active data processing, will come at higher cost than longer-term (archival) storage. It is therefore important to determine if there is sufficient temporary working storage available where you will be analyzing your data to accommodate the size of your project – or, more accurately, to accommodate the analysis of your project during an *acceptable time frame*. Having to analyze data in small increments to circumvent the limitations of restricted working storage space can result in large inefficiencies in the processing pipeline and greatly protract the time required for study completion.

Processing and Memory

Many biological algorithms are more dependent on local memory (RAM) than the mathematical and physics applications that have historically driven the design of academic high-performance computing centers. This is particularly the case for genome assembly, but is also applicable to many other aspects of high- throughput sequence processing, such as isoform discovery during expression analysis. Nodes with <16 GB RAM may struggle to accommodate some bioinformatics processing needs, and 24–96 GB RAM per multi-CPU compute node (or larger) is preferable when possible, depending on the number of CPUs and associated cores housed in the compute node. For some specific applications, such as whole-genome assembly, transcriptome assembly, and fusion analyses from RNA-seq data, even higher amounts of RAM are preferable, and it is becoming more common to find specialized compute nodes housing 256 GB to 1 TB of RAM dedicated for specific applications. Processing speed (i.e. CPU speed) is typically not a significant bottleneck for large data set analysis, particularly when using modern CPUs with adequate cache sizes and speeds in excess of 2.4 GHz. However, access to adequate *numbers* of compute nodes and processors to conduct analysis in parallel can represent a scarce resource impacting the time frame of analysis. Hence, it is important to investigate high-performance computing resources early on in the study design phase. As discussed below, commercial cloud services can help provide flexibility in this regard, but there are several limitations to their use that must also be considered.

Networking

Networking limitations are another common hindrance for next-generation sequencing studies, particularly when data are generated at a site remote from where it will be analyzed. Bottlenecks can occur at many possible points along the transfer path, including local network switches governing building network traffic. Testing expected transmission times can be challenging, as effective transfer rates can change over time due to increasing and decreasing network demand around the local area network. While efforts are being made across many institutions to improve their local networking capacity, in some instances, the most feasible means of moving data around remains shipping and carrying around hard drives of data from source to destination for upload via the so-called "sneaker net." Improvements in compression and transmission algorithms (e.g. IBM's UDP-based Aspera) have fortunately made this alternative an increasingly rare phenomenon. Nevertheless, it is good to be aware that physical media remains a viable option when network transfer proves either too slow or unreliable.

Genomics in the Cloud

Commercial cloud computing offerings, such as those provided by Amazon (Seattle, WA), Rackspace (San Antonio, TX), Microsoft (Redmond, Washington), and Google (Mountain View, CA) have become attractive solutions to address many of the infrastructure needs indicated above. Prices have fallen considerably, and it is now becoming feasible for academic researchers to use these services as an IT solution. Challenges remain, however. Chief among these is that institutional regulations often prohibit placing certain types of data, particular that pertaining to medical patients, in commercial cloud storage. While there has been a gradual relaxation of these prohibitions, institutional policies will continue to be an impediment to cloud computing until medical establishments are convinced that patient-derived data are adequately secured and that their liability is minimized. In addition to policy and regulatory concerns, commercial cloud offerings also do not necessarily resolve the data transfer challenges associated with large second-generation sequencing studies. Limitations in network capacity have the potential to make transferring data to remote cloud facilities a challenge. However, as indicated above, improvements in both physical network infrastructure and compression algorithms now largely mitigate these issues. To facilitate bioinformatics processing in the cloud, a number of biological sciences vendors provide cloud-based solutions tailored to sequence-based analyses. Two popular options include Illumina's BaseSpace and DNAnexus (https://www.dnanexus.com).

Processing and Analysis of Genomic Data

The nature of the platform used to acquire the data will often drive the manner of the analysis. This is particularly true for methods such as RNA-seq and ChIP-seq, where variations in experimental protocols can have a significant impact on the details of the acquired data and the type of analyses that might be performed. While the number of experimental platforms and the variety of experimental procedures for data generation have become too extensive for exhaustive overview, the following section overviews processing and analysis procedures for some of the more common platforms and strategies.

Array-Based Data

DNA Arrays and High-Throughput Genotyping

While the first genome-wide genetic studies were carried out with modest throughput technologies, such as microsatellites and single locus Taqman genotyping, it was the availability of relatively low-cost array-based assays that precipitated the explosion of genome scale studies, particularly genome-wide association studies (GWAS). These arrays provided the researcher with a large number of marker features to interrogate the genome. The array feature number, and hence the density of genomic coverage, has progressively increased in number over time while the cost per sample to process individuals on arrays has declined. The current state-of-the art chips readily accommodate millions of features. At present, the Illumina Omni chip offers up to five million features per sample, while the Affymetrix 6.0 genotyping platform offers 1.8 million, consisting of 906 600 single nucleotide polymorphisms (SNPs) and approximately 946 000 copy number variation probes. Typically, the markers included on these arrays consist of single nucleotide variants, but invariant loci are now frequently included as copy number probes to aid in the detection of structural variation. While vendors make an effort to include the set of variants that might prove useful to the widest user base, it is up to the researcher to determine which platform and array is best suited to their needs. Aspects to consider are how the markers are dispersed across the genome, as well as the populations from which the variant loci were derived. A recent trend among DNA array providers is to provide custom content arrays, such as the Exome chip and the Immunochip from Illumina. These arrays target a specific class of alleles meeting specified criteria for location and allele frequency. These could be known pathological variants, protein coding variants, or putative functional variants within a subset of genes. Other new arrays have focused on either (i) multiethnic populations, such as the Illumina Global Screening arrays or (ii) a genome-wide array with an enrichment of variants important for disease risk and pharmacogenomics such as the UK Biobank Axiom array. When dealing with custom chips, it is important to determine if the markers provided can satisfy all the various requirements of your study. For example, do they provide an appropriate set of polymorphic loci for determining underlying population structure? As many of these custom arrays include increasingly rare alleles, they may have limited capacity to detect disease alleles through association when the actual pathological allele itself is not included on the chip. Detailed consideration of all the factors involved here are beyond the scope of this chapter, but the researcher should make an effort to discern what diverse purposes the generated data may be expected to serve prior to selecting a specific targeted array.

Preprocessing and Initial Quality Control

Initial data acquisition and manipulation of Array data are handled through vendor-specific software provided with the instruments. Illumina's GenomeStudio and Affymetrix's Genotyping Console provide this functionality for the two most common platforms. Arrays are subject to processing effects and manufacturing flaws that can cause systematic signal intensity shifts at the intra- and inter-array level. These fluctuations can be the result of physical defects in the arrays themselves, usually causing localized anomalies. Other factors include reagent variation and/or exposure time differences that can impact signal intensities across entire samples. Preprocessing steps can identify and remove anomalous features, while normalization seeks to remove systematic effects on signals that are not the result of true biological differences. Unlike expression arrays, for which a wide range of normalization procedures are employed, preprocessing procedures for genotyping arrays are typically handled by the vendor Genome Studio and Genotyping Console for Illumina and Affymetrix, respectively. These preprocessing steps are sometimes integrated within downstream analysis steps themselves, as is the case with Illumina's Gentrain genotyping clustering algorithm. Among its QC options, Affymetrix Genotyping Console makes a per-sample Contrast QC metric available that indicates the amount of separation between homozygous and heterozygous peaks (following transformation). Experiments with >10% of samples yielding a contrast QC < 0.4 or an overall average at or below 1.7 are recommended to be examined for potential processing issues. Illumina likewise makes a number of QC metrics available in Genome Studio through the QC control board, which has information on sample-dependent and sample-independent controls incorporated into the assays. Batch effects, the introduction of systematic biases in array data that are not representative of underlying biological differences, are covered more fully in the expression array section (below). While not as commonly regarded as an issue in the context of genotyping, batch effects can nevertheless occur and results should be examined for their presence. Copy number inference from genotype array data is generally considered to be more sensitive to these effects than genotyping. For many of these effects, appropriate randomization can often minimize the impact on downstream biological inference. The researcher must determine if the effect is extreme enough to warrant an attempt at correction during analysis. There is always a risk with correction that true biological signal may be lost. Methods for genotype array copy number correction are available through the Bioconductor suite (Chierici et al. 2010; Scharpf et al. 2011). Discussion of batch effects in genotyping arrays can be found here (Miclaus et al. 2010).

Genotype Calling

The details of feature intensity extraction prior to genotype calling are dependent upon the specific platform generating the data and are typically handled by vendor-provided software. Although the underlying technologies are quite different, Affymetrix and Illumina both use fluorescence to measure the relative contributions of allele A and B at a given polymorphic locus. In an ideal world, a uniform optical fluorescent intensity would be registered for each allele when the chromosomes are in the homozygous state while the heterozygous state would yield and equal measure of contributions from alleles A and B. In reality, binding efficiencies of probes can differ both within and across assays, as well as across individuals, and the signals registered for a given assay can take assume a heterogeneous set of values. Hence, the relative contribution of alleles A and B across all individuals assayed in the experiment will exhibit a clustered set of values. Moreover, the shape of these clusters can change significantly from locus to locus, based on the hybridization properties of the specific oligonucleotides involved. While lower throughput experiments could rely on manual examination by humans to assign genotypes, the task has simply become untenable due to the size of modern array feature sets, necessitating the use of automated calling algorithms. (Although it should be noted that, as part of a quality control procedure, loci can be sorted for those with the least confident genotype scores, and these loci can be considered for manual curation.) The challenge of these genotyping algorithms is to determine whether the mixture of signals detected for a given locus corresponds to a homozygous A/A or B/B or the heterozygous state A/B. One key source of information is the use of a population training set to determine the shape of genotype clusters for each assay. Current vendor software packages provided by both Affymetrix and Illumina draw upon preexisting data from assayed populations (typically 1000 Genomes samples) to train calling algorithms for classifications. Options are also provided for users to train the calling algorithms with their own data sets. The Affymetrix calling algorithm for Affymetrix 6.0 Genotyping arrays is called Birdseed (currently v.2). Birdseed uses the Expectation Maximization (EM) algorithm to fit a two-dimensional Gaussian mixture model to A and B allele signal space (Korn et al. 2008). Illumina's Genome Studio uses its Gentrain algorithm to perform clustering, which can target either the users own data or publicly available 1000 Genomes sample data, or, alternatively, users can use their own data set for training. Gentrain employs an Artificial Neural Network for determining cluster centroids from polar transformed data. The calling algorithm, Gencall, subsequently classifies loci according to the closest matching cluster. Illumina recommends considering genotypes with Gencall (GC) scores of <0.15 as failed, and this can be used for the genotype calling efficiency metric. While the vendor provided genotype callers provide robust calling in many cases, particularly when an appropriate population training set is available, they can struggle significantly for rare variants. As the feature set of the arrays grows larger, the allele frequency spectrum of variants included on the arrays expands, and it is becoming common for arrays to include lower frequency alleles. This is particularly true for some of the custom-targeted arrays. The problem of rare alleles resides in the fact that, on account of their low frequency in the population, there are limited examples of the allele in the training and experimental data sets from which genotype clusters can be built. Multiple third-party algorithms have made efforts to increase accuracy for rarer alleles. For example, Opticall combines within and across-sample intensity data to attempt to improve calling accuracy (Shah et al. 2012). Additional third-party genotype calling options include GenoSNP, M(3), Sniper, ILLUMINUS (Hua et al. 2007; Giannoulatou et al. 2008; Li et al. 2012). BEAGLE incorporates known LD structure to improve accuracy (Browning and Yu 2009).

Call Efficiency

One key quality control metric is to assess the total fraction of loci called for each study sample. Typically, a study will set a minimal threshold for percentage of genotypes called for any one sample. The underlying premise is that samples failing to meet this threshold are likely to have inherent problems that may make their results questionable. A typical setting might be 95–98% call efficiency. However, note that this call efficiency stringency must be relaxed for Y chromosome markers, as these are expected to have no calls in females outside of the pseudoautosomal regions.

Data Cleaning and Additional Quality Control

Following genotype calling, there are several additional steps that can be taken to improve the quality of data prior to statistical analysis. For a more detailed discussion of data cleaning and marker pruning, see Chapter 9, *Genome Wide Association Studies*. Common quality control procedures include checking for mismatches between the sample manifest and sex of the samples and filtering for markers that deviate significantly from Hardy–Weinberg equilibrium. Where family structure is available, checking for Mendelian inconsistencies can aid in identifying problems with markers as well as potential sample labeling errors. Depending on the type of analysis being conducted, further pruning of markers based on linkage disequilibrium may be warranted. Several quality checks associated with post-genotyping GWAS data can be performed with the PLINK software suite (Purcell et al. 2007). Detailed discussion of quality control procedures for GWAS can be found in Laurie et al. (2010), Turner et al. (2011), and Zuvich et al. (2011).

Inferring Structural Variation From SNP-based Array Data

While array-based comparative genomic hybridization (aCGH; discussed below) is still considered superior to SNP-based array assays for the study of structural DNA variation, structural variants can nevertheless be accurately inferred from in many circumstances. This is particularly true as the size of the variants get larger, with the highest accuracies achieved for events of >500 kb that encompass 100 or more array markers. Both Affymetrix and Illumina arrays can be used for structural variant inference. One significant advantage of SNP arrays to aCGH arrays is the ability to discern the allelic state of the locus being assayed. Hence, it can readily detect stretches of homozygosity indicative of loss of heterozygosity, whether they are in a haploid or diploid state. In the case of duplications, SNP arrays can indicate that allele is being duplicated, which can have a direct impact on pathogenicity. Agilent has also begun to include single nucleotide variant options on its aCGH array chips to provide this information as well. Perhaps the largest advantage of using genotyping arrays for detecting structural variants is that the data are already produced in the course of genotyping studies; therefore, deriving additional structural variation information from the same data set represents an added value at no additional experimental cost. Each of the two major vendors, Illumina and Affymetrix, provides a copy calling algorithm as part of its analysis suite. Illumina's GenomeStudio offers the CNV Partition algorithm for copy number calling, which estimates the most likely of 14 allele copy states based on their expected intensity distributions. Affymetrix Genotyping Console provides the Canary algorithm to estimate the integer copy number at a given structural variant position (Korn et al. 2008). While both vendors provide algorithms for the analysis of structural variants, third-party algorithms are frequently preferred. PennCNV is a popular option; it uses a hidden Markov model and further incorporates family structure, where available, to augment CNV calling (Wang et al. 2007). Other academically developed options include QuantiSNP and BirdSuite (Colella et al. 2007; Korn et al. 2008). Ensemble approaches for copy number calling are often recommended, where several algorithms are used and variants are called where there is a consensus or majority of algorithms in agreement (see Kim et al. 2012 for discussion). Commercial solutions for copy number calling from genotyping arrays also exist, including Biodiscovery Nexus, GoldenHelix, and Partek.

A Note on Statistical Analysis and Interpretation of Results

The techniques for the downstream statistical analysis of data derived from genome-wide genotyping arrays are covered elsewhere in this volume (see Chapter 9). Once a set of prioritized loci of interest are obtained from statistical analysis, the next step is to identify a plausible biological target. In some instances, the significant SNP itself might constitute a potential variant. Having narrowed down to a genomic region of interest, strategies for the identification of plausible genes and variants are addressed in the third segment of this chapter.

Array-Based Analysis of Gene Expression

The first widespread use of array technology was in the context of gene expression assays. Today, expression arrays remain a powerful and low-cost method for surveying expression levels across biological and/or experimental conditions. Their primary limitation, as compared with newer RNA-Seq technologies, is their inability to detect novel transcripts. In order to assay a given genomic entity's expression, it has to have been previously known and incorporated on the array during manufacturing. In addition, the signal intensities of a microarray feature also have a restricted dynamic range in comparison to sequencing-based systems, which are based on discrete counts only limited by the amount of sequencing dedicated to the sample. An early developer of microarray expression chips, Affymetrix (Santa Clara, CA), uses a photolithographic manufacturing technique to achieve high feature density. The assay features typically consist of a fixed set of specifically designed oligonucleotides, some matching the target gene segment exactly and some containing intentional mismatches to provide background signal information. As with DNA arrays, expression array features have continued to expand over the years and now encompass most known genes, along with noncoding miRNAs. Typically, there are multiple array features per gene (collectively referred to as the gene's "probeset"), and specially designed arrays permit some degree of isoform abundance information to be obtained. In the assay, labeled cDNA from the experimental samples is hybridized to the oligonucleotide array and proprietary algorithms are used to extract intensity data from each feature. These signal intensities correspond with the relative expression of the gene or gene isoform targeted by the given feature. While chipsets from other commercial vendors (e.g. Illumina) use somewhat different technologies, common features include multiple probes per gene, assessments of background signals, and estimates of intensity data, all analyzed using proprietary algorithms.

Preprocessing and Quality Control As with DNA arrays, quality control for expression arrays occurs at multiple levels, beginning with experimental procedures in the laboratory. An important step in initial quality control is to examine array images for evidence of obvious deformations or scratching that can result in erroneous data points. These features should be flagged in order to be eliminated from downstream analysis. Strategies for the automated detection and elimination of aberrant features have been proposed (e.g. see Li and Wong 2001). Additional quality controls include ensuring that a similar percentage of genes are called across all chips, and the ratio of 3′ probes to 5′ probes for house-keeping genes, such as B-actin and GAPDH (Reviewed in Mohapatra and Krishnan 2011). Readers should also consult manufacturers' guidelines for quality control.

Batch Effects and Data Normalization

Batch effects occur when segments of the sample set are treated in some way that is systematically different than other subsets of the sample. If these differences are not distributed uniformly across sample classes, inferences concerning biological differences can be negatively impacted. The source of the effects can include the nature of the nucleic acid extraction process itself and need not necessarily occur in the lab during the experimental procedure. In the context of expression arrays, these systematic effects have been attributed to numerous factors, including the time of day that processing occurs and variance in ozone levels, which can affect labeling (Fare et al. 2003). These effects are particularly notorious in larger-scale projects, where laboratory processing may occur at different facilities. Sample sets processed over long periods of time are at greater risk, as there is increased likelihood that reagents and/or laboratory personnel may change during the processing. Systematic bias can potentially move pervasive in expression experiments compared to DNA genotyping experiments due to the increased number of steps involved in the process of RNA handling. A further problem is that RNA expression experiments are often conducted at a smaller scale than DNA array experiments, limiting the ability to detect and statistically compensate for systematic bias effects. On the positive side, these smaller experiments are often more likely to be conducted at the same time by the same technicians, restricting the number of opportunities for these effects to arise. The systematic differences caused by experimental artifacts can be eliminated by a variety of normalization procedures. See Luo et al. (2010) for a detailed empirically based comparison of strategies. The assumptions of normalization are usually based on constant expression level of internal housekeeping genes or external markers, the overall intensity levels, or the majority of genes being constantly expressed. Quantile normalization, currently a widely used normalization strategy, involves transforming the array data to a common distribution prior to later analyses (Bolstad et al. 2003). Combined approaches, such as Robust Multiarray Averaging (RMA), are now available, which provide summary statistics in addition to performing background correction and quantile normalization (Irizarry et al. 2003).

Differential Expression

Following normalization, frequently the next goal of analysis is the search for subsets of genes that are differentially expressed across various experimental or biological conditions. Several criteria for establishing a threshold for differentially expressed genes have been proposed, including a threshold level of a twofold difference. However, simple thresholds fail to account for inter-array variance. Extensions of the analysis of variance (ANOVA) and Student's *t*-test provide one means for taking variance into account (Tusher et al. 2001). The moderated *t*-test has been implemented in the Bioconductor package *limma* (Ritchie et al. 2006). The downside of these approaches is that they require multitest correction, and there is limited agreement on what correction method is most appropriate. While Bonferroni correction measures have been applied, they are often considered to be overly conservative. Another approach that has been used is to generate a ranked list of genes with $p < 0.05$. This list might then be used as the set of differentially expressed genes for downstream analysis, such as pathway enrichment analysis (described below). Further filtering this ranked list to include only those genes above a set fold difference threshold has been shown to increase replicability across expression experiments (Consortium et al. 2006), although sensitivity to true positives would be expected to be impacted. While practical and readily calculated, the overall value of ranked lists based on alpha-value thresholds is dependent upon the number of hypotheses rejected in the experiment. In some scenarios, expressed gene lists composed in this way may consist mostly of false positives. Due to these limitations, alternative approaches are now commonly used, such as controlling for the false discovery rate (FDR) (Benjamini and Yosef 1995); several extensions of this approach have since been developed. Methods that control FDR incorporate the notion of *p*-value ranking while also taking into account the alpha level and total number of hypotheses being tested. In addition to questions about multiple test correction methods, some researchers have questioned the appropriateness of parametric tests for differential expression altogether, due to violation of parametric assumptions exhibited by the data. Several nonparametric tests have thus been suggested, including the Fisher-Pitman rank sum test and the Wilcoxon rank sum test. Neuhauser and Senske (2004) proposed the Baumgartner–Weiss–Schindler test and demonstrate it to be both less conservative and more powerful, particularly in cases of asymmetric distributions or distributions that are heavily tailed (Neuhauser and Senske 2004).

Classification and Clustering Methods

Other common analysis tasks performed on microarray expression data are *classification* and *clustering*. Consider, for example, the scenario where a tumor tissue biopsy is taken from a patient and subjected to a microarray expression assay. Given a training set of expression data from malignant and nonmalignant tumors previously classified by experts, the task of a classification algorithm might be to accurately assign the unknown sample to the benign or malignant categories. Several supervised machine learning approaches have been applied to this and similar problems, ranging from partitional clustering (e.g. K nearest neighbors) to Artificial Neural Networks, Random Forests, Support Vector Machines, and a variety of hierarchical clustering approaches (reviewed in Tarca et al. 2007). In addition to supervised learning, which is based on training data sets, unsupervised learning approaches are also commonly applied to expression array data to identify data groupings. In unsupervised learning, an attempt is made to find underlying structure in the data without the use of prior knowledge or training data sets. Unsupervised clustering strategies include hierarchical clustering, self-organizing maps (SOM), and fuzzy clustering. The clusters that are discovered through the unsupervised learning process may yield meaningful biological categories that are not immediately evident from traditional analysis methods or anticipated based on existing biological knowledge. Again returning to the cancer analogy, if an unsupervised clustering analysis can identify two subgroups of tumors based on patterns of gene expression, a number of additional testable hypotheses naturally follow, such as how these various subgroups respond to specific pharmacological treatments.

All clustering approaches begin by first defining an appropriate measure of distance between the objects being clustered. Common choices for distance metrics include Euclidean distance and the Pearson coefficient. Readers can find many of the clustering strategies used in molecular biology outlined and explained in the following reviews (Kerr et al. 2008; Nugent and Meila 2010; Xu and Wunsch 2010). Here, a brief summary of hierarchical clustering is provided, as it is among the most common clustering strategies used by molecular biologists and implementations can be found in most popular software packages. Hierarchical clustering can be divided into two main categories, *Agglomerative clustering* and *Divisive clustering*. Agglomerative clustering takes a bottom-up approach, beginning with each object, typically a gene but sometimes an experimental grouping, in its own cluster. Next, individual objects are joined into higher level clusters with the nearest object to them, as defined by the specified distance metric. This process is repeated, forming increasingly coarse-grained clusters, until all objects finally belong to a single cluster representing the root of the cluster tree (pictorially represented as the root of the dendrogram). Divisive clustering (less frequently used in the context of microarray analysis) takes a top-down approach, starting with one cluster containing all objects and then systematically divides them into increasingly more refined clusters, with the process ending when each object forms its own cluster. The results of hierarchical clustering algorithms are normally displayed as dendrograms associated with heat maps (described in next section). Researchers have a wide range of software options for conducting the more common machine learning and clustering methods; robust implementations can be found in a number of commercial and academic software packages, including Biodiscovery, Genespring, Partek, and a host of R and SAS packages. Once clusters are established, it is natural to inquire as to how meaningful or robust they are. Several strategies for cluster validation exist. These include the use of external label information (where available) to validate the cluster results. Resampling techniques, such as bootstrapping, have also been employed to determine the stability of clusters (Felsenstein 1985). A wide variety of cluster validation strategies are helpfully reviewed in (Handl et al. 2005).

Visualization of Expression Data

Combining clustering results with heat maps can provide an intuitive representation of how genes are clustered according to their expression values. As indicated above, the dendrograms are often the product of hierarchical clustering routines. Classic Venn diagrams are also used to display set relationships among genes expressed across multiple biological conditions or treatments. Depending on the type of algorithm used for analysis, a natural visualization technique may arise directly from the algorithmic output. For example, self-organizing maps also naturally generate a visualization scheme for array data (see, for example Torkolla et al. 2001). One of the advantages of commercial software packages, such as Partek and Genespring, is that visualization options applicable to algorithmic outputs are readily accessible and do not require separate software or learning the syntax of a new graphics library. Rarely, however, will commercial software provide the full range of flexibility available within open source graphics libraries, such as R's *ggplot2* or Python's *matplotlib*. More recently, polyglot libraries, such as *plotly* (https://plotly.com/graphing-libraries), provide visualization APIs for multiple scripting languages, removing additional barriers to adopting these increasingly versatile tool sets. More recently, visualization techniques based on dimensional reduction, such as tsne and UMAP, have become increasingly popular methods for visualizing cluster results from a variety of data types, including array data (Figure 10.1).

Pathway and Network Analyses

Another common set of analyses performed on expression array data and results are enrichment analyses targeted gene ontologies, pathways, or networks. These analyses are applicable to gene sets derived from a large number of assay types. Pathway and/or Network Analysis are names often

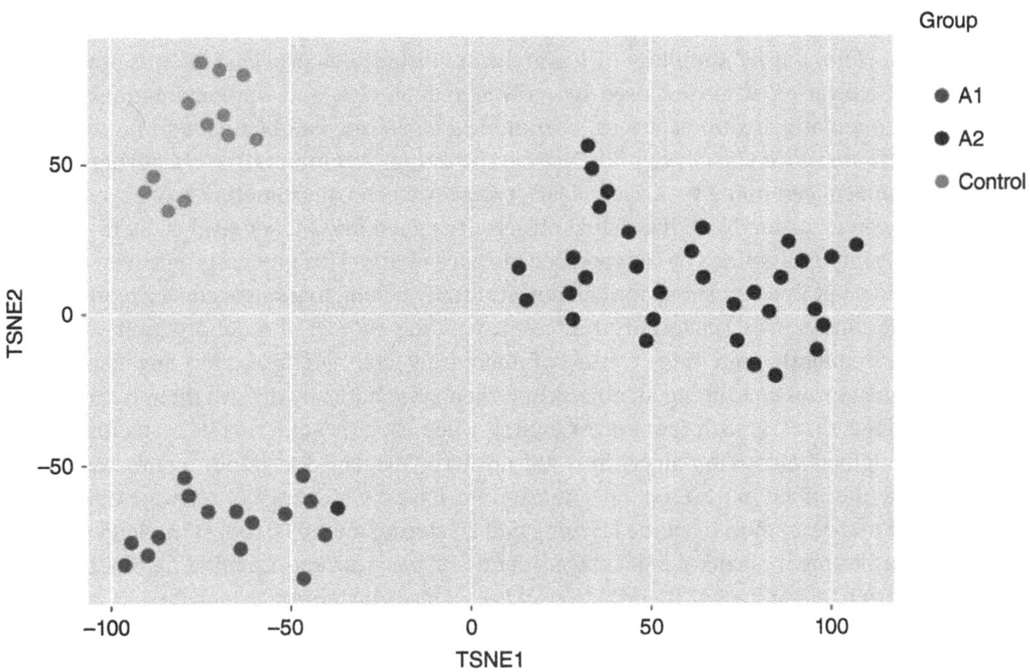

Figure 10.1 Tsne visualization of methylation array data clusters for pediatric adrenocortical tumors. *Source:* Data for plot obtained from (PMID: 32923859). Images courtesy of Brent Orr and Quynh Tran, St. Jude Children's Research Hospital.

used in reference to a diverse set of strategies for examined the re. Typically the term "pathway" is restricted to a succession of biological interactions, where the constituents have a physical relationship. Network analysis, on the other hand, is a more comprehensive term and can include topologies where no known physical contact among constituents need exist. Network analysis can apply just as readily to text mining as it can to a protein interaction context. Although more broad than pathways, ontological analysis of genes is frequently grouped in this category as well, as there is significant overlap among strategies. A gene ontology is a formal language used in an attempt to categorize gene products in a consistent manner. Multiple ontologies can coexist, with individual entities. For example, there are ontologies based on participation in pathways as well as ontologies based on physical location with respect to the cell, biochemical activity, or broader biological process with defined beginnings and endpoints that are directly relevant to the function of biological organisms. Querying whether particular ontological terms are overrepresented among your results can aid in the identification of biological systems or functions relevant to disease states. For example, the set of prioritized differentially expressed genes may correspond to the activation of a particular DNA repair pathway. Alternatively, the set of results from a genome-wide association analysis may reveal an over-abundance of genes involved in synaptic junctions. These queries can help identify biological systems relevant to the disease of interest, and sometimes these can be quite different from what may have been predicted a priori. Unfortunately, "network," "pathway," and less commonly "ontological" analysis are often used interchangeably to refer to quite different things. In the context of statistical genetics, for example, pathway analysis may include incorporating existing biological knowledge to examine epistatic effects of variants within a particular biological pathway (e.g. folate pathway). In the context of bioinformatics, it is most often associated with determining whether a particular set of genes, often significant candidates from a previous expression or association analysis, are overrepresented among a subset of pathways or ontologies. Still another group of methods sometimes referred to as "pathway analysis" deals with discerning gene interactions from expression data, such as the action of transcription factor networks. Here, we restrict our discussion to the class of category overrepresentation analyses most commonly associated with bioinformatics.

Several software packages and online resources are available for pathway analysis for data from gene expression and other sources. These include IPA, Genespring, Partek, and Genego. Popular free academic options include DAVID, Webgestalt, and PARIS (Zhang et al. 2005; da Huang et al. 2007; Sherman et al. 2007; Yaspan et al. 2011). Characterizing features of these analyses include controlling for the relative size (i.e., gene membership count) of the categories under consideration, how relative ranking of gene lists are considered, as well as if and how the presence hierarchical structure among the categories is accommodated. The genes input to enrichment analysis tools are ultimately the outcomes of upstream analyses, such as differential gene expression. It remains incumbent upon the users of gene set enrichment software tools to determine how best to generate and filter the set of significant gene results used as input. Needless to say, methodology and parameterization choices made during the production of input gene lists can significantly impact the biological significance of results.

Direct Counting and Other Expression Assay Procedures

Although technically not array platforms, it should be noted that a technology for directly counting RNA molecules via fluorescent tagging and optical microscopy was introduced by Nanostring (Geiss et al. 2008). The lack of a need to first convert RNA molecules to cDNA confers many advantages to the experimental process by removing potential vectors for bias introduction, and the higher-throughput capacity of the system provides for an attractive alternative over traditional quantitative PCR. However, since RNA counts are conducted under an optical field of restricted size, the results of the direct counting method at low levels of RNA will often not match those obtained from more sensitive quantitative PCR in the same samples. That is, while the benefits of direct counting remove several potential biases of PCR that can undermine accurate quantification, the overall sensitivity of qPCR for the detection of small number remains superior. As the data yielded by the Nanostring instrument are discrete count data, analysis is relatively straightforward. Sample counts are readily normalized using control features included in the protocol, and quantification and differential analysis procedures similar to those described below for RNA-seq experiments can be easily adopted. Approaches for normalization and differential expression analysis are available as R software packages (Brumbaugh et al. 2011; Waggott et al. 2012).

Additional Uses for Oligonucleotide Arrays

While genotyping and expression analysis remain the two primary uses for array technologies, they have also have important applications other analyses examining additional genomic features, such as protein binding and methylation marks. A subset of this broader group of array-based methodologies is briefly described.

ChIP-chip Oligonucleotide arrays have been used to assay the products of chromatin immunoprecipitation (ChIP) reactions. Here, antibodies targeting specific proteins known or suspected to be associated with DNA are used to pull the associated DNA fragments out of solution. These fragments are then hybridized to the oligonucleotide array, which typically has oligonucleotide features spanning the genome. The signal intensity generated by the DNA fragment binding to the corresponding oligonucleotides on the array indicates which genomic DNA segments are potentially associated with the targeted protein. There are two primary questions that are asked of the data that is generated from the ChIP-chip experimental procedure. The first regards what genomic regions are associated with the protein of interest, and the second concerns the set of genomic regions exhibiting change across biological and experimental conditions. As with other array analysis methods, normalization procedures on raw intensity values are necessary prior to biological inference. Furthermore, due to the repetitive nature of the genome, care must be taken to correct for over-represented motifs that might otherwise yield false peaks. Determining which regions are associated with the bound protein is essentially a peak-finding task, and there are several available software options for analyzing the data, several of which are listed below in the discussion of Methyl-chip data. A key factor in the analysis of ChIP-chip data concerns the shape of the peak generated. Some proteins, such as histones, are observed to generate broader peaks, while transcription factors can generate sharper peaks. Particular software may be more robust for one or the other peak type or otherwise require parameter adjustment from defaults to better accommodate broader or sharper peaks. The second common analysis step involves testing for differential peak sizes across biological conditions. Strategies similar to those used for gene expression can be used. Numerous software packages are available that handle these ChIP-chip data. Considerations for the design and analysis of ChIP-chip data are reviewed in (Buck and Lieb 2004).

Methyl-chip One approach to assay methylation levels by means of chips is to use an immunoprecipitation reaction with an antibody targeting 5-methyl-cytosine (MeDip) (Gebhard et al. 2006). Alternatively, the methyl-binding MBD domain of the MBD protein family has also been used to a similar effect. In both scenarios, bound DNA fragments that are pulled out of solution are then extracted and hybridized to an oligonucleotide array (Ballestar et al. 2003).

Analysis, in large part, consists of peak finding strategies and methodology is similar to ChIP-chip described above. Software for both types of analyses is available in the open-source R Bioconductor package and is likewise accommodated by many commercial package providers, including Biodiscovery, Partek, and GoldenHelix.

Site-specific Methylation Arrays Another methylation assay based on array technology is the site-specific array, which uses a combination of bisulfite conversion and site-specific DNA probes to measure the relative level of methylation at specific loci. Currently, Illumina is the most common provider of these arrays, originally producing a 27k feature version that primarily covered CpG islands and largely targeted cancer research. Expanded versions containing 450 and 850k sites are now available from Illumina, which has expanded the feature set substantially to include additional methylation sites outside the of CpG islands, including those within exons. While Illumina does include a methylation module in its Genome Studio, it is common to use third-party alternatives for analysis. Third-party analysis options include *lumi* (Du et al. 2008), *minfi* (PMID: 24478339), RnBeads (PMID: 30871603), and ChAMP (PMID: 24478339) provide integrated R package solutions. Performance of these and other approaches are reviewed in (PMID: 25233806). The most commonly targeted metric for analysis is the beta value, which is the ratio between the measured methylated probe intensities vs. the total (methylated and nonmethylated probe intensities). This produces a value in the range of 0–1, which can readily be interpreted as a methylation level for the locus. However, Du et al. noted that the beta value exhibits extreme heteroscedasticity at highest and lowest methylation ranges and has thus recommended the M value, the log2 ratio of methylated probe intensity vs. unmethylated probe intensity, as an alternative metric for analysis (Du et al. 2010). They consider the behavior of the M value to be more statistically suitable for the range of downstream analyses, such as analysis of differential methylation. Use of the M-value has been incorporated into the MethylumiM class of the *Methylumi* package. After performing appropriate normalization and transformation of data from methylation arrays, examination of differential expression is similar to that for expression arrays, and the same techniques can be performed.

High-Throughput Sequencing Methods for Genomics

Introduction

High-throughput sequencing represented a fundamental shift in the way nucleotide sequence data were generated for genomic research. As opposed to the Sanger sequencing strategy, which focused on data from a single locus being generated for each independent sequencing reaction conducted, the second generation of sequencers represent a massive parallelization of the base detection process, with data being acquired from millions of separate DNA fragments independently. To accomplish this "high-throughput sequencing," clonal copies of DNA molecules from a sequence fragment library are arranged into clusters across a 2D surface using multiple strategies, including bead-based and bridge amplification of tethered DNA fragments. These clonal "colonies" allow sufficient signal to be generated for detection. In the case of the widely used Illumina sequencing instruments, this is optical signal. For Ion Torrent and Ion Proton systems, the signal is generated through the release of hydrogen ions, which could be detected by pH-sensitive silicon wafers. The density at which these signal generating molecular clusters can be produced and assayed has grown dramatically over the past decade and the cost per base of sequencing declined precipitously. One of the primary changes from the type of data generated previously, however, is that data from second-generation sequencing systems come in the form of short sequence read fragments, typically ranging from 36 to 500 bp. The algorithms that previously worked for Sanger sequencing are no longer always adequate, in some cases due to the size of the reads, but in other cases, it was a consequence of the overall sizes of the data sets themselves. Many of the analysis strategies described below center around how to efficiently and effectively derive meaning from the millions of DNA fragments produced by the high-throughput sequencing process. In the case of human genetic disease studies and for common model organisms, the task is simplified greatly by the presence of a reference genome build that can serve as a scaffold for alignment of these short reads. Although not without its challenges due to both the computational requirements and the high levels of repetition and homology in mammalian genomes, the fragments generated from sequencing can be effectively mapped to reference genomes, generating various degrees of depth of coverage, depending on the amount of sequence produced for the targeted library. As alluded to in the first section of this chapter, the advent of high-throughput sequencing also brought with it an increase in demand for computational power. It necessitated new algorithmic approaches for data analysis, some of which are highlighted below. One key feature of several of the next generation platforms is the availability of "paired end" sequence, where sequence reads from both sides of a DNA fragment could be obtained, typically with a "dark" region in the center that remains unsequenced. Informatically, these two ends can be associated together, and they add a new dimension to analysis that is capitalized upon for mapping specificity, structural variation inference as well as other more customized methods. In the event that one end of the pair lands in a repetitive region such that it could not otherwise be uniquely placed on the reference genome, the paired end can serve as an anchor, allowing the other read (and its associated coverage) to be salvaged where it may previously have been discarded.

High-Throughput Sequencing for Genotype Inference

A primary use for high-throughput sequencing is to infer genotypes for downstream statistical analysis, such as the linkage and association methods described throughout this volume. The genomic target of high-throughput sequencing can be as all-encompassing as whole-genome sequencing or it could be the selective sequencing of a subset of genomic regions. Typically, this targeting occurs by hybridization-based selection with biotinylated oligonucleotides, although other PCR-based strategies also exist (e.g. droplet-based PCR). In either case, the processes described below are generally applicable, although it should be noted that some quality improvement procedures (such as base quality recalibration) require sufficient amount of sequence complexity to work effectively. While the workflow described here will frequently allude to the GATK software suite (DePristo et al. 2011), a popular and freely available analysis package developed at the Broad Institute, it should be noted that several alternative solutions exist, and new algorithms and software suites continue to be produced by both academic and commercial groups. Notably, for those researchers less comfortable with command line-based Linux/Unix environments, the commercially available CLC Bio software suite is a popular choice, integrating analysis and visualization options. Partek and GoldenHelix also provide robust software options that implement a variety of statistical methods covering most phases of genomic data analysis. The appropriate analysis software for a given research group will depend upon several factors, including the scale of the operation and desired level of automation, as well as which specific sequencing platform is being used to generate the data. Commercial software is often the more attractive solution when analysis must be conducted on a stand-alone high-end computer workstation. Alternatively, if a high-performance computing environment (local cluster or cloud) and appropriate computational expertise are available, software frameworks such as GATK can be deployed to more readily accommodate higher volume of analysis in a timely fashion.

Base Calling Base calling is the process of converting signal intensity data from the sequencing instrument to nucleotide bases and associated quality scores. This foundational step in the sequence analysis process is typically handled by the vendor provided software. In the case of Illumina, the calling algorithm "Bustard" is now handled on-instrument by their Real Time Analysis (RTA) software. While several efforts were made early on to improve upon vendor provided calling algorithms (e.g. Altacyclic, Erlich et al. 2008), these have largely failed to gain traction among the wider research community. In part, this is due to the fact that third-party software options for early-stage analysis (such as base calling or conversion of optical images to intensities) can often run counter to vendor's efforts to simplify and streamline the analysis process for end users. Obtaining access to the appropriate intensity input files and running third party calling options often entails side-stepping data processing automation features that are incorporated into vendor software. It is, generally speaking, only at the base quality recalibration or alignment stage (see below) that many users find that intervening on the otherwise convenient vendor workflows can have a significant enough impact on results to justify the added effort. All this notwithstanding, efforts continue to be made to improve upon the efficiency and/or accuracy of vendor-provided base calling methods. For example, the ParticleCall method adopts a Hidden Markov model approach and has been shown to improve accuracy under certain conditions (Shen and Vikalo 2012).

Base Quality Recalibration The procedure of base calling produces associated phred-scaled quality values that can be factored in to downstream alignment and genotype calling. One common procedure following base calling is to recalibrate the initial base qualities reported by the base calling algorithm (e.g. Bustard, in the case of Illumina). Initial quality phred-scaled quality scores are often found to be inflated upon empirical evaluation, and examining the covariance of key sequence attributes (e.g. machine cycle, dinucleotide, etc.) can allow for scores to be adjusted to more accurately reflect empirical results. Note that the empirical accuracy is determined by examining the frequency at which bases from a certain feature class mismatch the reference after excluding locations that are known to vary in the population. A function for performing this recalibration is made available through the GATK package, helpfully named BaseRecalibrator. Base quality calibration, however, is one of the areas where targeted sequencing can sometimes require a different treatment than whole genome or exome scale sequencing. For example, insufficient sequence complexity among targeted regions can have a negative impact upon results. Users must also carefully consider at which level data should be combined for calibration purposes (lane level, sample level, etc.) in order to minimize the potential for batch effects during analysis. These and other considerations are addressed in the GATK documentation. As was the case with base calling, several authors have now questioned whether the accuracy improvements obtained through base quality recalibration justify the added CPU cycles required. Nevertheless, the recalibration step remains within GATK's current best practice guidelines; users are encouraged to weigh the impact of these improvement gains in the context of their specific study goals.

Alignment To accommodate the vast amount of short read data, new alignment algorithms were necessitated. Early alignment algorithms targeting high-throughput sequencing data, including MAQ and ELAND, used a spaced seed-based search strategy (Li et al. 2008). The length of the short read fragment was first divided into multiple shorter seed sequences. These seeds were then mapped to the genome for perfect matches, and matches were examined for possible larger alignments using traditional local alignment strategies, such as the Smith-Waterman algorithm. Two subsequent aligners, BWA and Bowtie, employed the Burroughs-Wheeler transform

(Langmead et al. 2009; Li and Durbin 2009, 2010), originally developed in the context of data compression, to represent genetic data in a compact fashion that was amenable to rapid, more computationally efficient searching. BWA has become the most commonly used alignment strategy for DNA-based tasks, but a host of other alignment algorithms with different strengths and weaknesses are available. For RNA-seq data in particular, however, the STAR aligner is now broadly adopted on account of its combination of speed and robustness for downstream applications (PMID: 23104886). For specialized applications, such as the detection of fusion events from RNA-seq data, the supplementation of STAR with alternate alignment strategies, has been showed to yield improvements in sensitivity. Fusioncatcher, for example, internally employs an ensemble alignment strategy to increase robustness (see bioRxiv, Nov. 2014, DOI:10.1101/011650). For cases where RNA-seq is being assessed for quantitation purposes, a variety of "pseudo-aligner" options are now available (e.g., Kallisto) that use kmer-based approaches to quantify RNA-seq data in a fraction of time a full aligner typically require (PMID: 27504780) For structural variant detection within DNA, the Mr. and Mrs. Fast aligners rely on the earlier spaced seed based strategy, but they have features that provide advantages when analyzing structural variation (Alkan et al. 2009; Hach et al. 2010). Exhaustive head to head comparisons of modern sequence aligners are challenging due to the numerous parameters options available and the variety of potential data sets. With these caveats in mind, there are a number of review articles available that compare alignment algorithms (e.g. Li and Homer 2010; Ruffalo et al. 2011; Pereira et al. 2020). There are typically tradeoffs necessitated between speed and accuracy and sensitivity vs. specificity. Efficiency trade-offs may necessitate additional data cleaning procedures that are less necessary (although still recommended!) with slower, but more stringent, algorithms. When selecting an aligner, one should always factor in its strengths/weaknesses for the planned downstream analyses. Although it computationally expensive, it sometimes necessary to align the same data set with multiple algorithms, each targeting some specific aspect of downstream data analysis.

Variant Detection and Local Realignment Around Insertion/Deletions The process of variant detection is that of determining nucleotide locations in the alignment where there are alleles that are different from the reference genome that has been targeted. Regions where such variants are detected will then be subjected to genotype calling, as described below. However, many of these variant locations will be red herrings, often the result of faulty alignments or sequencing error. An important step in improving the quality and efficiency of the analysis process is to weed out those potentially erroneous locations. One of the primary causes of incorrect alignments is the presence of insertion/deletion polymorphisms with respect to the reference genome. Local realignment around suspected insertion/deletion sites can greatly reduce the number of false-positives attributable to alignment issues. The GATK suite includes a set of modules for this purpose. The first, RealignerTargetCreator, takes in a file detailing known polymorphic indel locations and attempts to find all suspicious alignment regions indicative of indel-related alignment error. The second module, IndelRealigner, conducts the local realignment around the targeted regions and produces a new, cleaned BAM file for subsequent analysis. More recent versions of GATK use haplotype information to make these separate operations unnecessary, but they remain available in the event earlier calling algorithms prove more suitable for the task at hand. Many clinical diagnostic labs, for instance, will maintain use a given algorithm version for extended periods of time, as the process of validating workflow changes can be costly in terms of both time and sample materials.

Genotype Calling Recall that the nature of a high-throughput sequencing data set, even for a single individual at a single nucleotide locus, will generally consist of multiple short read fragments mapped to the corresponding reference location. As mammalian genomes, except for the mitochondrial and sex chromosomes, are predominantly diploid, the goal of genotype calling is to infer the correct allelic state for both diploid chromosomes from the sampling of alleles obtained after alignment of the fragments to the locus. Due to the nature of high-throughput sequencing, the two chromosomal states at a given locus are effectively randomly sampled, and sampling effects can have large impacts on genotype inference. For example, consider the case where we happen to know from prior experimental data that the allelic state of the two chromosomes in an individual at a given autosomal position are A and T. If there are five sequence fragment reads randomly selected from the chromosomes at this position, it is unlikely but still quite possible to sample five A-alleles or five-T-alleles and thereby falsely infer a homozygous status. While this scenario has a relatively low probability at any given locus, consider that this same procedure is potentially being carried out at millions of polymorphic loci across the genome. The number of falsely inferred homozygous states will be high if our read depth average is five, which is one of the reasons a much higher average read depth is recommended, typically in the range of 20–30X for any given locus. In order to achieve this level of coverage depth across the largest possible number of sequenced sites, given stochastic variation in depth and other factors impacting uniformity of coverage depth, a considerably higher level of depth on average should be targeted. In the case of exome sequencing, a targeted average depth of 80–120X is frequently recommended, although financial considerations often necessitate lower levels. For WGS, depending on the downstream uses for the data, target coverage ranging from 30 to 60X is now common, with expectations for targeted sequencing coverage being many-fold higher. In addition to the false-negative scenario, which can impact sensitivity for variant detection and negatively impact genotype quality, the false-positive allele calls can present more of a problem, as they can result in costly amounts of confirmation testing and/or drown out true biological signals. Consider the scenario where we know the state of two chromosomes at a particular position to be homozygous for the A allele. It is nevertheless quite possible to observe 15 A alleles and 5 T alleles. The nonreference alleles in this case are attributable to sequencing error. Although error rates of high- throughput sequencing are quite low, due to the volume of the reads being sequenced the number of falsely inferred variants can be substantial if no consideration of error is incorporated within genotype calling. Errors can come from several sources; common error sources include alignment error, misincorporation of bases by polymerases during sequencing library preparation, and phasing effects that impact base call quality at the ends of reads. Some of the factors that correlate with error, such as sequencing run cycle, are incorporated in the base quality recalibration procedure previously described. This allows the quality scores for those sequenced bases most at "risk" for error to be down modulated accordingly. Early calling strategies dealt with error by setting minimum thresholds for which the minor allele component must be present for a locus to be deemed heterozygous. One pitfall of this approach is that it did not require factoring in the total number of reads at a site when calculating a calling threshold, and the relative reliability of calls was highly subject to sampling effects. Furthermore, it did not require accounting for base quality scores in determining allele calls. More recent genotyping strategies attempt to factor in both the sampling effect and underlying error model (Reviewed in Nielsen et al. 2011). These callers incorporate sample population data and haplotype information to improve accuracy. In general, these later genotype calling approaches attempt to determine the most likely underlying chromosome state to have generated the observed read distribution, and to do this they use both sample data and other information sources to estimate error rates and allele frequencies, which influence the likelihood of one genotype vs. another. A benefit of these strategies, aside from increases in accuracy, is that they yield both a genotype inference (where there is adequate coverage data to do so) and an associated confidence score for the genotype call, which can later be factored in to genotype quality assessment for prioritization. For some low cutoff threshold of read depth, it may be determined that there is not sufficient confidence for a genotype call to be made. Depending on the level of coverage achieved in the experiment, many researchers will limit genotype inference to loci with eight or more reads, although when the motivation for the experiment is primarily variant detection, as opposed to genotyping, thresholds are routinely lowered below this (e.g. 3X coverage). For WGS, this approach can have drawbacks, however, due to the fact that regions with the poorest coverage can also be the regions with the highest error rates, due extreme GC content, homology, or other problems. In addition to the GATK, a popular calling algorithm is SAMTOOLS mpileup (Li et al. 2009which is conveniently implemented within the easily accessible GALAXY platform, described elsewhere in this chapter (https://usegalaxy.org).

Indel Calling Small insertion–deletion variants, also known as indels or microindels, are a widespread source of variation in the human genome. They can be either directly associated with pathology or used for association and/or linkage analyses. While somewhat more straightforward than inferring larger structural variation, the calling of short indels currently remains less robust than SNV calls. One of the most widely used algorithmic strategies is that of Dindel (Albers et al. 2011). Dindel employs "probalistic realignment" to predict the presence of an indel variant, whereby alternate haplotypes to and experimental data are proposed, and the experimental data aligned against these alternate reference sequences. The posterior probability of the alignment, taking into account factors such as base quality and read depth, is then evaluated to determine if an indel will be called at the location. Indel calling is of particular importance for 454 and related technologies, such as Ion Torrent and Ion Proton, where runs of identical nucleotides can be difficult to correctly enumerate, leading to insertion–deletion errors. While software packages are available that can compensate for these events to some extent, when data are derived from these instruments, the researcher needs to ensure that the software being used is able to mitigate these false-positive indels to some extent. Another important aspect of indel calls is that indels can impact alignment and add error to variant calling. As previously discussed, it is generally advisable to conduct a more rigorous local realignment around the vicinity of known or suspected indels prior to genotype calling in order to rule out false positives. The except would be cases where equivalent indel correction strategies are directly incorporated within the genotype caller employed (e.g. GATK Haplocaller).

Quality Assessment of Genotypes Not all genotypes are created equal. As indicated earlier, modern genotype calling algorithms not only assign a genotype but they also provide a measure of the confidence in the call that was made. This confidence score can then be used to estimate the probability that a given genotype (e.g. the presence of a rare damaging allele) will be empirically confirmed through a separate molecular test. The ability to triage variants in this manner is exceedingly important for high-throughput sequencing studies, as the volume of variants potentially needing follow-up after an experiment can be quite large. This confidence score alone, however, is not the only source of information for estimating the probability that a given genotype call will be confirmed molecularly. Additional sources of information include the mapping qualities, position of reference and nonreference reads within the sequence reads, and the number of haplotypes required to explain the level of variation a region, which can be an indicator of alignment problems due to homology. Various machine learning approaches, such as support vector machines, can be used to delineate high-quality from low-quality calls. In GATK, this process is variant quality score recalibration (VQSR), and it is available for both genotype and indel calls.

For genotypes, a Gaussian mixture model approach is used to learn characteristics of good and bad calls from the experimental data set. Techniques are rapidly evolving, however, and the underlying learning algorithms used in GATK and other software can be expected to shift over time. The basic need to evaluate genotype quality in a separate stage will likely remain, however, unless these procedures are more seamlessly integrated with upstream processing phases.

Sequence Annotation The topic of annotation is a large and diverse one and is addressed in multiple contexts throughout this chapter. In the immediate context of high-throughput sequencing to obtain genotypes, the term "sequence annotation" typically refers to obtaining several key biological attributes about the variant and its location that can aid with prioritizing its particular functional significance. Perhaps the most basic question one might begin with is whether the variant is located within a gene. There are then a series of follow on questions that immediately arise. Does the variant lie within a protein coding sequence within the gene, and does the nucleotide change represent a protein coding change in the gene? From there, one might further inquire if the amino acid change is of a conservative or nonconservative in nature with respect to its charge. To dig still deeper, we might want further want to know if the particular position has been conserved over evolutionary history or has been inclined to change frequently. If the variant is outside of the coding sequence, there are still a number of attributes that can be consulted to determine its potential to be functional, including whether it is at a splice junction or part of a wide diversity of regulatory motifs. The above are but a brief list of the type of nested questions that might be applied to a particular variant location, some of which are addressed in the third segment of this chapter in the context of variation and annotation. Fortunately for researchers, there are a number of excellent software packages that facilitate the basic annotation process for next generation sequencing data, many of which can be further extended to include custom annotations provided by the researcher or other outside groups. Among the more popular options include the SeattleSeq web server (http://snp.gs.washington.edu), which accumulates a variety of information from several databases and algorithms. Annotations include evolutionary conservation and predicted effect of the variant on protein function. Other popular options include VEP (http://useast.ensembl.org/info/docs/tools/vep/index.html) and ANNOVAR (http://annovar.openbioinformatics.org/en/latest) (Wang et al. 2010), both of which can be extended with custom annotation criteria. An additional option is Biofilter, which incorporates a diverse set of sources, including GWAS and eQTL study results, to aggregate available genetic knowledge about a genetic position (Pendergrass et al. 2013). Further details regarding annotation of sequencing data are provided in the third segment of this chapter under annotation.

Structural Variation Inference from High-Throughput Sequencing Data In addition to changes at the single nucleotide level, our genomes are also subject to larger structural alterations. These include deletions, insertions, inversions, transpositions, and a variety of more complex rearrangements whose causal mechanisms are only just beginning to be elucidated. While cytogenetic analysis via light microscopy has long allowed us to appreciate that large chromosomal aberrations were associated with genomic disorders, modern genomic technologies have revealed that these structural variants are far more pervasive in the genome than initially appreciated. Comparative genomic hybridization arrays and similar technologies opened a window into a range of variation that was below the resolution of light microscopy but too large to be readily discerned from capillary sequencing. These structural variants have a wide size distribution and fitness spectrum; they can be completely neutral, pathological, and even occasionally confer positive benefits. As a consequence, detection of these structural variants has become an important component of modern genetic studies. While detection of these variants via array technologies remains an important strategy, there is an increasing focus on the use of high-throughput sequencing for structural variation discovery. One key advantage of the sequencing-based approach is that it has the potential, depending on the specific methodology being used, to directly reveal the junction points of structural variants. Often the precise location of the variant junction can be a key determination of the potential pathogenicity of a structural mutation or polymorphism. The technical complexity associated with structural variant detection has long made it a challenge and sensitivity and specificity for structural events lags behind that associated with single nucleotide variation. Regions with copy number variation often occur in the vicinity of homologous DNA regions, which cause problems with cross -hybridization (in the array context) and nonspecific fragment mapping (with sequencing approaches). Nevertheless, considerable progress has been made in structural variation detection. Due the diversity in high-throughput sequencing protocols, there are several potential ways in which sequencing data can be used. These strategies are briefly discussed below, along with some of the common software platforms used for structural variant detection.

Conceptually, the use of sequence fragment counts for estimating locus copy number might at first appear relatively straightforward. However, several complicating factors make sole reliance on copy number data for structural variation detection a poor strategy in practice. (Not the least of which is that not all types of structural variations involve copy number changes [e.g. translocations, inversions].) First, there is stochastic variation in coverage depth across the genome, resulting from the sampling process. Second, there is bias in the level of coverage across regions of the genome, often resulting from extremes in CG content among other factors. Another complication is that the homologous sequence that is present both within and in the vicinity of many structural variants makes the placement of read fragments ambiguous. All these factors combine to undermine the reliability of raw fragment counts as the sole proxy for copy number. A second strategy for assessing copy number has been to use paired-end and mate paired reads to extract structural information. As mentioned previously, paired-end reads are a set of short reads that are known (due to the underlying technology) to come from opposite ends of a single DNA fragment. Mate pair reads are formed from a somewhat more complex circularization strategy and allow the physical fragment to be of a larger size. Using the known size of the DNA fragments being sequenced, we know approximately how far apart and in which orientation any two paired (or mated) sequence fragments should map to the reference genome. Appreciable deviations in mapping distance can be used to infer that a structural aberration relative to the reference has occurred. Note that it may well be the case that the reference genome itself is altered with respect to the experimental sample. To accurately determine which sequence is the ancestral state and which is the derived state, a third (ancestral) sample is required for comparison. For humans, this might be a chimpanzee or gorilla sample. The relative orientation of aligned fragments can be used to infer inversion events. Likewise, if one end of the pair aligns on one chromosome, and the other maps to a separate chromosome, this could indicate a translocation event. The final category of information involves reads spanning the actual breakpoint junction of the structurally variant region, appropriately called "split-reads" because the alignments of the sequencing reads are "split" across two noncontiguous genomic regions due to the event which produced the variation. In order to detect these split reads, information about partial alignments must be maintained by the aligner and subsequently analyzed, which is one of the reasons that Mr./Mrs. Fast are well suited for this purpose (Alkan et al. 2009; Hach et al. 2010).

In addition to inferring the existence of a structural event, another aspect of analysis is to determine the integer copy number of the variant region. Magnitude of copy number can have direct impacts on the pathology (or lack thereof) of an individual's chromosomal state. In the context of inferring copy number, read count data are most informative. While the difference between zero and two copies of a locus is usually easy to infer, inferring copy number for higher multiplicities can be more challenging, as the relative difference in expected counts diminishes, and a variety of algorithms have been developed toward this end (e.g. Boeva et al. 2011). As with copy number analysis from array data, many authors recommend using a combination of multiple algorithms and then only retaining variant calls where there is complete or majority consensus. Such ensemble approaches have gained increasing popularity in cases, such as structural variants, where no consensus has formed regarding a single algorithm or tool.

Software options for inferring structural variation. There are several software options available for inferring structural variation and associated copy number changes. Note that most of these packages assume that whole genome or otherwise contiguous genomic sequencing has been conducted, as there are several additional factors that must be taken into account when assessing sequencing data derived from targeted enrichment strategies (e.g., exome sequencing). While the focus of the current volume is on complex genetic disease, it is important to note that a large fraction of available structural variation detection software targets somatic variation, typically in the context of cancer. These tools often require paired samples of cancerous and normal tissue, making their applicability to the germline context limited. For general purpose, structural variant calling, Illumina's open source manta (https://github.com/Illumina/manta) and delly (https://github.com/dellytools/delly) are popular options with good performance profiles (PMID: 22962449).

A separate subset of structural variant software is aimed at sequencing from target-enriched libraries. Data from enrichment have additional coverage biases introduced and reads a discontinuously distributed among "islands" of targeted regions across the genome. Software options for sequencing analysis of target-enriched data include Conifer and SplitRead, among others (Fromer et al. 2012; Karakoc et al. 2012; Krumm et al. 2012). More recently, strategies involving a hybrid of alignment and local de novo assembly have also been introduced (Nijkamp et al. 2012).

A Brief Note on Whole-Genome Assembly Currently, human genetic studies rely heavily upon using alignment algorithms to map read fragments back to available reference genomes. However, when a genome is not available for a given model organism, or it is determined that added information (e.g. structural variant inference) may be obtained by assembling a nonreference individual of a species, de novo assembly may be warranted. Genome assembly is computationally demanding, highly dependent upon the technology or combination of technologies used for generating the data, and frequently necessitates additional experiments to fill in information gaps, depending on the level of resolution required. To date, assembly and annotation of mammalian sized genomes has been undertaken by large research collaborations. Proper treatment of the many issues surrounding the development of an accurate genome assembly is well beyond the scope of this chapter, but a few of the major strategies for modern genome assembly are briefly highlighted below. The predominant assembly strategy prior to the high-throughput sequencing revolution was the overlap-layout-consensus (OLC) approach, wherein a graph of sequence contig relationships was constructed such that each sequenced contig was represented by a node in the graph and overlapping contig sequence was represented by edges. Finding the best assembly was a matter of finding the Hamiltonian path through the graph, with each node being visited once. While effective for contigs, this strategy was ultimately not feasible for very short read sequence assembly. In 2001, Pevzner, Tang, and Waterman introduced a new method of sequence assembly based on de Bruijn graphs (Pevzner et al. 2001). This powerful method, implemented in their EULER software, began with the rather unintuitive approach of breaking the short reads down into even shorter K-mers. Later modifications of the de Bruijn graph approach also better handled the repetitive regions that are characteristic of mammalian and plant genomes far more efficiently and accurate that the OLC strategy. In 2008, Zerbino further expanded upon the de Bruijn graph approach with more efficient algorithms and a strategy for handling repetitive DNA (Zerbino and Birney 2008). Their strategy was implemented in the VELVET software package, which became one of the first assemblers to effectively handle the very short read data generated from the newly developed high-throughput sequencing platforms. Simpson and Durbin introduced a more memory efficient approach, String Graph Assembly (SGA), that draws upon compression algorithms and String Graph assembly (Simpson and Durbin 2012). In addition to the above strategies, there are several short read de novo assemblers currently in use, each with particular strengths and weaknesses depending on the context of the genome being assembled. These include SoapDenovo and Newbler (Reviewed in Lin et al. 2011). See Miller et al. for a review of de novo assembler strategies and their relative performance (Miller et al. 2010; Sohn and Nam 2018).

Expression Analysis from High-Throughput Sequencing Data – RNA-Seq

While many aspects of the underlying alignment algorithms used for RNA analysis are similar to those used in expression array analysis, there notable differences as well, including increased diversity in the experimental protocols generating RNA sequencing data. These details have a direct impact on the nature of what can effectively be analyzed in the data and also the type and magnitude of biases that might be present. For example, depending on the experimental details, nonpolyadenylated noncoding RNAs may or may not have been captured for sequencing. As a result, the process of RNA expression analysis has the potential to be more involved and include more decision points than the basic analysis procedure for DNA sequencing. Partly as a consequence, there is presently no clear consensus on what represents the state of the art in RNA-seq analysis methods, and analysis techniques continue to evolve. For the purposes of this chapter, the collection of high-throughput sequencing protocols that yield RNA sequencing results are referred to as RNA-seq. Both modifications of the standard RNA-seq procedure, as well as specialized experimental protocols directly targeting subcategories of RNA for generation, also exist, and their analysis may require a more tailored approach than the general strategy outlined in this chapter. It is incumbent upon the person analyzing the data to become sufficiently familiar with the experimental techniques involved so that they have an understanding of the nature of the library that has been sequenced. With this understanding, it is more likely that the appropriate analysis protocol will be used and that the investigators awaiting the results (assuming they are not the same as those doing the analysis) have appropriate expectations concerning what can be discerned through analysis. For example, it should be clear that mature miRNA cannot generally be obtained without a specialized library preparation, and a lack of miRNAs in a set of differentially expressed loci from a standard RNA-seq library is not indicative that they are not in fact differentially expressed. Further, it should be noted that except in very specialized cases involving single molecule sequencing techniques (generally third-generation sequencers), RNA-seq does not actually involve the direct sequencing of RNA. Rather, RNA is first converted into cDNA.

An ever-expanding list of analysis methodologies have been proposed for RNA-seq, as analysis of high-throughput sequence data remains a fertile ground for method development. There are a number of reviews available that cover the diversity of methodologies (see Oshlack et al. 2010; Garber et al. 2011; Stegle et al. 2015). What follows is a general discussion of the basic concepts and strategies used in RNA-seq analysis. Tools and analysis frameworks are mentioned in the following discussion, where relevant, but an exhaustive enumeration of all current strategies is not attempted.

Ongoing changes in both technology and analysis methods mean that even the most widely adopted software tools may be rendered obsolete over relatively short time frames.

Base-Calling and Foundational Preprocessing Steps Prior to alignment of sequencing reads to a reference genome, the informatics processing for RNA expression data, including generation of intensity signals and base calling, is handled similarly to that for DNA sequencing, and the reader should reference the high-throughput DNA sequencing above for a general description.

Alignment As with DNA sequencing data, most analysis procedures will begin with alignment to an appropriate reference genome. While techniques exist for de novo assembly of transcriptomes in situations where either it is the case that no reference genome exists, or only a distantly related reference genome is available, most analyses will ultimately proceed from the point of alignment to a reference. Alignment for RlNA-seq data is complicated by the fact that the vast majority of expressed gene transcripts in mammalian genomes, such as human, are interrupted by introns. Thus expressed sequence is not contiguous when mapped against the genome; some sequence reads will span intron/exon junction points and thus be discontinuous in genome sequence space. The alignment process for RNA sequence data must take this phenomenon into account in order to accurately represent expressed data and extract the full amount of information from RNA sequencing reads. Two basic alignment strategies have been used, unspliced and spliced alignment (Reviewed in Garber et al. 2011). Unspliced alignment is the most straightforward approach; examples include both the seed and extend and the Burroughs Wheeler transform methods discussed above in the DNA alignment section. Unspliced aligners effectively ignore the splicing phenomenon and map expressed sequence fragments to the genome in the same fashion as they would genomic DNA fragments. Specialized reference target files that include known exons and splice junctions are used in this process, maximizing the amount of information that could be obtained compared with simply aligning to the entire genome, where discontiguously mapped transcripts would result in the failure to map many more reads. Nevertheless, unspliced aligners are limited to the quantitation of known exon junctions unless further analysis is undertaken.

Spliced aligners are divided into two main approaches, termed "exon first" and "seed and extend." Exon first is the strategy employed in tools such as the Tophat suite, and the mapping process occurs in two fundamental stages. Reads are first mapped to the *entire* genome using an unspliced aligner. In the case of Tophat, the Bowtie aligner is used. After the initial alignment process is complete, read fragments are next divided into smaller fragments, which are mapped again against the entire genome. Subreads that map to multiple discontinuous locations are then examined. This approach is very efficient when a relatively small subset of the reads does not unambiguously align to exons or known junctions, and hence investigators do not need to employ the computationally expensive second phase of analysis. The second spliced alignment strategy is called the "seed-extend" method, an example of which is the software GSNAP (Wu and Nacu 2010). These methods take a more rigorous approach of breaking sequence reads into smaller k-mer sized fragments, and conducting mapping and splice junction searches simultaneously. The seed and extend approach has increased sensitivity in instances where pseudogenes exist, but they have the disadvantage of being significantly more costly in computational time.

Reconstructing the Units of Transcription Following alignment, the number of sequence fragments associated with a given genome feature must be quantitated. A set of known features, often provided to analysis software in the form of a .GTF format file, can be used to provide an initial list of transcription units. These predefined annotations also serve the purpose of allowing the researcher to relate transcriptional units in the current experiment, to those described and/or used in previous work. However, in order to detect novel expressed regions, units of transcription must also be inferred from the experimental data itself. This process of transcription unit reconstruction has been conducted using two fundamental strategies, genome-guided and genome-independent reconstruction. As their names suggest, they differ by whether or not a genomic reference is used for reconstruction of the units of transcription. Cufflinks employs a genome guided assembly method, making use of a graph topology derived from sequence read fragments that have been mapped to a known genome (Trapnell et al. 2010, 2012). Spliced reads bridging transcripts are represented by edges. Each path through the graph represents a possible transcription unit. Cufflinks attempts to find the minimum number of isoforms that can explain the observed data, and when there is more than one minimal path set available, the number of reads covering each path is used to infer the most likely minimal set. Unlike Scripture (Guttman et al. 2010), which uses a similar graph, Cufflinks maximizes specificity over sensitivity. More recently, methods such as RSEM, which does not require a reference genome, have gained widespread adoption (PMID: 21816040).

Counting Sequence Reads and Normalization On the surface, the read counts resulting from RNA-Seq are discrete and readily straightforward to correlate with transcript abundance. However, several features of transcriptome biology undermine this simple treatment and necessitate added complexity to the analysis. Initial efforts to quantify the level of expression from gene transcripts involved directly counting the number of read fragments mapping to the individual exons of a gene and subsequently performing a scaling procedure to account for differences in the amount of read data gathered per sample and the size of the genes being measured. Two standard scaling procedures have been most commonly used for the basic scaling step, both of which attempt to simultaneously correct for differences in the number of reads across experimental samples and differences in the length of genes. While the need to correct for different amounts of sequencing for each sample is fairly transparent, what may be slightly less obvious that the larger gene transcripts could potentially show increased expression if transcript size is not adequately addressed. A very common approach used for scaling is termed Reads Per Kilobase of Transcripts per Million Rapped Reads (RPKM). A slightly different variant of RPKM, called Fragments per 1 k Megabases (FPKM), accommodates paired-end sequence reads and is used by Cufflinks for quantitation. Here, the number of DNA *molecules* mapping to the transcription unit is counted as opposed to the "reads," of which more than one may be connected to the same physical DNA molecule. There are limitations with the "raw count method" approach, namely for scenarios where expression levels change at the isoform level without correspondingly changing at the whole transcript level and vice versa (Trapnell et al. 2012). As a result, the strategy of direct counting does not always accurately reflect the underlying biological changes taking place. This issue is further exacerbated by the fact that many sequence reads cannot be uniquely assigned to one isoform vs another, since many exons will be included in multiple splice forms of a gene. This problem with multiplicity of read locations is generally termed "count uncertainty" (Reviewed in Garber et al. 2011). A set of strategies, termed "isoform deconvolution" try to better model the inherent complexity of alternatively spliced genes.

Differential Expression Analysis After the units of transcription being studied have been enumerated and their boundaries defined and after the reads mapping to them have been appropriately quantified and normalized, the next step in analysis is often to look for differentially expressed genes across biological and/or experimental conditions. While conceptually similar to the approaches used for expression array data, there are distinct differences due to the fact that data from NGS is ultimately discrete in nature, requiring a different underlying statistical model. Early approaches to analysis of RNA-seq gene expression focused on the use of a Poisson distribution and negative binomial distribution to model sampling variance and detect statistical differences between conditions (Reviewed in Garber et al. 2011). However, fragments counts from NGS data often exhibitmore variance than expected under a simple Poisson distribution, a problem that has been referred to as "count overdispersion." The problem is exacerbated by the issue of "count uncertainty," which results from the fact that specific reads cannot always be uniquely assigned to a single isoform of a gene. Both of these phenomena increase the level of error and variance in the data. Cufflinks2 makes an effort to address both of these problems during differential expression analysis by using a more accurate estimate of count error and variance; it mixes a negative binomial distribution, which captures the uncertainty of fragment counts, and a beta distribution, which captures the overdispersion of counts to model variance. The combination of the two distributions is termed a "beta negative binomial distribution," which encompasses both count uncertainty and count overdispersion. Another added benefit of Cufflinks 2, as opposed to many other differential expression detection methods, is that it can also be used to detect statistically significant differences is isoform abundance across samples. While the Tophat/Cufflinks suite is a popular analysis pipeline, a plethora of methods for the analysis of differential expression are now available, many of which are accessible through the R modules, including the Bioconductor package. In addition to these and other freely available academic software options, several popular commercial software options exist. These include Partek, Genego, and Genespring. Commercial options tend to be more graphically oriented and can be readily used successfully by researchers who are less familiar with the Linux/Unix and associated command line environment.

Classification, Clustering, and Tertiary Analyses Classification and clustering procedures, as discussed in the context of expression arrays, are equally applicable to data derived from RNA-seq experiments. The challenge in the context of sequencing is that of scale and increased dimensionality of the data. Additional dimensional reduction steps may be necessary so that existing supervised and unsupervised algorithms can be directly applied to the data.

Additional Assays and Analysis based on High-Throughput Sequencing Data While the primary uses of sequencing have been described above, modern sequencing platforms have become the basis of a wide variety of molecular techniques, which have rapidly achieved an even greater diversity than that of array-based assays. While there are common themes found in their analysis (e.g. alignment), each is subject to its own subtleties. The variety of these approaches precludes exhaustive overview, but below several of these strategies are mentioned, along with a brief mention of associated analysis methods.

ChIP-Seq and Methylation-based Sequences

ChIP-seq and MeDip-seq ChIP-seq and MeDip-seq are the sequencing-based analogs to ChIP-chip and Methyl-chip analysis described earlier. The only difference on the experimental side is that the DNA fragments that are pulled out of solution are used to form a sequencing library instead of being hybridized to an oligonucleotide array. The fundamental algorithmic goal in the initial phase of these analyses is peak finding. The next task, after peak locations have been established, is typically differential analysis of peaks. Several software options are available for both peak finding and differential expression, which is often the outcome of interest for loci that demonstrate differential binding to specific factors (histones, transcription factors) or differential methylation (Reviewed in Kim et al. 2011).

Methyl-Seq A separate class of experiments involves bisulfite treatment of DNA prior to sequencing. While this can be followed by whole-genome sequencing, it is more common due to cost considerations and the need for increased sequence depth to use restriction enzymes to select a subset of the genome to target for sequencing. Due to the loss of sequence information resulting from the conversion of unmethylated cytosines to uracil, which are read as thymines, alignment of bisulfite treated DNA is more difficult than nontreated DNA, but the problem has been greatly mitigated by the expanded length of sequence reads from the latest versions of high-throughput sequencers. Furthermore, due to the repetitive nature of the genome, care must be taken to correct for overrepresented motifs that might otherwise yield false peaks.

Bioinformatics Resources

In the section above, and elsewhere in this volume, a multitude of data generation platforms have been discussed. The volume of data currently being generated from various sources is enormous, and in order to address the challenge of turning all of this data into useful curated information, a great diversity of online web resources and other tools have been produced. These range from databases housing biochemical pathway relationships to genotype and sequence read warehouses. In addition to databases, some sites directly provide analysis services, such as the SeattleSeq web server, which offers basic annotations for variant loci. These diverse resources contain various amounts of manually and/or computationally curated data or types of services that can be drawn upon to both generate and assess biological hypotheses. Collectively, these various bioinformatics offerings form a continually shifting ecology of websites and downloadable resources. They tend to follow the patterns of business lifespans, with startups arriving and initially offer a new product or service. Sometimes there is high demand, and the sites flourish or its particular functionality gets subsumed within larger site. Others end up as ghosts, no longer being maintained, as the original demand they aimed to serve either is no longer relevant or becomes replaced by a superior service. Certain web resource presences have longer lasting presences, particularly those run by government agencies (e.g. NCBI and ENSEMBL) or large academic institutions (e.g. UCSC Genome Browser). These larger entities tend to aggregate a diverse set of data sets, annotations, and services. These sites serve an important aggregation function in what can often be a bewildering excess of available bioinformatics resources with wildly varying quality and usability. Generally, these larger sites are referred to in this chapter as bioinformatics portals. UCSC Genome Browser serves as both a browser and a bioinformatics gateway. The National Center for Biotechnology and Innovation (NCBI) is another such portal, making a variety of important databases available. These include the Entrez database, which represents the primary repository of DNA sequences from diverse organisms. Also available is dbSNP, which is a compendium of single nucleotide and small indel variation derived from a multitude of past and ongoing sequencing efforts. These represent but the tip of the iceberg in terms of what is available through the NCBI portal. It is worthwhile to make an effort to familiarize oneself with the full scope of available resources for genomic and proteomic data. The European Bioinformatics Institute provides the ENSEMBL site, which is an alternative portal for a similar array of bioinformatics tools and data resources. Of particular importance to human genetic studies are sites that provide information regarding population allele frequencies. These include the Exome Variant Server (http://evs.gs.washington.edu/EVS), the International Genome Sample Resource (formerly the Thousand Genomes Project portal) (www.internationalgenome.org), the Genome Aggregation Database (gnomAD) https://gnomad.broadinstitute.org, which has whole exome sequence of over 125000 individuals and whole-genome sequence data from over 15 000 individuals from multiple populations. It is important, however, when relying on these or any other databases as a source for population frequency data, that the researcher understand how the individuals sequenced were selected for sequencing and whether they represent a true control population with respect to your study cohort.

Annotation of Genomic Data

Annotation in genomics, in a very general sense, is the process of associating or mapping one or more biologically relevant features to a set of chromosomal coordinates. The process of annotation can occur at many different scales, and hence its discussion arises in several different contexts in bioinformatics.

Annotations can be derived from in silico analyses, wherein computational methods, sometimes based on empirically derived data sets, are used to predict the presence and boundaries of genomic features. An example of this would be RepeatMasker annotation, which is performed by the RepeatMasker program and is available as an annotation track for current genome builds from various browser websites. RepeatMasker provides the locations of repetitive elements in the genome, knowledge of which can be useful for a number of technical procedures (frequently to avoid them). Other common features that are predicted are transcription factor binding sites, nucleosome formation (and exclusion) sites, and laminar binding regions. Still other annotation data sets are inferred from direct experimental evidence, such as Chip-Seq or gene expression arrays. Among the most useful in silico analyses for human genetic disease studies are those indicating evolutionary conservation, typically derived from multiple alignment of several species. The tendency of a region to resist change at the nucleotide or protein level is generally indicative of functional significance, although there can be exceptions, such as when regions have experienced frequent positive selection events. In these latter cases, an excess of change compared to neutrality may be observed.

Rather than enumerate the increasingly diverse annotation options available, this section will instead focus on the general strategies for annotation. The context in which one may need to derive functional information from a genomic region can vary significantly, from gene mapping to pharmacogenomics. A large number of annotation data sets are available for download through the bioinformatics portals previously mentioned, and others are frequently provided as supplemental components of published papers. Typically, these annotation sets consist of tabular data that includes the start and stop coordinates for the annotation feature of interest. It is, of course, important to know and maintain documentation on the particular genome assembly from which the coordinates are derived; this information is generally provided by the source of the annotation data. In the case of most annotation, a simple intersection of a user's region of interest with the annotation coordinates will suffice. Such intersections can be performed without knowledge of programming using tools available through both Galaxy (described below) and the UCSC Genome Browser table viewer section. Some annotations, however, require additional processing of user data before the annotation can be made. Consider, for example, the case of a nucleotide variant discovered during the course of a sequencing study, which may require annotations indicating if the change is alters an amino acid, and if that amino acid change is conservative or nonconservative. While on the one hand, an exhaustive list of all possible mutations and their consequences could be precomputed and provided through a table, this may not always be the most efficient solution. In these situations, annotation software packages may compute the annotations on a site by site basis.

In the context of human genetic studies, the annotated genome is often referenced to determine the loci most likely to be subject to disruption that can result in a phenotype. For example, if several variants are discovered across a region of interest, it is generally those that intersect with functional annotations that will be most likely to have phenotypic effect. Not all annotations are created equal, however, and the researcher must determine which particular class of annotations are most likely related to the phenotype of interest. If, for example, a certain disease is typified by abnormally low expression of a gene, then among the top candidates for disrupted loci might include the set of transcription factor binding sites associated with the gene. Among the most generically useful annotations are those of evolutionary conservation, typically derived from multispecies alignment (e.g., phastCons and GERP). These measures can be used for both coding and noncoding sequence. When multiple annotation tools are employed, as is frequently the case among different research groups and institutions, conflicting annotation results should be anticipated. Such differences can stem from a number of factors, including differences in modeling assumptions for predictions, as well as differences within the source information used for those predictions. Hence when differences arise among study findings, such as, for example, the frequency of observed variants in transcription factor binding sites, or "promoter mutations," the range of assumptions involved in designating these entities within the respective studies should be carefully assessed.

Genome Browsers as Versatile Tools

A genome browser is an end-user-oriented tool for visualizing regions of the genome. Numerous such browsers have been developed over the years by commercial and academic entities. Among the most popular are the UCSC Genome Browser, ENSEMBL, and IGV. Modern genome browser websites, however, such as UCSC Genome Browser and ENSEMBL, can represent much more than a way to examine the sequence at a given set of genomic coordinates. They have evolved into a rich software framework to overlay many common annotation sets provided by a host of research groups over the years. These include methylation data, recombination rates, SNV locations from multiple sources (e.g., dbSNP and 1000 Genomes), structural variation data, and the list goes on. These annotation layers can bring together information from numerous databases, saving the researcher from having to navigate to numerous sites to integrate information. Even more powerfully, using custom annotation features, the user can display their own data sets in conjunction with public data. The availability to readily accumulate and display annotation tracks over the genome has made the modern genomic browser a powerful tool in the visualization toolkit (Figure 10.2). One is generally offered many means for navigating the structure of the genome within browsers, including directly typing in genome coordinates or gene names when they are known. When only sequence is available, alignment tools, such as BLAST and BLAT, allow the user to directly query the genome for matching or related sequence. Alternatively, when one wants to traverse the genome based on some particular annotation set, table-based strategies are made available, where various annotation intersections can be performed in order to restrict the genomic space to that which meets certain desired criteria. In addition to these functions, the browser and their associated websites also serve as storehouses for a wide variety of genomic annotation data, which can be downloaded directly from the UCSC, ENSEMBL, and NCBI websites. They provide direct access to underlying database tables, and for those without working knowledge of MySQL or similar database systems, a fair amount of filtering and selection of data subsets can nevertheless be accomplished by way of user-friendly graphical user interfaces. To facilitate sharing, browser instances including personalized settings and data can be shared with colleagues for collaboration. Suffice it to say, the utility of modern genome browsers cannot be overstated, and it is

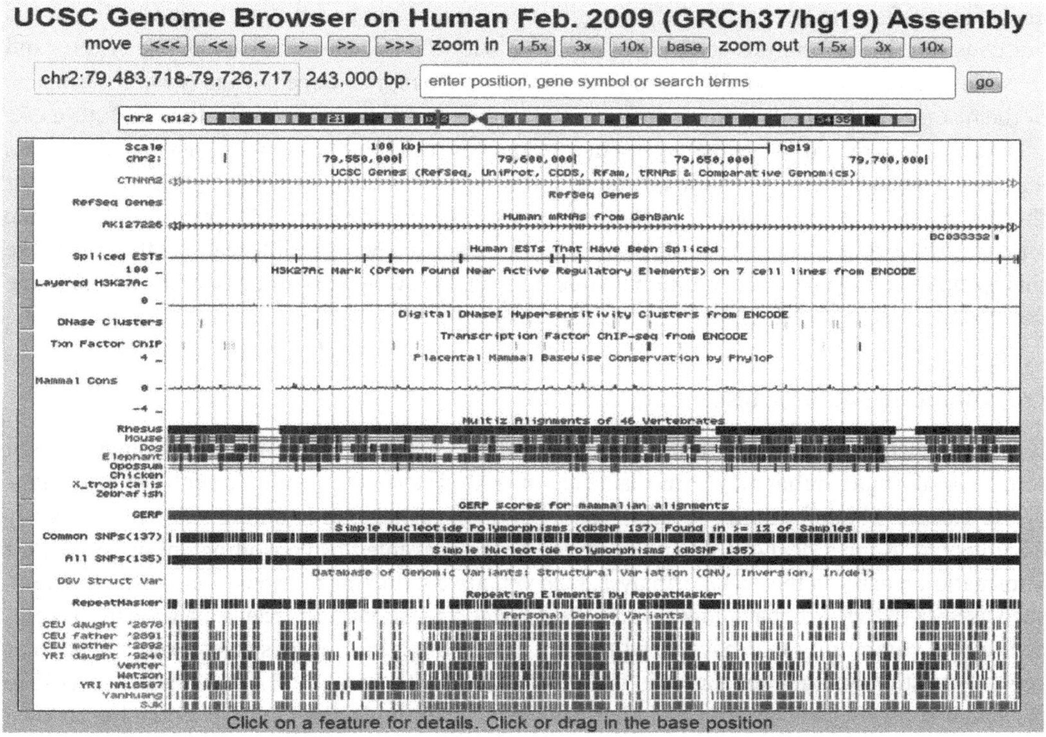

Figure 10.2 View from the UCSC genome browser of a segment of the *CTNNA2* gene with several overlayed annotations.

important for bioinformatics researchers to become familiar with the expanding set of features they make available.

Bioinformatics Frameworks and Workflows

In the realm of bioinformatics, the varieties of software packages, versions, and parameters available to the end user can be daunting. In the course of any given project, data analysis may be attempted with multiple versions of software operating under different parameter choices. Not only does this present a challenge for the researcher but it also can lead to difficulties generating quality documentation, and, as a result, difficulties for other groups (or even the same researchers!) to repeat experiments exactly as previously carried out. Replication of experiments is a cornerstone of the scientific process; one goal of the Galaxy framework was to capture the procedures performed in a given bioinformatics solution so that all parameters and software versions could be captured (Lazarus et al. 2008; Blankenberg et al. 2010; Goecks et al. 2010). Galaxy is a modular framework with a straightforward graphically oriented user interface. It can be accessed via the web, or users can download their own instances of the software and deploy them either on their own computer hardware or even the cloud, through provided virtual machine images. Several common software analysis tools and generic data handling utilities have been encapsulated within Galaxy, and users with their own installed instances can customize it to work with their own tools, using a relatively straightforward XML template. Galaxy further facilitates transactions between software tools by also performing appropriate formatting conversions. Importantly, Galaxy allows analysis workflows to be constructed, wherein analysis steps can be arranged in order to be carried out consistently. All analysis actions are captured within a database that tracks the history, and these data can be drawn upon like a laboratory notebook to replicate specific analyses as necessary, or the history of a particular series of actions can be directly converted to a workflow for future use.

Note that several workflow management systems now exist that can encompass bioinformatics analysis protocols. Along with Galaxy, these can represent the equivalent of an LIMS system for the processing and data analysis portion of a study. While existing LIMS are typically flexible enough to be extended to analysis pipelines, they sometimes have inflexible features that can make this process cumbersome. Additional freely available workflow frameworks for pipeline development include PEGASUS (https://pegasus.isi.edu).

Crowdsourcing and Troubleshooting

It cannot be emphasized enough that in the age of social media, blogs, and wikis, often the best source of information that may be available about analysis tools may come through online forums. While academic bioinformatics software abounds, these packages are notorious for being sparsely documented. A great deal of trial and error can be circumvented by consulting online forums for information regarding the specific analysis type you are conducting. Various websites have arisen, such as SEQANSWERS, that allow users to pose-specific questions and receive answers from members of the community that have faced similar issues (http://seqanswers.com). Another popular general question and answer forum is Biostars (http://biostars.org). Still other forums specific to particular software packages can be helpful as well. For example, the GATK package has its own forum at http://getsatisfaction.com. Due to the rapid pace of change in bioinformatics, particularly in areas related to high-throughput sequencing technologies, use of these "live" resources, as opposed periodically released review articles or even software manuals themselves, which are often sparse and out of date, has become essential for coming up to date with state of the art procedures. A word of advice before posing your question on one of these forums: search the frequently asked questions (FAQ) file first! Chances are very good your question has already been answered.

Data Sharing

Funding agencies and journals are placing increasing emphasis on the need for research data to be made publicly available. Hence, another aspect of bioinformatics has to encompass the proper understanding of the sites and resources available for sharing data and any programming source code generated during the study. For certain varieties of projects, the NIH demands that genotype data be included in its dbGaP server, a restricted access database to which researchers can gain access after meeting a set of public requirements. Data sharing requirements pose interesting challenges for the research community, as the resources for storing the ever-increasing amount of data being generated are limited. Microarray data, including array-CGH, can be archived at the Gene Expression Omnibus site. For short read sequencing data, NCBI provides the Short Read Archive. Whether or not these archive resources scale to meet increasing amount of data being generated remains to be seen. Some projects, such as the Alzheimer Disease Sequencing Project (ADSP), are depositing whole-genome sequence data for sharing with other investigators via a custom portal established by the NIH. Increasingly, funding agencies are also requiring research to be published in open access journals or otherwise provide public access to manuscripts. To accommodate this trend, some commercial journals are providing open access options. Several options for providing open access to research source code are available. These include SourceForge (http://sourceforge.net) and Github (http://github.com), among others.

References

Albers, C.A., Lunter, G., MacArthur, D.G. et al. (2011). Dindel: accurate indel calls from short-read data. *Genome Res* 21 (6): 961–973.

Alkan, C., Kidd, J.M., Marques-Bonet, T. et al. (2009). Personalized copy number and segmental duplication maps using next-generation sequencing. *Nat Genet* 41 (10): 1061–1067.

Ballestar, E., Paz, M.F., Valle, L. et al. (2003). Methyl-CpG binding proteins identify novel sites of epigenetic inactivation in human cancer. *EMBO J* 22 (23): 6335–6345.

Benjamini, Y.H. and Yosef, H. (1995). Controlling the false discovery rate: a practical and powerful approach to multiple testing. *J R Stat Soc Ser B* 57 (1): 289–300.

Blankenberg, D., G. Von Kuster, Nathaniel Coraor et al. (2010). "Galaxy: a web-based genome analysis tool for experimentalists." *Curr Protoc Mol Biol* Chapter 19: Unit 19 10 11–21.

Boeva, V., Zinovyev, A., Bleakley, K. et al. (2011). Control-free calling of copy number alterations in deep-sequencing data using GC-content normalization. *Bioinformatics* 27 (2): 268–269.

Bolstad, B.M., Irizarry, R.A., Åstrand, M., and Speed, T.P. (2003). A comparison of normalization methods for high density oligonucleotide array data based on variance and bias. *Bioinformatics* 19 (2): 185–193.

Browning, B.L. and Yu, Z. (2009). Simultaneous genotype calling and haplotype phasing improves genotype accuracy and reduces false-positive associations for genome-wide association studies. *Am J Hum Genet* 85 (6): 847–861.

Brumbaugh, C.D., Kim, H.J., Giovacchini, M., and Pourmand, N. (2011). NanoStriDE: normalization and differential expression analysis of NanoString nCounter data. *BMC Bioinform* 12: 479.

Buck, M.J. and Lieb, J.D. (2004). ChIP-chip: considerations for the design, analysis, and application of genome-wide chromatin immunoprecipitation experiments. *Genomics* 83 (3): 349–360.

Chierici, M., Miclaus, K., Vega, S., and Furlanello, C. (2010). An interactive effect of batch size and composition contributes to discordant results in GWAS with the CHIAMO genotyping algorithm. *Pharmacogenomics J* 10 (4): 355–363.

Cochrane, G., Alako, B., Amid, C. et al. (2013). Facing growth in the European nucleotide archive. *Nucleic Acids Res* 41 (D1): D30–D35.

Colella, S., Yau, C., Taylor, J.M. et al. (2007). QuantiSNP: an objective Bayes hidden-Markov model to detect and accurately map copy number variation using SNP genotyping data. *Nucleic Acids Res* 35 (6): 2013–2025.

Consortium, M., Shi, L., Reid, L.H. et al. (2006). The MicroArray quality control (MAQC) project shows inter- and intraplatform reproducibility of gene expression measurements. *Nat Biotechnol* 24 (9): 1151–1161.

DePristo, M.A., Banks, E., Poplin, R. et al. (2011). A framework for variation discovery and genotyping using next-generation DNA sequencing data. *Nat Genet* 43 (5): 491–498.

Du, P., Kibbe, W.A., and Lin, S.M. (2008). lumi: a pipeline for processing Illumina microarray. *Bioinformatics* 24 (13): 1547–1548.

Du, P., Zhang, X., Huang, C.-C. et al. (2010). Comparison of Beta-value and M-value methods for quantifying methylation levels by microarray analysis. *BMC Bioinform* 11: 587.

Erlich, Y., Mitra, P.P., delaBastide, M. et al. (2008). Alta-cyclic: a self-optimizing base caller for next-generation sequencing. *Nat Methods* 5 (8): 679–682.

Fare, T.L., Coffey, E.M., Dai, H. et al. (2003). Effects of atmospheric ozone on microarray data quality. *Anal Chem* 75 (17): 4672–4675.

Felsenstein (1985). Confidence limits on phylogenies: an approach using the bootstrap. *Evolution* 39: 783–781.

Fromer, M., Moran, J.L., Chambert, K. et al. (2012). Discovery and statistical genotyping of copy-number variation from whole-exome sequencing depth. *Am J Hum Genet* 91 (4): 597–607.

Garber, M., Grabherr, M.G., Guttman, M., and Trapnell, C. (2011). Computational methods for transcriptome annotation and quantification using RNA-seq. *Nat Methods* 8 (6): 469–477.

Gebhard, C., Schwarzfischer, L., Pham, T.-H. et al. (2006). Genome-wide profiling of CpG methylation identifies novel targets of aberrant hypermethylation in myeloid leukemia. *Cancer Res* 66 (12): 6118–6128.

Geiss, G.K., Bumgarner, R.E., Birditt, B. et al. (2008). Direct multiplexed measurement of gene expression with color-coded probe pairs. *Nat Biotechnol* 26 (3): 317–325.

Giannoulatou, E., Yau, C., Colella, S. et al. (2008). GenoSNP: a variational Bayes within-sample SNP genotyping algorithm that does not require a reference population. *Bioinformatics* 24 (19): 2209–2214.

Goecks, J., Nekrutenko, A., Taylor, J. et al. (2010). Galaxy: a comprehensive approach for supporting accessible, reproducible, and transparent computational research in the life sciences. *Genome Biol* 11 (8): R86.

Guttman, M., Garber, M., Levin, J.Z. et al. (2010). Ab initio reconstruction of cell type-specific transcriptomes in mouse reveals the conserved multi-exonic structure of lincRNAs. *Nat Biotechnol* 28 (5): 503–510.

Hach, F., Hormozdiari, F., Alkan, C. et al. (2010). mrsFAST: a cache-oblivious algorithm for short-read mapping. *Nat Methods* 7 (8): 576–577.

Handl, J., Knowles, J., and Kell, D.B. (2005). Computational cluster validation in post-genomic data analysis. *Bioinformatics* 21 (15): 3201–3212.

Hua, J., Craig, D.W., Brun, M. et al. (2007). SNiPer-HD: improved genotype calling accuracy by an expectation-maximization algorithm for high-density SNP arrays. *Bioinformatics* 23 (1): 57–63.

da Huang, W., Sherman, B.T., Tan, Q. et al. (2007). DAVID bioinformatics resources: expanded annotation database and novel algorithms to better extract biology from large gene lists. *Nucleic Acids Res* 35 (Web Server issue): W169–W175.

Irizarry, R.A., Hobbs, B., Collin, F. et al. (2003). Exploration, normalization, and summaries of high density oligonucleotide array probe level data. *Biostatistics* 4 (2): 249–264.

Karakoc, E., Alkan, C., O'Roak, B.J. et al. (2012). Detection of structural variants and indels within exome data. *Nat Methods* 9 (2): 176–178.

Kerr, G., Ruskin, H.J., Crane, M., and Doolan, P. (2008). Techniques for clustering gene expression data. *Comput Biol Med* 38 (3): 283–293.

Kim, H., Kim, J., Selby, H. et al. (2011). A short survey of computational analysis methods in analysing ChIP-seq data. *Hum Genomics* 5 (2): 117–123.

Kim, S.Y., Kim, J.H., Chung, Y.-J. et al. (2012). Effect of combining multiple CNV defining algorithms on the reliability of CNV calls from SNP genotyping data. *Genomics Inform* 10 (3): 194–199.

Korn, J.M., Kuruvilla, F.G., McCarroll, S.A. et al. (2008). Integrated genotype calling and association analysis of SNPs, common copy number polymorphisms and rare CNVs. *Nat Genet* 40 (10): 1253–1260.

Krumm, N., Sudmant, P.H., Ko, A. et al. (2012). Copy number variation detection and genotyping from exome sequence data. *Genome Res* 22 (8): 1525–1532.

Langmead, B., Trapnell, C., Pop, M. et al. (2009). Ultrafast and memory-efficient alignment of short DNA sequences to the human genome. *Genome Biol* 10 (3): R25.

Laurie, C.C., Doheny, K.F., Mirel, D.B. et al. (2010). Quality control and quality assurance in genotypic data for genome-wide association studies. *Genet Epidemiol* 34 (6): 591–602. https://doi.org/10.1002/gepi.20516.

Lazarus, R., Taylor, J., Qiu, W. et al. (2008). Toward the commoditization of translational genomic research: design and implementation features of the Galaxy genomic workbench. *Summit Translat Bioinforma* 2008: 56–60.

Li, H. and Durbin, R. (2009). Fast and accurate short read alignment with Burrows-Wheeler transform. *Bioinformatics* 25 (14): 1754–1760.

Li, H. and Durbin, R. (2010). Fast and accurate long-read alignment with Burrows-Wheeler transform. *Bioinformatics* 26 (5): 589–595.

Li, H. and Homer, N. (2010). A survey of sequence alignment algorithms for next-generation sequencing. *Brief Bioinform* 11 (5): 473–483.

Li, C. and Wong, W.H. (2001). Model-based analysis of oligonucleotide arrays: expression index computation and outlier detection. *Proc Natl Acad Sci U S A* 98 (1): 31–36.

Li, H., Ruan, J., and Durbin, R. (2008). Mapping short DNA sequencing reads and calling variants using mapping quality scores. *Genome Res* 18 (11): 1851–1858.

Li, H., Handsaker, B., Wysoker, A. et al. (2009). The sequence alignment/map (SAM) format and SAMtools. *Bioinformatics* 25: 2078–2079. [PMID: 19505943].

Li, G., Gelernter, J., Kranzler, H.R. et al. (2012). M(3): an improved SNP calling algorithm for Illumina BeadArray data. *Bioinformatics* 28 (3): 358–365.

Lin, Y., Li, J., Shen, H. et al. (2011). Comparative studies of de novo assembly tools for next-generation sequencing technologies. *Bioinformatics* 27 (15): 2031–2037.

Luo, J., Schumacher, M., Scherer, A. et al. (2010). A comparison of batch effect removal methods for enhancement of prediction performance using MAQC-II microarray gene expression data. *Pharmacogenomics J* 10 (4): 278–291.

Mangalam, H. (2002). The Bio* toolkits--a brief overview. *Brief Bioinform* 3 (3): 296–302.

Miclaus, K., Wolfinger, R., Vega, S. et al. (2010). Batch effects in the BRLMM genotype calling algorithm influence GWAS results for the Affymetrix 500K array. *Pharmacogenomics J* 10 (4): 336–346.

Miller, J.R., Koren, S., and Sutton, G. (2010). Assembly algorithms for next-generation sequencing data. *Genomics* 95 (6): 315–327.

Mohapatra, S.K. and Krishnan, A. (2011). Microarray data analysis. *Methods Mol Biol* 678: 27–43.

Neuhauser, M. and Senske, R. (2004). The Baumgartner-Weiss-Schindler test for the detection of differentially expressed genes in replicated microarray experiments. *Bioinformatics* 20 (18): 3553–3564.

Nielsen, R., Paul, J.S., Albrechtsen, A. et al. (2011). Genotype and SNP calling from next-generation sequencing data. *Nat Rev Genet* 12 (6): 443–451.

Nijkamp, J.F., van den Broek, M.A., Geertman, J.M. et al. (2012). De novo detection of copy number variation by co-assembly. *Bioinformatics (Oxford, England)* 28 (24): 3195–3202.

Nugent, R. and Meila, M. (2010). An overview of clustering applied to molecular biology. *Methods Mol Biol* 620: 369–404.

Oshlack, A., Robinson, M.D., and Young, M.D. (2010). From RNA-seq reads to differential expression results. *Genome Biol* 11: 220.

Pendergrass, S.A., Frase, A., Wallace, J. et al. (2013). Genomic analyses with biofilter 2.0: knowledge driven filtering, annotation, and model development. *BioData Min* 6 (1): 25.

Pereira, R., Oliveira, J., and Souda, M. (2020). Bioinformatic and computational tools for next-generation sequencing analysis in clinical genetics. *J Clin Med* 9: 132.

Pevzner, P.A., Tang, H., Waterman, M.S. et al. (2001). An Eulerian path approach to DNA fragment assembly. *Proc Natl Acad Sci U S A* 98 (17): 9748–9753.

Purcell, S., Neale, B., Todd-Brown, K. et al. (2007). PLINK: a tool set for whole-genome association and population-based linkage analyses. *Am J Hum Genet* 81 (3): 559–575.

Ritchie, M.E., Diyagama, D. et al. (2006). Empirical array quality weights in the analysis of microarray data. *BMC Bioinform* 7: 261.

Ruffalo, M., LaFramboise, T., Koyutürk, M. et al. (2011). Comparative analysis of algorithms for next-generation sequencing read alignment. *Bioinformatics* 27 (20): 2790–2796.

Scharpf, R.B., Ruczinski, I., Carvalho, B. et al. (2011). A multilevel model to address batch effects in copy number estimation using SNP arrays. *Biostatistics* 12 (1): 33–50.

Shah, T.S., Liu, J.Z., Floyd, J.A.B. et al. (2012). optiCall: a robust genotype-calling algorithm for rare, low-frequency and common variants. *Bioinformatics* 28 (12): 1598–1603.

Shen, X. and Vikalo, H. (2012). ParticleCall: a particle filter for base calling in next-generation sequencing systems. *BMC Bioinform* 13: 160.

Sherman, B.T., da Huang, W., Tan, Q. et al. (2007). DAVID knowledgebase: a gene-centered database integrating heterogeneous gene annotation resources to facilitate high-throughput gene functional analysis. *BMC Bioinform* 8: 426.

Simpson, J.T. and Durbin, R. (2012). Efficient de novo assembly of large genomes using compressed data structures. *Genome Res* 22 (3): 549–556.

Sohn, J.I. and Nam, J.W. (2018). The present and future of de novo whole-genome assembly. *Brief Bioinform* 19: 23–40.

Stajich, J.E., Block, D., Boulez, K. et al. (2002). The Bioperl toolkit: Perl modules for the life sciences. *Genome Res* 12 (10): 1611–1618.

Stegle, O., Teichmann, S.A., and Marioni, J.C. (2015). Computational and analytical challenges in single-cell transcriptomics. *Nat Rev Genet* 16: 133–145.

Tarca, A.L., Carey, V.J., Chen, X.-w. et al. (2007). Machine learning and its applications to biology. *PLoS Comput Biol* 3 (6): e116.

Torkolla, K., Gardner, R.M., Kaysser-Kranich, T., and Ma, C. (2001). Self-organizing maps in mining gene expression data. *Inform Sci* 139: 79–96.

Trapnell, C., Williams, B.A., Pertea, G. et al. (2010). Transcript assembly and quantification by RNA-Seq reveals unannotated transcripts and isoform switching during cell differentiation. *Nat Biotechnol* 28 (5): 511–515.

Trapnell, C., Roberts, A., Goff, L. et al. (2012). Differential gene and transcript expression analysis of RNA-seq experiments with TopHat and Cufflinks. *Nat Protoc* 7 (3): 562–578.

Turner, S., Armstrong, L.L., Bradford, Y. et al. (2011; Chapter 1:Unit1.19). Quality control procedures for genome-wide association studies. *Curr Protoc Hum Genet* doi: 10.1002/0471142905.hg0119s68.

Tusher, V.G., Tibshirani, R., and Chu, G. (2001). Significance analysis of microarrays applied to the ionizing radiation response. *Proc Natl Acad Sci U S A* 98 (9): 5116–5121.

Waggott, D., Chu, K., Yin, S. et al. (2012). NanoStringNorm: an extensible R package for the pre-processing of NanoString mRNA and miRNA data. *Bioinformatics* 28 (11): 1546–1548.

Wang, K., Li, M., Hadley, D. et al. (2007). PennCNV: an integrated hidden Markov model designed for high-resolution copy number variation detection in whole-genome SNP genotyping data. *Genome Res* 17 (11): 1665–1674.

Wang, K., Li, M., and Hakonarson, H. (2010). ANNOVAR: functional annotation of genetic variants from high-throughput sequencing data. *Nucleic Acids Res* 38 (16): e164.

Wu, T.D. and Nacu, S. (2010). Fast and SNP-tolerant detection of complex variants and splicing in short reads. *Bioinformatics* 26 (7): 873–881.

Xu, R. and Wunsch, D.C., 2nd (2010). Clustering algorithms in biomedical research: a review. *IEEE Rev Biomed Eng* 3: 120–154.

Yaspan, B.L., Bush, W.S., Torstenson, E.S. et al. (2011). Genetic analysis of biological pathway data through genomic randomization. *Hum Genet* 129 (5): 563–571. https://doi.org/10.1007/s00439-011-0956-2.

Zerbino, D.R. and Birney, E. (2008). Velvet: algorithms for de novo short read assembly using de Bruijn graphs. *Genome Res* 18 (5): 821–829.

Zhang, B., Kirov, S., and Snoddy, J. (2005). WebGestalt: an integrated system for exploring gene sets in various biological contexts. *Nucleic Acids Res* 33 (Web Server issue): W741–W748.

Zuvich, R.L., Armstrong, L.L., Bielinski, S.J. et al. (2011). Pitfalls of merging GWAS data: lessons learned in the eMERGE network and quality control procedures to maintain high data quality. *Genet Epidemiol* 35 (8): 887–898. https://doi.org/10.1002/gepi.20639.

11

Complex Genetic Interactions/Data Mining/Dimensionality Reduction

William S. Bush and Stephen D. Turner

[1]*Department of Population and Quantitative Health Sciences, Case Western Reserve University, Cleveland, OH, USA*
[2]*Signature Science, LLC, Charlottesville, VA, USA*

Human Diseases Are Complex

The first human genetic disease was identified by Archibald Garrod, the father of human genetics. As a physician, Garrod related the patterns of autosomal recessive inheritance discovered by Gregor Mendel to families with the metabolic disorder alkaptonuria (Garrod 1903). A few years later, James Herrick and his intern Ernest Irons documented the first described case of sickle cell anemia after examining a patient's blood smear, and in 1949 Linus Pauling and colleagues demonstrated that this genetic disease was the result of a mutation in a specific protein complex, hemoglobin (Pauling and Itano 1949). This seminal work began a scientific race to identify the damaged gene for every known genetic disorder, which was exemplified with the identification of mutations in the cystic fibrosis transmembrane conductance regulator (CFTR) gene that result in autosomal recessive cystic fibrosis (Kerem et al. 1989).

The discovery of primary genetic factors for diseases like sickle cell anemia, Huntington's disease, and cystic fibrosis dramatically changed our understanding of and diagnosis of these diseases, and due to their simple and clear inheritance patterns these became known as "Mendelian diseases." Despite the influence of a single strong genetic factor, much variability in the presentation, onset, and severity of these diseases is unexplained by known genetic and environmental factors.

Multiple distinct mutations can lead to sickle cell anemia, and the severity of symptoms can vary from individual to individual. Several additional genetic factors modify the effect of the primary mutation to alter severity and progression of the disease. Additional variants in the α-thalassaemia gene, the 158C->T mutation upstream of the β-globin-γ gene, and yet unknown genes in the female population all reduce disease severity (Stuart and Nagel 2004). After accounting for the primary genetic cause of sickle cell anemia in an individual, remaining phenotypic variation includes the life span and density of red blood cells, adhesion of sickle cells to the interior of blood vessels, changes in urine concentration, and altered activity of anion transport proteins within the red blood cell membranes (Nagel 2001).

Likewise, there is variation in the hallmark lung phenotype of cystic fibrosis that is not explained by mutations in the CFTR gene alone. The primary mutations for cystic fibrosis do predict other organ symptoms, such as pancreatic exocrine insufficiency (Zielenski 2000), but there is a full range of lung disease severity among the most frequent CFTR mutations, even after adjusting for

Genetic Analysis of Complex Diseases, Third Edition. Edited by William K. Scott and Marylyn D. Ritchie.

environmental factors typically associated with lung disease, such as tobacco use. There are several genes suspected to contribute to the lung phenotype independent of CFTR mutation, and it is hypothesized that alleles of these genes may act in combination to contribute to the lung phenotype. Furthermore, mouse knockouts of the CFTR gene develop an intestinal obstruction shortly after birth, a similar finding present in 20% of CF newborns, and genes in both humans and mice were found to modify its development and lethality (Merlo and Boyle 2003).

These examples of variation in sickle cell anemia and cystic fibrosis phenotypes provide compelling evidence that even diseases with nearly deterministic genetic components following clear Mendelian inheritance are ultimately quite complex and involve multiple genetic factors. Conversely, some more "complex" disorders have strong Mendelian components or subtypes. Mutations in *BRCA1* and *BRCA2* were discovered in families with higher rates of breast cancer than the general population using linkage analysis (Miki et al. 1994; Thorlacius et al. 1997). Individuals with these mutations have an earlier onset and a different clinical course. Similarly, a relatively common allele of the *APOE* gene was strongly associated to late-onset Alzheimer's disease (Corder et al. 1993), though early-onset familial forms are caused by mutations acting in an autosomal dominant fashion (reviewed in Barbier et al. 2018). More recently, a common allele of the *CFH* gene was found to increase risk for age-related macular degeneration (Edwards et al. 2005; Haines et al. 2005).

Despite these successes, the traditional linkage-based gene mapping approaches showed sequential failure at identifying single genetic causes for many common human diseases (Altmuller et al. 2001). There are many possible explanations for the failure of linkage analysis, but chief among them is that it assumes the Mendelian transmission of the same moderately influential disease alleles within a single genetic locus in affected families (Risch 2000).

As explored in other chapters, the observation of disease is a mathematical function of disease prevalence, phenotype definition, disease allele frequency, modifier allele effects, and environmental influences. Rather than classifying human diseases as either simple or complex, consider that diseases vary along a spectrum of these two extremes. The properties of a given disease can make discovering susceptibility genes easier or more difficult depending on the tools and study design used. Whatever the disease being studied, it is now generally assumed that all human phenotypes will exhibit great genetic complexity in one form or another.

Complexity of Biological Systems

The central dogma of biology describes a hierarchical structure that governs all biological systems: DNA is transcribed to RNA which is translated to amino acid sequences that fold into functional proteins. This dogma has developed and matured over many years to account for noted exceptions, such as functional RNA sequences. The general hierarchy could be extended to describe higher level biological structure as well. Proteins work in concert to form the fundamental structural and chemical building blocks of the cell, providing channels, pores, receptors, and other molecular transport systems, and complex metabolic and regulatory pathways, eventually creating specialized organelles within the cell. The functional structure and output of cells delineate cell types; aggregations of multiple cell types form tissues; multiple tissues combine to form organs and organ systems; and ultimately an entire organism is formed that interacts with the environment through defined behaviors. These layers of biological complexity exist between nucleotide and structural alterations at the DNA level and fully expressed human traits. Genetic variation can contribute to subtle changes in each of these layers. Collections of genetic variation likely function in concert to alter all of these layers, changing probability of disease along the way.

The complexity of human phenotypes is nicely illustrated by the plethora of genetic associations generated from genome-wide association studies (GWAS) (Manolio et al. 2008; Buniello et al. 2019). For the vast majority of phenotypes studied, multiple genetic variants have been identified, most of which have a very modest influence on the phenotype (Hindorff et al. 2009). Variants associated with complex traits tend not to change protein sequence but change gene regulation by altering sequences bound by transcriptional machinery (Nicolae et al. 2010; Schaub et al. 2012). The extreme complexity discovered through GWAS is well illustrated by autoimmune phenotypes such as multiple sclerosis and Crohn's disease, each of which have over 50 replicating genetic loci involved in disease (Franke et al. 2010; Gourraud et al. 2012). Furthermore, many of these loci influence other autoimmune phenotypes through common pleiotropic mechanisms (Ramos et al. 2011).

As evidenced by genetic studies to date, human phenotypes can be riddled with complexity in a number of ways. A review by Thornton-Wells et al. nicely categorizes the types of complexity identified in typical genetic studies (Thornton-Wells et al. 2004), which includes allelic heterogeneity, locus heterogeneity, gene–gene interaction, and gene–environment interaction. There is also variability in the genetic endpoint, including trait heterogeneity, phenocopy, and pleiotropy. Each of these facets of the true genetic architecture of a disease has an impact on study design, statistical analysis, and interpretation of results.

Genetic Heterogeneity

Allelic heterogeneity occurs when two or more alleles influencing a single genetic mechanism are independently associated with disease risk. The most striking examples of allelic heterogeneity occur in Mendelian disorders such as cystic fibrosis, where over 1000 variants in the *CFTR* gene have been identified (Kulczycki et al. 2003). This phenomenon is by no means isolated to Mendelian disorders. Several studies of complex disease have identified multiple variants influencing the same genetic mechanism, such as multiple sclerosis (Zuvich et al. 2011). Examining the NHGRI GWAS catalog reveals multiple allelic associations to the same trait within the same region; however, it is unclear if these are due to independent signals, or if they are an artifact of studies conducted across multiple genotyping platforms that ultimately represent a single genetic signal (Hindorff et al. 2009; Buniello et al. 2019). Allelic heterogeneity, notably, is readily detected by linkage analysis as multiple alleles within the same chromosomal region can be detected co-segregating with the trait across families.

Locus heterogeneity occurs when two or more genetic variations with different genetic functions both contribute independently to disease risk. A striking example of locus heterogeneity is tuberous sclerosis, characterized by the formation of benign focal tumors in multiple organ systems. Tuberous sclerosis is caused by mutations in the *TSC1* gene in 10–30% of cases, with mutations in the *TSC2* gene accounting for the majority of cases (and 15–20% with no identifiable mutations) (Povey et al. 1994). Notably, *TSC1* and *TSC2* encode for proteins (hamartin and tuberin, respectively) that physically interact to form a heterodimer that is thought to function as a tumor suppressor (Crino et al. 2006; Leung et al. 2007). Locus heterogeneity is prolific among common diseases. The vast majority of traits studied by GWAS have identified at least two associated variants either located on different chromosomes or influencing different genetic mechanisms.

Given the complexity of biological processes, it is intuitive to assume that combinations of alleles across multiple genes interact to influence a trait – a phenomenon called epistasis or gene–gene interaction. As a hypothetical example, assume that there are multiple receptors that trigger a downstream response; variants that reduce the expression of one receptor alone may not influence a trait to a strong degree, but individuals with variants altering multiple receptors may have a significantly different biological function resulting in a phenotype. This phenomenon was identified early in the field of genetics by Lancefield and Castle who discovered combinations of variants that influence Drosophila bristle and eye phenotypes. Yeast and other model organisms exhibit epistasis for a broad number of phenotypes. Notable examples include serial crosses of *Saccharomyces cerevisiae*, which identified epistatic interactions responsible for chemical resistance and mitochondrial function (Ehrenreich et al. 2010), and response to thermal environment in *C. elegans*, controlled by a complex network of interacting genetic variants (Gaertner et al. 2012). Robust examples of epistasis in humans are more difficult to point out. Carrasquillo et al. identified mutations in the *endothelin B* gene and *RET* gene that interact to produce severe Hirschprung's disease (Carrasquillo et al. 2002), and several other modifier effects have been reported for this phenotype (Wallace and Anderson 2011). Also, in multiple sclerosis, alleles of the MHC region demonstrate functional epistasis (Gregersen et al. 2006).

Trait heterogeneity occurs when samples are ascertained based on a superficial case definition that is derived from a combination of multiple underlying traits. A hallmark example of trait heterogeneity is autism spectrum disorders. The diagnosis of "autism" has changed a variety of times since studies of the disease began, and now the phenotypic characteristics associated with autism are divided into multiple domains, many of which are present in other similar disorders (Geschwind 2009). As a result, most studies now conducted to study these subphenotypes together – the autism spectrum disorders (Veenstra-Vanderweele et al. 2004). As mentioned before, early and late onset forms of Alzheimer's disease have distinct genetic influences. Similarly, the notable *BRCA1* and *BRCA2* mutations identified for breast cancer are moderately penetrant variants but ultimately account for a small proportion of overall breast cancer cases (Walsh et al. 2006). These variants are generally identified in *familial* breast cancer, so named due to earlier onset and increased relative risk in families of individuals with breast cancer ("Pathology of Familial Breast Cancer: Differences between Breast Cancers in Carriers of *BRCA1* or *BRCA2* Mutations and Sporadic Cases. Breast Cancer Linkage Consortium" Stratton 1997).

Statistical and Mathematical Concepts of Complex Genetic Models

When examining the effect of a single genetic variation on a phenotype, statistical procedures are used to associate an individual's genotypic state to the outcome. As an individual has three possible genotype states (AA, Aa, aa), there are multiple ways to perform this analysis (Figure 11.1). Assuming a dominant effect of the locus, we might group the AA and Aa genotypes together into one class versus individuals with the aa genotype. Alternatively, we might assume an additive dose-like effect of each A allele, encoding the genotypes as 0(aa), 1(Aa), and 2(AA). We could also assume a multiplicative effect of the A allele, encoding the genotypes as 0(aa), 4(Aa), and 16(AA). It is accepted practice to analyze genotype data with additive encodings, as this has been shown to have reasonable power to detect all of these models (Lettre et al. 2007). Each of these models, however, is supported by an intuitive underlying biological mechanism.

Multi-locus modeling becomes far more complex, and the terminology is far less precise. Consider two variants A and B with genotypes (AA, Aa, aa, and BB, Bb, bb) where both have an additive effect on the phenotype. When put together, the simplest case is that the effects of A and B alleles are completely independent of one another, and their cumulative effects can be estimated

Figure 11.1 Genotype models. Given a single biallelic variant, there are three possible genotypes which are generally encoded as one of four possible genetic models. Increasing cell shading indicates increased effect on the phenotype.

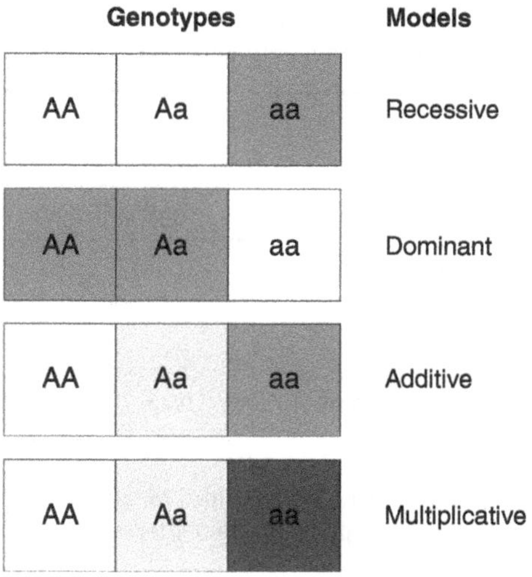

by adding them together. This would be a *multi-locus additive model*. The combination of alleles from A and B could have synergistic effects however, where additional copies of the A and B alleles *together* increase risk beyond what is expected under the multi-locus additive model. There are a variety of ways in which these models can manifest; as a *multi-locus multiplicative model*, as a *multi-locus dominant model*, or some intermediate form. Multi-locus models may also be *antagonistic*, where additional copies of the A and B alleles *together* decrease risk below what is expected under the multi-locus additive model. These models are also referred to as *genetic heterogeneity* models.

Some multi-locus models may not contain strong independent effects of either variant A or B, and where the effect of these variants on a phenotype manifests only within certain multi-locus combinations of alleles. These models are often called *purely epistatic models*, and because they contain no independent or *main effects* of the two variants, the variants cannot be detected by single-locus analyses. Examples of these functions are readily found (Moore et al. 2004), including the diagonal model (Table 11.1) and XOR model (Table 11.2). Despite their mathematical possibility, these models are challenging to explain in biological terms using traditional dominant or additive marginal effects of variants (Frankel and Schork 1996).

Similar complex models are possible when considering gene–environment interactions. When environmental variables are measured as discrete (exposure yes/no) or ordinal (severity 1–5), the problems with multi-locus modeling exist here as well. We might assume a particular model, such as an additive genotype by ordinal severity interaction, where more copies of the minor allele in concert with more severe exposure levels exponentially increase risk. Continuous environmental variables can be similarly modeled, where genotypes alter the change in risk per unit. It is important to note that these types of interactions could also easily be influenced by dominant or multiplicative genotype effects.

A variety of machine learning and statistical approaches can be applied to detect complex genetic models. An important point to consider with these techniques is that they are all influenced by the encoding of the data. Using the intuitive additive encoding (0[aa], 1[Aa], and 2[AA]) forces constraints on the modeling strategy, forcing a dose–response effect per allele, unless the approach can explicitly correct for such assumptions. In some cases, dummy-encoding the genotype states or using allelic encodings allow more flexibility in the modeling.

Table 11.1 Penetrance values for combinations of genotypes from two SNPs exhibiting interactions in the absence of independent main effects.

	Table penetrance			
	AA (0.25)	Aa (0.50)	aa (0.25)	Margin penetrance
BB (0.25)	0	0	1	0.25
Bb (0.50)	0	0.50	0	0.25
bb (0.25)	1	0	0	0.25
Margin penetrance	0.25	0.25	0.25	

Table 11.2 Penetrance values for combinations of genotypes from two SNPs exhibiting interactions in the absence of independent main effects.

	Table penetrance			
	AA (0.25)	Aa (0.50)	aa (0.25)	Margin penetrance
BB (0.25)	0	1	0	0.5
Bb (0.50)	1	0	1	0.5
bb (0.25)	0	1	0	0.5
Margin penetrance	0.5	0.5	0.5	

Analytic Approaches to the Detection of Complex Interactions

Linkage Analysis/Genomic Sharing

Linkage analysis has been applied extensively to multiple complex traits with the goal of identifying genomic regions inherited with disease status in families (Bush and Haines 2010). The approach is robust to allelic heterogeneity but relies upon the basic premise that a common, single genomic region influences disease in a family. Extensions of linkage analysis have been developed to allow conditional analyses that can model genetic interactions.

Linkage analysis can also be performed within strata of the original dataset (see Chapter 6). For example, Hall et al. stratified familial breast cancer families based on age of onset (≤45 and >45 years), which refined the *BRCA1* linkage from 50- to an 8-cM region (Hall et al. 1992). When such a cutoff is not evident from clinical criteria, computational approaches can be used to explore possible stratifications. Orderedsubset analysis was developed to automatically subset families based on various phenotypic or genetic criteria with the hypothesis that these subgroups of families are more genetically homogenous (Chung et al. 2008). Linkage analysis is then performed on each of the subgroups to identify influential chromosomal regions. As such, this approach can identify interactions by determining loci that co-segregate in families that also harbor additional susceptibility alleles.

Association Analysis

Far more techniques for examining genetic interactions have been developed for and applied to population-based genetic data, typically in the form of case–control studies. Prior to 2005, these studies were usually designed by selecting candidate genes based on biomedical literature and genotyping less than 1000 single nucleotide polymorphism (SNP)s within those genes. The limited number of variables in such a dataset allowed the application of computationally intensive procedures, including multiple machine learning techniques.

Case-only analysis exploits the idea that a true interaction among genotypes, or genotypes and environmental variables, would require an enrichment of certain multi-locus genotype classes relative to others in affected individuals. For example, if a compound homozygote of the major allele (AABB) had dramatically increased risk, cases would have higher proportions of the AABB multi-locus genotype than we would expect given the A and B allele frequencies. As such, a case-only analysis of gene–gene interaction looks for deviations from expected multi-locus genotype frequencies can be conducted (as its name implies) using only the cases of a dataset. This analysis is generally conducted using a chi-square test, similar in principal to tests for Hardy–Weinberg equilibrium. Extensions of this general approach have been incorporated into PLINK software (Ueki and Cordell 2012).

Artificial neural networks attempt to model complex genetic effects using what is essentially complex hierarchical logistic regression. SNP genotypes are mapped to input nodes, which weight the influence of each SNP and pass this value to one or more layers of hidden nodes, which in turn combine and weight these values and pass this result forward through a function (typically a logistic function). The end result is a prediction of disease status or a prediction of a quantitative trait. The weights for the connections of the network can be fit using a variety of approaches, most commonly backpropagation. Neural networks have been extensively evaluated for use in genetic epidemiology (Motsinger-Reif and Ritchie 2008) and have been applied to the cardiovascular disease (Costello et al. 2003), type 2 diabetes (North et al. 2003), Crohn's disease (Giachino et al. 2007), pharmacogenomic traits (Serretti and Smeraldi 2004; Lin et al. 2006), and asthma (Tomita et al. 2004).

Classification and Regression Trees (CART) and its variants have been applied to multiple phenotypes. CART works by sequentially partitioning the cases and controls into subsets using the input variables. Subsets (and the variables used to generate them) are selected by optimizing some function (usually a Gini index or something similar) which measures the accuracy of the resulting prediction. CART has been applied in genetic studies of hypertension (Huang et al. 2004), pharmacogenomic traits (Zaykin and Stanley Young 2005), and various other aspects of genetic study design (Yu et al. 2005). Similarly, basic rule-mining approaches can efficiently discover complex models when genotype encodings are properly specified (Bush et al. 2007).

Multivariate Adaptive Regression Splines (MARS) is an extension of CART that dynamically fits regression splines within each subset of the data (rather than traditional linear or discriminant functions). MARS has been applied to prostate cancer (Lin et al. 2008) and ischemic stroke (Cook et al. 2004).

Multifactor dimensionality reduction (MDR) is another CART-like approach that examines patterns of multi-locus genotypes (assigning them to high- and low-risk categories) to estimate complex genetic association models. Rather than relying on an optimization function (like the Gini index), MDR performs an exhaustive search of all multi-locus combinations. MDR is one of the most widely applied methods for detecting gene–gene interactions with applications to asthma and allergy (Chan et al. 2006; Millstein et al. 2006; Park et al. 2007), autoimmune diseases (Julià et al. 2007; Beretta et al. 2008, 2008; Okazaki et al. 2008), and cancer (Andrew et al. 2006; Chen et al. 2007; Huang et al. 2007; Cao et al. 2008) among many others.

Random Forests are another machine learning variant that constructs multiple decision trees. Random Forests create an ensemble of decision trees to predict an outcome for each sample by averaging all individual tree predictions over the entire ensemble. The Random Forest algorithm begins by randomly drawing a bootstrap sample of n individuals with replacement. For each bootstrap sample, a regression tree is grown using a modified CART algorithm; it draws a random subset of input variables instead of using all predictors, and the resulting trees are not pruned or simplified in any way. In this way, the speed of training classification trees is greatly improved, and the nonselected (out-of-bag) samples from the bootstrapping procedure are used to measure the predictive ability of the ensemble. Random Forests have successfully been used for many genetic data analysis applications, including candidate gene studies (Bureau et al. 2005) and interaction analysis in large-scale association studies (Lunetta et al. 2004).

Genome-Wide Association Analysis

Since 2005, the field of human genetics has generated millions of genotypes for thousands of genetic traits using GWAS (see Chapter 9). The majority of approaches applied to candidate gene studies are overwhelmed by the scale of these datasets, which often contain hundreds of thousands of SNPs and tens of thousands of cases and controls. To address this challenge, multiple techniques have been applied to reduce the set of input SNPs for interaction methods. These loosely fall into analytical and biological filtering approaches.

The most basic analytical filtering approach is to evaluate interactions among the most significant marginal SNP associations (e.g. the top 10% of GWAS results). This approach will only detect interactions where there are marginal effects strong enough to fall into this top tier of results. Other algorithmic approaches may also be applied that identify potentially influential SNPs that do not rely on marginal effects. For example, the Relief algorithm and its various extensions identify potential interactions through a procedure similar to clustering (Kononenko 1994). Good attributes are those that have the best ability to discriminate among cases and controls. ReliefF is the most commonly used extension of Relief and is included in the MDR software package as a filtering option prior to MDR analysis (Moore et al. 2006).

A fundamentally different strategy is to use biological information to extract subsets of SNPs for interaction analysis. The underlying hypothesis for this method is that variants within genes whose products physically interact (or function together in the same system) are more likely to statistically interact to influence a phenotype. This approach can also have the virtue of focusing the hypothesis on a core set of biological mechanisms, improving the interpretability of the genetic model. Two software packages have been developed to automate this procedure; Biofilter (Bush et al. 2009; Pendergrass et al. 2013, 2) and INTERSNP (Herold et al. 2009). These packages couple the bioinformatics SNP selection to various statistical methods for interaction analysis.

By far the most applied approach to interaction analysis in GWAS-scale data has been logistic and linear regression. Generally, by coupling regression to either biological or analytical filters, pairwise interactions are assessed using logistic or linear regression. The model is typically specified using additive main effects with a multiplicative interaction term (Kraft et al. 2007). Also, a variant of Random Forests has been developed specifically to scale to genome-wide SNP data. This method, called Random Jungle, has been successfully used on GWAS, where input predictor variables can number in the millions, and where few are associated with the disease outcome (Schwarz et al. 2010).

GWAS have known confounding factors, most notably the issue of *population stratification*. This issue arises when differences in allele frequency between ethnicities in a sample are mistakenly associated due to differences in disease prevalence (which is discussed in more detail in Chapter 8). Population stratification may be especially problematic when examining multi-locus combinations of SNPs, as individuals of a similar genetic ancestry are more likely to share multi-locus genotypes. Also, many machine learning methods are prone to over-fitting, especially in datasets with large numbers of predictors. As such, the most robust approach for confirming a complex genetic association is replicating this effect in an additional dataset.

Conclusion

Biological systems are inherently complex, and genetic and environmental factors can influence these systems in various ways to influence a phenotype. The rapid, cost-effective generation of genetic data has ushered in a new era of human genetics, where the comprehensive analysis of entire biological mechanisms is now possible. The new challenge is not in data collection, but rather in understanding how to interconnect pieces of the genetic puzzle underlying complex phenotypes. New approaches to genetic analysis that can account for this known complexity are needed to map genetic variation to phenotypes of interest. As such, in the coming years, statistical and data mining techniques for genetic data will become ever more important.

References

Altmuller, J., Palmer, L.J., Fischer, G. et al. (2001). Genomewide scans of complex human diseases: true linkage is hard to find. *American Journal of Human Genetics* 69 (5): 936–950.

Andrew, A.S., Nelson, H.H., Kelsey, K.T. et al. (2006). Concordance of multiple analytical approaches demonstrates a complex relationship between DNA repair gene SNPs, smoking and bladder cancer susceptibility. *Carcinogenesis* 27 (5): 1030–1037. https://doi.org/10.1093/carcin/bgi284.

Barbier, M., Wallon, D., and Le Ber, I. (2018). Monogenic inheritance in early-onset dementia: illustration in Alzheimer's disease and frontotemporal lobar dementia. *Geriatrie Et Psychologie Neuropsychiatrie Du Vieillissement* 16 (3): 289–297. https://doi.org/10.1684/pnv.2018.0744.

Beretta, L., Cappiello, F., Moore, J.H. et al. (2008). Ability of epistatic interactions of cytokine single-nucleotide polymorphisms to predict susceptibility to disease subsets in systemic sclerosis patients. *Arthritis and Rheumatism* 59 (7): 974–983. https://doi.org/10.1002/art.23836.

Beretta, L., Cappiello, F., Moore, J.H., and Scorza, R. (2008). Interleukin-1 gene complex single nucleotide polymorphisms in systemic sclerosis: a further step ahead. *Human Immunology* 69 (3): 187–192. https://doi.org/10.1016/j.humimm.2007.12.006.

Buniello, A., MacArthur, J.A.L., Cerezo, M. et al. (2019). The NHGRI-EBI GWAS catalog of published genome-wide association studies, targeted arrays and summary statistics 2019. *Nucleic Acids Research* 47 (D1): D1005–D1012. https://doi.org/10.1093/nar/gky1120.

Bureau, A., Dupuis, J., Falls, K. et al. (2005). Identifying SNPs predictive of phenotype using random forests. *Genetic Epidemiology* 28 (2): 171–182. https://doi.org/10.1002/gepi.20041.

Bush, W.S. and Haines, J. (2010). Overview of linkage analysis in complex traits. *Current Protocols in Human Genetics* Chapter 1 (January): Unit 1.9.1–18. doi: https://doi.org/10.1002/0471142905.hg0109s64.

Bush, W.S., Thornton-Wells, T.A., and Ritchie, M.D. (2007). Association rule discovery has the ability to model complex genetic effects. *IEEE Symposium on Computational Intelligence and Data Mining IEEE Symposium on Computational Intelligence and Data Mining*, 2007, pp. 624–629. doi:https://doi.org/10.1109/CIDM.2007.368934.

Bush, W.S., Dudek, S.M., and Ritchie, M.D. (2009). Biofilter: a knowledge-integration system for the multi-locus analysis of genome-wide association studies. *Pacific Symposium on Biocomputing* 2009: 368–379.

Cao, G., Lu, H., Feng, J. et al. (2008). Lung cancer risk associated with Thr495Pro polymorphism of GHR in Chinese population. *Japanese Journal of Clinical Oncology* 38 (4): 308–316. https://doi.org/10.1093/jjco/hyn007.

Carrasquillo, M.M., McCallion, A.S., Puffenberger, E.G. et al. (2002). Genome-wide association study and mouse model identify interaction between RET and EDNRB pathways in Hirschsprung disease. *Nature Genetics* 32 (2): 237–244. https://doi.org/10.1038/ng998.

Chan, I.H.S., Leung, T.F., Tang, N.L.S. et al. (2006). Gene-gene interactions for asthma and plasma total IgE concentration in Chinese children. *The Journal of Allergy and Clinical Immunology* 117 (1): 127–133. https://doi.org/10.1016/j.jaci.2005.09.031.

Chen, M., Kamat, A.M., Maosheng, H. et al. (2007). High-order interactions among genetic polymorphisms in nucleotide excision repair pathway genes and smoking in modulating bladder cancer risk. *Carcinogenesis* 28 (10): 2160–2165. https://doi.org/10.1093/carcin/bgm167.

Chung, R.-H., Schmidt, S., Martin, E.R., and Hauser, E.R. (2008). Ordered-subset analysis (OSA) for family-based association mapping of complex traits. *Genetic Epidemiology* 32 (7): 627–637. https://doi.org/10.1002/gepi.20340.

Cook, N.R., Zee, R.Y.L., and Ridker, P.M. (2004). Tree and spline based association analysis of gene-gene interaction models for ischemic stroke. *Statistics in Medicine* 23 (9): 1439–1453. https://doi.org/10.1002/sim.1749.

Corder, E.H., Saunders, A.M., Strittmatter, W.J. et al. (1993). Gene dose of apolipoprotein E type 4 allele and the risk of Alzheimer's disease in late onset families. *Science* 261 (5123): 921–923.

Costello, T.J., Falk, C.T., and Kenny, Q.Y. (2003). Data mining and computationally intensive methods: summary of group 7 contributions to genetic analysis workshop 13. *Genetic Epidemiology* 25 (Suppl. 1): S57–S63. https://doi.org/10.1002/gepi.10285.

Crino, P.B., Nathanson, K.L., and Henske, E.P. (2006). The tuberous sclerosis complex. *The New England Journal of Medicine* 355 (13): 1345–1356. https://doi.org/10.1056/NEJMra055323.

Edwards, A.O., Ritter, R., Abel, K.J. et al. (2005). Complement factor H polymorphism and age-related macular degeneration. *Science (New York, N.Y.)* 308 (5720): 421–424. https://doi.org/10.1126/science.1110189.

Ehrenreich, I.M., Torabi, N., Jia, Y. et al. (2010). Dissection of genetically complex traits with extremely large pools of yeast segregants. *Nature* 464 (7291): 1039–1042. https://doi.org/10.1038/nature08923.

Franke, A., McGovern, D.P.B., Barrett, J.C. et al. (2010). Genome-wide meta-analysis increases to 71 the number of confirmed Crohn's disease susceptibility loci. *Nature Genetics* 42 (12): 1118–1125. https://doi.org/10.1038/ng.717.

Frankel, W.N. and Schork, N.J. (1996). Who's afraid of epistasis? *Nature Genetics* 14: 371–373.

Gaertner, B.E., Parmenter, M.D., Rockman, M.V. et al. (2012). More than the sum of its parts: a complex epistatic network underlies natural variation in thermal preference behavior in *Caenorhabditis elegans*. *Genetics* 192 (4): 1533–1542. https://doi.org/10.1534/genetics.112.142877.

Garrod, A.E. (1903). Some further observations on the reaction of urochrome with acetaldehyde. *The Journal of Physiology* 29 (4–5): 335–340. https://doi.org/10.1113/jphysiol.1903.sp000959.

Geschwind, D.H. (2009). Advances in autism. *Annual Review of Medicine* 60: 367–380. https://doi.org/10.1146/annurev.med.60.053107.121225.

Giachino, D.F., Regazzoni, S., Bardessono, M. et al. (2007). Modeling the role of genetic factors in characterizing extra-intestinal manifestations in Crohn's disease patients: does this improve outcome predictions? *Current Medical Research and Opinion* 23 (7): 1657–1665. https://doi.org/10.1185/030079907x210471.

Gourraud, P.-A., Harbo, H.F., Hauser, S.L., and Baranzini, S.E. (2012). The genetics of multiple sclerosis: an up-to-date review. *Immunological Reviews* 248 (1): 87–103. https://doi.org/10.1111/j.1600-065X.2012.01134.x.

Gregersen, J.W., Kranc, K.R., Ke, X. et al. (2006). Functional epistasis on a common MHC haplotype associated with multiple sclerosis. *Nature* 443 (7111): 574–577. https://doi.org/10.1038/nature05133.

Haines, J.L., Hauser, M.A., Schmidt, S. et al. (2005). Complement factor H variant increases the risk of age-related macular degeneration. *Science* 308 (5720): 419–421.

Hall, J.M., Friedman, L., Guenther, C. et al. (1992). Closing in on a breast cancer gene on chromosome 17q. *American Journal of Human Genetics* 50 (6): 1235–1242.

Herold, C., Steffens, M., Brockschmidt, F.F. et al. (2009). INTERSNP: genome-wide interaction analysis guided by a priori information. *Bioinformatics (Oxford, England)* 25 (24): 3275–3281. https://doi.org/10.1093/bioinformatics/btp596.

Hindorff, L.A., Sethupathy, P., Junkins, H.A. et al. (2009). Potential etiologic and functional implications of genome-wide association loci for human diseases and traits. *Proceedings of the National Academy of Sciences of the United States of America* 106 (23): 9362–9367. https://doi.org/10.1073/pnas.0903103106.

Huang, J., Lin, A., Narasimhan, B. et al. (2004). Tree-structured supervised learning and the genetics of hypertension. *Proceedings of the National Academy of Sciences of the United States of America* 101 (29): 10529–10534. https://doi.org/10.1073/pnas.0403794101.

Huang, M., Dinney, C.P., Lin, X. et al. (2007). High-order interactions among genetic variants in DNA base excision repair pathway genes and smoking in bladder cancer susceptibility. *Cancer Epidemiology, Biomarkers & Prevention: A Publication of the American Association for Cancer Research, cosponsored by the American Society of Preventive Oncology* 16 (1): 84–91. https://doi.org/10.1158/1055-9965.EPI-06-0712.

Julià, A., Moore, J., Miquel, L. et al. (2007). Identification of a two-loci epistatic interaction associated with susceptibility to rheumatoid arthritis through reverse engineering and multifactor dimensionality reduction. *Genomics* 90 (1): 6–13. https://doi.org/10.1016/j.ygeno.2007.03.011.

Kerem, B., Rommens, J.M., Buchanan, J.A. et al. (1989). Identification of the cystic fibrosis gene: genetic analysis. *Science (New York, N.Y.)* 245 (4922): 1073–1080. https://doi.org/10.1126/science.2570460.

Kononenko, I. (1994). Estimating attributes: analysis and extensions of RELIEF. In: *Proceedings of the 7th European Conference on Machine Learning*. ECML'94. Berlin, Heidelberg: Springer-Verlag, pp. 171–182. doi: https://doi.org/10.1007/3-540-57868-4_57.

Kraft, P., Yen, Y.-C., Stram, D.O. et al. (2007). Exploiting gene-environment interaction to detect genetic associations. *Human Heredity* 63 (2): 111–119. https://doi.org/10.1159/000099183.

Kulczycki, L.L., Kostuch, M., and Bellanti, J.A. (2003). A clinical perspective of cystic fibrosis and new genetic findings: relationship of CFTR mutations to genotype-phenotype manifestations. *American Journal of Medical Genetics. Part A* 116A (3): 262–267. https://doi.org/10.1002/ajmg.a.10886.

Lettre, G., Lange, C., and Hirschhorn, J.N. (2007). Genetic model testing and statistical power in population-based association studies of quantitative traits. *Genetic Epidemiology* 31 (4): 358–362. https://doi.org/10.1002/gepi.20217.

Leung, A.K.C., Lane, W., and Robson, M. (2007). Tuberous sclerosis complex: a review. *Journal of Pediatric Health Care: Official Publication of National Association of Pediatric Nurse Associates & Practitioners* 21 (2): 108–114. https://doi.org/10.1016/j.pedhc.2006.05.004.

Lin, E., Hwang, Y., Wang, S.-C. et al. (2006). An artificial neural network approach to the drug efficacy of interferon treatments. *Pharmacogenomics* 7 (7): 1017–1024. https://doi.org/10.2217/14622416.7.7.1017.

Lin, H.-Y., Wang, W., Liu, Y.-H. et al. (2008). Comparison of multivariate adaptive regression splines and logistic regression in detecting SNP-SNP interactions and their application in prostate cancer. *Journal of Human Genetics* 53 (9): 802–811. https://doi.org/10.1007/s10038-008-0313-z.

Lunetta, K.L., Brooke Hayward, L., Segal, J., and Van Eerdewegh, P. (2004). Screening large-scale association study data: exploiting interactions using random forests. *BMC Genetics* 5 (December): 32. https://doi.org/10.1186/1471-2156-5-32.

Manolio, T.A., Brooks, L.D., and Collins, F.S. (2008). A HapMap harvest of insights into the genetics of common disease. *The Journal of Clinical Investigation* 118 (5): 1590–1605. https://doi.org/10.1172/JCI34772.

Merlo, C.A. and Boyle, M.P. (2003). Modifier genes in cystic fibrosis lung disease. *The Journal of Laboratory and Clinical Medicine* 141 (4): 237–241. https://doi.org/10.1067/mlc.2003.29.

Miki, Y., Swensen, J., Shattuck-Eidens, D. et al. (1994). A strong candidate for the breast and ovarian cancer susceptibility gene BRCA1. *Science (New York, N.Y.)* 266 (5182): 66–71. https://doi.org/10.1126/science.7545954.

Millstein, J., Conti, D.V., Gilliland, F.D., and James Gauderman, W. (2006). A testing framework for identifying susceptibility genes in the presence of epistasis. *American Journal of Human Genetics* 78 (1): 15–27. https://doi.org/10.1086/498850.

Moore, J.H., Hahn, L.W., Ritchie, M.D. et al. (2004). Routine discovery of complex genetic models using genetic algorithms. *Applied Soft Computing* 4 (1): 79–86. https://doi.org/10.1016/j.asoc.2003.08.003.

Moore, J.H., Gilbert, J.C., Tsai, C.-T. et al. (2006). A flexible computational framework for detecting, characterizing, and interpreting statistical patterns of epistasis in genetic studies of human disease susceptibility. *Journal of Theoretical Biology* 241 (2): 252–261. https://doi.org/10.1016/j.jtbi.2005.11.036.

Motsinger-Reif, A.A. and Ritchie, M.D. (2008). Neural networks for genetic epidemiology: past, present, and future. *BioData Mining* 1 (1): 3. https://doi.org/10.1186/1756-0381-1-3.

Nagel, R.L. (2001). Pleiotropic and epistatic effects in sickle cell anemia. *Current Opinion in Hematology* 8 (2): 105–110. https://doi.org/10.1097/00062752-200103000-00008.

Nicolae, D.L., Gamazon, E., Zhang, W. et al. (2010). Trait-associated SNPs are more likely to be EQTLs: annotation to enhance discovery from GWAS. *PLoS Genetics* 6 (4): e1000888. https://doi.org/10.1371/journal.pgen.1000888.

North, B.V., Curtis, D., Cassell, P.G. et al. (2003). Assessing optimal neural network architecture for identifying disease-associated multi-marker genotypes using a permutation test, and application to Calpain 10 polymorphisms associated with diabetes. *Annals of Human Genetics* 67 (Pt 4): 348–356.

Okazaki, T., Wang, M.-H., Rawsthorne, P. et al. (2008). Contributions of IBD5, IL23R, ATG16L1, and NOD2 to Crohn's disease risk in a population-based case-control study: evidence of gene-gene interactions. *Inflammatory Bowel Diseases* 14 (11): 1528–1541. https://doi.org/10.1002/ibd.20512.

Park, H.-W., Shin, E.-S., Lee, J.-E. et al. (2007). Multilocus analysis of atopy in Korean children using multifactor-dimensionality reduction. *Thorax* 62 (3): 265–269. https://doi.org/10.1136/thx.2006.065482.

Pauling, L. and Itano, H.A. (1949). Sickle cell anemia a molecular disease. *Science (New York, N.Y.)* 110 (2865): 543–548. https://doi.org/10.1126/science.110.2865.543.

Pendergrass, S.A., Frase, A., Wallace, J. et al. (2013). Genomic analyses with Biofilter 2.0: knowledge driven filtering, annotation, and model development. *BioData Mining* 6 (1): 25. https://doi.org/10.1186/1756-0381-6-25.

Povey, S., Burley, M.W., Attwood, J. et al. (1994). Two loci for tuberous sclerosis: one on 9q34 and one on 16p13. *Annals of Human Genetics* 58 (2): 107–127. https://doi.org/10.1111/j.1469-1809.1994.tb01881.x.

Ramos, P.S., Criswell, L.A., Moser, K.L. et al. (2011). A comprehensive analysis of shared loci between systemic lupus erythematosus (SLE) and sixteen autoimmune diseases reveals limited genetic overlap. *PLoS Genetics* 7 (12): e1002406. https://doi.org/10.1371/journal.pgen.1002406.

Risch, N.J. (2000). Searching for genetic determinants in the new millennium. *Nature* 405 (6788): 847–856.

Schaub, M.A., Boyle, A.P., Kundaje, A. et al. (2012). Linking disease associations with regulatory information in the human genome. *Genome Research* 22 (9): 1748–1759. https://doi.org/10.1101/gr.136127.111.

Schwarz, D.F., König, I.R., and Ziegler, A. (2010). On safari to random jungle: a fast implementation of random forests for high-dimensional data. *Bioinformatics (Oxford, England)* 26 (14): 1752–1758. https://doi.org/10.1093/bioinformatics/btq257.

Serretti, A. and Smeraldi, E. (2004). Neural network analysis in pharmacogenetics of mood disorders. *BMC Medical Genetics* 5 (December): 27. https://doi.org/10.1186/1471-2350-5-27.

Stratton, M.R. (1997). Pathology of familial breast cancer: differences between breast cancers in carriers of BRCA1 or BRCA2 mutations and sporadic cases. Breast cancer linkage consortium. *Lancet (London, England)* 349 (9064): 1505–1510.

Stuart, M.J. and Nagel, R.L. (2004). Sickle-cell disease. *Lancet (London, England)* 364 (9442): 1343–1360. https://doi.org/10.1016/S0140-6736(04)17192-4.

Thorlacius, S., Sigurdsson, S., Bjarnadottir, H. et al. (1997). Study of a single BRCA2 mutation with high carrier frequency in a small population. *American Journal of Human Genetics* 60 (5): 1079–1084.

Thornton-Wells, T.A., Moore, J.H., and Haines, J.L. (2004). Genetics, statistics and human disease: analytical retooling for complexity. *Trends in Genetics* 20 (12): 640–647.

Tomita, Y., Tomida, S., Hasegawa, Y. et al. (2004). Artificial neural network approach for selection of susceptible single nucleotide polymorphisms and construction of prediction model on childhood allergic asthma. *BMC Bioinformatics* 5 (September): 120. https://doi.org/10.1186/1471-2105-5-120.

Ueki, M. and Cordell, H.J. (2012). Improved statistics for genome-wide interaction analysis. *PLoS Genetics* 8 (4): e1002625. https://doi.org/10.1371/journal.pgen.1002625.

Veenstra-Vanderweele, J., Christian, S.L., and Cook, E.H. (2004). Autism as a paradigmatic complex genetic disorder. *Annual Review of Genomics and Human Genetics* 5: 379–405. https://doi.org/10.1146/annurev.genom.5.061903.180050.

Wallace, A.S. and Anderson, R.B. (2011). Genetic interactions and modifier genes in Hirschsprung's disease. *World Journal of Gastroenterology* 17 (45): 4937–4944. https://doi.org/10.3748/wjg.v17.i45.4937.

Walsh, T., Casadei, S., Coats, K.H. et al. (2006). Spectrum of mutations in BRCA1, BRCA2, CHEK2, and TP53 in families at high risk of breast cancer. *JAMA* 295 (12): 1379–1388. https://doi.org/10.1001/jama.295.12.1379.

Yu, K., Xu, J., Rao, D.C., and Province, M. (2005). Using tree-based recursive partitioning methods to group haplotypes for increased power in association studies. *Annals of Human Genetics* 69 (Pt 5): 577–589. https://doi.org/10.1111/j.1529-8817.2005.00193.x.

Zaykin, D.V. and Stanley Young, S. (2005). Large recursive partitioning analysis of complex disease pharmacogenetic studies. II. Statistical considerations. *Pharmacogenomics* 6 (1): 77–89. https://doi.org/10.1517/14622416.6.1.77.

Zielenski, J. (2000). Genotype and phenotype in cystic fibrosis. *Respiration; International Review of Thoracic Diseases* 67 (2): 117–133. https://doi.org/10.1159/000029497.

Zuvich, R.L., Bush, W.S., McCauley, J.L. et al. (2011). Interrogating the complex role of chromosome 16p13.13 in multiple sclerosis susceptibility: independent genetic signals in the CIITA-CLEC16A-SOCS1 gene complex. *Human Molecular Genetics* 20 (17): 3517–3524. https://doi.org/10.1093/hmg/ddr250.

12

Sample Size, Power, and Data Simulation

Sarah A. Pendergrass[1] *and Marylyn D. Ritchie*[2]

[1] *Human Genetics, Genentech, San Francisco, CA, USA*
[2] *Department of Genetics, Perelman School of Medicine, University of Pennsylvania, Philadelphia, PA, USA*

To call in the statistician after the experiment is done may be no more than asking him to perform a post-mortem examination: he may be able to say what the experiment died of.

—Ronald Fisher

Introduction

The goal of genetic association studies is to determine if there is a relationship between genetic variation and some complex and multifactorial trait, in a defined selection of individuals, from a population of interest. The goal of linkage analysis is to determine, using pedigree information, the gene or genes underlying a trait of interest. With either, the genetic variation data, phenotypes/traits/outcomes of interest, sampling method, and statistical tests can vary widely between studies. For instance, genetic variation data can include genome-wide single-nucleotide polymorphisms (SNPs) or single-nucleotide variants (SNVs) from independent or related individuals, or other forms of genetic variation including copy number variants (CNVs). In both kinds of studies, phenotypes, traits, and/or outcome information can include quantitative phenotypes, such as measurements of lipid levels among individuals, grouped case–control or categorical data based on a phenotypic measurement, or outcome information such as response to a drug or treatment.

An important consideration for all studies, regardless of variability in overall study design, is designing an experiment with adequate *power* to answer a question of interest with confidence. In this chapter, we describe some of the broader considerations for obtaining adequate power in designing a genetic association test for single genetic variants of common frequency (with a minor allele frequency of 1% or greater) or a linkage analysis and some of the software available for evaluating the power a study may have over a range of potential parameters. These approaches may also be applied to the study of individual rare variants for association with complex traits; as the discussion will show, such studies require larger sample sizes due to the rare exposure. Power considerations for multivariant (gene- or set-based) methods for rare-variant association studies (minor allele

This revised chapter contains material contributed by Yi-Ju Li and Susan Shao in the previous edition of this book.

Genetic Analysis of Complex Diseases, Third Edition. Edited by William K. Scott and Marylyn D. Ritchie.

frequency $< 1\%$) are not discussed within this chapter as such studies have an even larger number of assumptions that need to be considered in a power estimation and few tools currently exist for such estimates.

Sample Size and Power

In designing an experiment, a sample from the population must be selected to answer the question of interest. For example, in a study of the relationship between SNPs and lipid levels in Europeans, it is not possible to collect all genotypic and lipid-level measurements across all of Europe. Thus, a sample of individuals must be carefully selected in order to properly investigate the relationship between genetic variation and the phenotypic trait. Further, a decision needs to be made regarding having highly related individuals in the study and thus using a family-based testing association method or choosing a population-based study approach using unrelated individuals. For linkage analysis, it is critical to know whether the available pedigree information or intended study design is sufficient for the detection of the gene(s) underlying the trait of interest. The number of individuals in a sample, known as the **sample size**, is a critical component of determining the power of a study.

Following a statistical test, the null hypothesis is either rejected or not. Four possible outcomes of an experiment exist. If the null hypothesis is rejected, the investigator may be correct or incorrect. Similarly, if the null hypothesis is not rejected, the investigator may be correct or incorrect. In statistical testing, the null hypothesis is never "accepted"; evidence can only support or fail to support the alternative hypothesis.

The four possible outcomes of an experiment are shown in Table 12.1. A Type I error, quantified as α, occurs when the null hypothesis is incorrectly rejected, also known as a false-positive outcome. A Type II error, quantified as β, occurs when a false null hypothesis is not rejected, also known as a false-negative outcome. The power of the study, or the probability of detecting a correct positive result (e.g. linkage or association), is quantified by $1-\beta$. An underpowered study, without a large enough sample size to account for multiple considerations and factors of a study, can lead to inconclusive results or a false failure to reject the null hypothesis (false negative), potentially wasting time, energy, and funds. Similarly, a study in a small sample size may lead to a false-positive result that is not robust, whereby well-powered replication attempts fail to reproduce the initial finding.

Having a larger sample size than needed, while evading power issues, will use more time and resources than necessary. A larger than necessary sample size can also be an ethical consideration

Table 12.1 The four possible outcomes of an experiment.

Decision	True state of null hypothesis Null hypothesis true H_0	 Null hypothesis false H_a
Fail to reject H_0	$1-\alpha$	β (Type 2 error)
Reject H_0	α (Type 1 error)	$1-\beta$

in a pharmacological or intervention study, when it may be important to avoid enrolling more individuals than necessary in a treatment study. There are multiple practical considerations when determining sample size, including time, resources, heritability of the trait, and error in the measurement of the outcome or phenotype of interest. These are outlined in Table 12.2 and described further here.

Heritability is one important consideration when designing a genetic association study or linkage study. Is the disease, condition, outcome, or response heritable? If there is little evidence of heritability for the trait of interest, then very little can be done to adequately power a study. If the disease has low heritability (10–20%), then sample sizes may need to be very large to find a significant association or linkage. Recall that there are more details about how to determine the heritability of a trait in Chapter 3.

In genetic association and linkage studies, the potential **effect size** of a variant, or the extent of how a variant affects a trait of interest, is another consideration for determining the sample size for a study. If an effect size is likely to be large, then the sample size can be smaller but still have adequate power to detect a significant relationship between genetic variation and a phenotype. In genetic association studies, there is also a trade-off between the effect size of the variants of interest and the **allele frequency** of the variants of interest. Is the allele frequency likely to be common or rare? If the allele frequencies are likely to be very low, and the effect size of variants are modest,

Table 12.2 Practical considerations when determining sample size.

Factor	Effect
Heritability	The lower the heritability of a trait of interest, the more a sample size may need to be increased to adequately investigate potential associations
Effect size	The degree to which genetic variation impacts a trait of interest also impacts the study sample size necessary
Causative or tag SNP	In a genome-wide association study, whether or not a SNP is considered to be causative or a tag SNP, and the LD structure between causative genetic variation and a tag SNP, will affect sample size considerations
Allele frequency	If the allele frequencies are likely to be very low and the effect size of variants will be modest, then a much larger sample size will be necessary, compared to a low-frequency variant with a large effect size
Choosing individuals to contribute to the sample	If multiple factors contribute to a phenotype outcome, careful choices about who to obtain the phenotypic measurement from can improve the power of a study
Defining the phenotype	Decisions about the phenotype that will be used, including how clearly that phenotype manifests in individuals in relationship to genetic variation, is an important consideration. Phenotypic and/or genetic heterogeneity may be increased or decreased dependent on the phenotype chosen for a study
Measurement of the phenotype	How the phenotype is measured may increase or decrease the amount of error, and thus power, associated with using a specific phenotype
Recombination factor between trait and marker locus	In a linkage study, the further the marker locus is from the trait locus, the more families are required to detect a significant result

then a much larger sample size will be necessary than a low-frequency variant with a large effect size. If the genetic risks are likely to be modest, a population-based study will likely require thousands of individuals to detect those modest genetic risks.

In a genetic association study, are the variants being investigated considered to be *tag SNPs* or actual **causative variants**? The linkage disequilibrium structure between causative variants and the variants genotyped for a study will also impact power. If tag SNPs are being used, and linkage disequilibrium (LD) is likely high with causative variants, then potentially a smaller sample size is necessary to be adequately powered. The LD structure across race/ethnicity/ancestry is therefore an important consideration to take into account when designing a study, as LD structure varies within different populations, and will be affected by admixture. More details about linkage disequilibrium can be found in Chapter 8. Similarly, in a linkage study, the **recombination fraction** between a trait and marker locus affects power. More details about the recombination fraction can be found in Chapter 6.

In **choosing the sample** for a defined phenotype, if multiple factors (such as genetic heterogeneity, or nongenetic factors such as environment) potentially contribute to a phenotype outcome and are not considered in study design, the power of a study can be affected (Edwards et al. 2009; Thornton-Wells et al. 2004). Making decisions about limiting and excluding individuals to limit the impact of heterogeneity can increase power of a study to detect specific variants related to a disease. Stratification of a sample is another approach that can be used for addressing this issue, subdividing a sample based on genetic, environmental, or demographic factors. It is important to note that this may affect power through the resulting smaller sample size for each subsample after stratification. The power of a study may be increased by choosing individuals with a family history of a condition, or by selecting multiple cases across related individuals, and correcting for relatedness or using a family-based association test. However, it is important to remember that adding related individuals to a study may also decrease power if the study is unintentionally picking up more of a shared causative environmental factor.

Defining the phenotype to use in a study is an important consideration that affects power in a study. The more poorly defined a phenotype is, the more likely the genetic, environmental, and other heterogeneity will increase and subsequently decrease the power of the study (Thornton-Wells et al. 2004). One example is a study using a case or control definition for type 2 diabetes based on the answer to the question "has your doctor ever told you that you have type 2 diabetes." For patients correctly reporting their diagnosis of type 2 diabetes, there are additional factors contributing to individual-level disease such as varying environmental exposure and other conditions like high body mass index, resulting in greater heterogeneity in those included in the sample. In addition, there is a greater chance of individuals being incorrectly placed into a case or control group with this measurement of type 2 diabetes status compared to type 2 diabetes case–control status based on the clinical measurement of glycated hemoglobin levels. Thus, a greater sample size would be necessary with the first metric of type 2 diabetes to account for heterogeneity introduced with a broadly defined phenotype, as well as the greater likelihood of misclassification of patients compared to a more reliable clinical measurement of type 2 diabetes. If it is a phenotype with wide variability between individuals, the phenotype itself (although accurately measured) may require a larger sample size in order to obtain a group large enough to represent the distribution of the phenotype within the population.

The **measurement of the phenotype** itself is also a consideration. If the phenotypic measurement incorporates considerable error or bias into a study, a larger sample size may be required. On the other hand, if the measurement of the phenotype is costly or difficult and would result in extremely low sample size despite being a high-quality phenotype, using a less stringent phenotypic measurement or a combined set of phenotypic measurements may be essential in order to achieve a high enough study sample size. With the increase in studies that utilize data across multiple research centers, challenges may be faced in harmonizing phenotypes across studies, with a resultant impact on sample size through exclusion of some individuals from contributing studies to properly match phenotype definition.

Power Calculations and Simulation

To address some of the many study design considerations that have been described thus far, power calculations can be made to estimate the sample size necessary for a study with various genotypic and phenotypic parameters. Software packages that use a probabilistic framework exist for power calculations for genetic association and linkage analysis studies. However, simulation work is another viable approach for determining the power a given dataset will have. All of the parameters for a study are not likely to be known, so using in silico experiments to investigate a variety of study designs and scenarios can provide a way to optimize a robust study design, through considering some of the best and worst-case scenarios that may occur within the real data.

Power Studies for Association Analysis

If implementing a frequently used statistical approach (i.e. chi-square test or regression) or study design for a genetic association study, there are multiple user-friendly probabilistic power calculation options available for power estimates and sample size determination. However, it is important to consider simulating data for power calculations even when using a common study design, as data simulation can better reflect the complexity of the study data and possible outcomes of a study being planned. With the move to high-throughput sequencing and exploration of rare variants, simulation has been an important approach for determining what methods and study design show the most promise for an effective rare-variant study. We briefly describe here a few specific types of association studies, and in the next section detail various software tools for power calculation and simulation for association analysis for a variety of study designs.

Candidate polymorphism studies focus on a specific SNP or SNPs hypothesized to be associated with disease. This approach stems from prior evidence of an association between this SNP or SNPs and a phenotype/trait, thus there is likely to be considerable information the parameters necessary for a power calculation. This form of study is very amenable to the use of probabilistic software for power calculation.

Candidate gene studies can be considered a type of candidate polymorphism study, specifically focused on the investigation of the association between SNPs within a *single gene or genes*, and a trait of interest. Prior knowledge of the gene (or gene family), such as function, location, or evidence from a linkage study may be the driving force behind this kind of study. As with candidate polymorphism studies, there may be enough information existing in the literature about the SNPs of the study to aid in setting parameters for use with probabilistic software for power calculation.

In **genome-wide association studies** (GWAS), hundreds of thousands to several million SNPs are preselected for coverage across the genome using specific commercially available array or chip technology for genotyping. These studies take advantage of the linkage disequilibrium structure (LD) that exists between SNPs, assuming that the genotyped SNPs will be in LD with causative variants.

With the complexity of GWAS data, due to the large number of SNPs, LD, potential interactions, and potential for pleiotropy (the effect of the variation of one SNP or gene on multiple phenotypes), these studies are highly amenable to the approach of data simulation for study design and estimates of power. Simulations of various study scenarios can be used as an in silico set of pilot studies to determine the power a study would have for various variables, including parameters investigated across various ranges.

In a **Fine-mapping** study, a candidate region in the genome is chosen for high-resolution SNP genotyping. The study may include several hundred SNPs for the investigation of association. These studies are typically pursued after an initial association study that used tag SNPs, such as GWAS. The goal of the fine-mapping study is to elucidate causative loci around previously discovered tag SNPs. Power calculation software exists for estimating sample size for these kinds of studies. Again however, simulation studies are a way to better account for the complexity of the data of a fine-mapping study, and a way to explore a range of parameters for the study.

An approach combining candidate SNP, candidate gene, and the fine-mapping approach is the use of **custom chips,** developed for investigating the association between genetic variation and outcome across specific genes and regions (Cortes and Brown 2011). The design of these chips is based on previous GWAS knowledge for SNP selection with the inclusion of fine-mapping regions. With the combination of data on these chips, simulation data to account for the complexity of the data is likely the best approach.

While **population-based** approaches focus on collecting an unrelated sample of individuals in a population, **family-based** genetic association studies are another relevant method. For the studies of association using the family-based approach, the transmission disequilibrium test (TDT), or one of its derivatives for different pedigree shapes and sizes, can be used to analyze this form of data. The TDT tests for linkage in the presence of association. The affected sibpair (ASP) test is another test for family-based data. These association study designs were discussed in great detail in Chapter 8. Below, we describe existing software for calculating power for family-based genetic association studies. It is important to note, unlike population-based data, family-based data can also be used for linkage analyses.

Software for Calculating Power for Association Studies, Family- or Population-Based

A series of software packages are listed here that use probability models getting estimates of power for various kinds of studies. Some of these probability-based models can be used to create in silico families from which to draw datasets or provide parameter sweeps across various potential study designs for population-based studies. While these tools are easier to use than the dataset simulation software discussed in more detail later in this chapter, it is important to remember that with the way these models are generated, they may not sufficiently model the degree of noise, outliers, and other complexity that may exist within datasets.

PGA: Power for Genetic Association Analyses (Menashe et al. 2008)

This software is used for the power and sample size calculation for various types of genetic association studies such as candidate gene, fine mapping, and GWAS studies, with case–control based data. The software was developed for multiple possible study designs: candidate gene studies, fine-mapping studies, and genome-wide scans.

http://dceg.cancer.gov/tools/design/pga

Fine-Mapping Power Calculator (Udler et al. 2010)

This program uses a likelihood-based approach to estimate sample size for detecting a minimal set of potentially causal variants, based on effect size and degree of LD under different scenarios.
http://apps.ccge.medschl.cam.ac.uk/ocac/FineMappingPowerCalculator.html

Quanto (Gauderman 2002, 2003)

This program allows the user to determine sample size, or power at a given sample size. Study designs include matched case–control, case-sibling, case-parent, case-only, and quantitative traits, with the ability to specify gene–gene and gene–environment interactions. https://preventivemedicine.usc.edu/download-quanto

PAWE: Power for Association with Errors (Gordon et al. 2002, 2003)

This software is used for computing power for case–control association studies when genotyping errors are present. Results include power or sample size values dependent on input parameters and error level.
http://compgen.rutgers.edu/pawe

PAWE-3D (Gordon et al. 2005)

This is the related software to PAWE that is a webtool, where results (depending on parameters) are plotted as graphs, contour plots, and histograms.
https://github.com/gaow/genetic-analysis-software/blob/master/pages/PAWE-3D.md

GPC: Genetic Power Calculator (Purcell et al. 2003)

This power calculation software is for data/analysis types such as case–control association studies, quantitative trait locus (QTL) association studies for sibships and singletons, TDT for discrete traits, and epistasis.
http://zzz.bwh.harvard.edu/gpc

CaTS (Skol et al. 2006)

This software is for calculating power for case–control genetic association studies, including power calculations for two-stage genetic association studies.
http://www.sph.umich.edu/csg/abecasis/CaTS/tour.html

INPower (Park et al. 2010)

R package is available for performing power calculations for up to three-stage GWAS.
https://cran.r-project.org/web/packages/powerGWASinteraction/powerGWASinteraction.pdf

Software for Calculating Power for Transmission Disequilibrium Testing (TDT) and Affected Sib-Pair Testing (ASP)

GPC: Genetic Power Calculator (Purcell et al. 2003)

This power calculation software is for data/analysis types such as case–control association studies, QTL association studies for sibships and singletons, TDT for discrete traits, and epistasis.
http://zzz.bwh.harvard.edu/gpc

TDT-PC: Transmission Disequilibrium Test Power Calculator (Chen and Deng 2001)

This software is for computing the statistical power of TDT tests under a variety of study scenarios. There is also a built-in simulation module that can be used to simulate empirical power and Type I error rates.
http://people.virginia.edu/~wc9c/TDTPC/Download.htm

TDTASP (Brown 2004)

This software is for power and sample-size calculation for TDT and ASP.

https://biostatistics.mdanderson.org/SoftwareDownload/SingleSoftware.aspx?Software_Id=20

TDTPOWER (Ferreira et al. 2007)

This software estimates the asymptotic power of the TDT to detect association between a bi-allelic marker locus and disease status while accounting for the effects of untested loci and shared environmental factors.

https://genepi.qimr.edu.au/staff/manuelF/power_tdt/main.html

ASP/ASPSHARE

ASP is for calculating power for gene mapping using a sibpair design.

http://www.uni-kiel.de/medinfo/mitarbeiter/krawczak/download/index.html

Simulation Software for Association Study Power Assessment

There are multiple considerations when choosing a simulation method. One consideration is the quantity of genetic sequence data desired, as some methods are only useful for generating a small length of genetic sequence data. Also, the degree of computational intensity will vary dependent on the design of the simulation software and choice of model. This in turn will impact the time it takes to do a simulation, as well as the number of replications possible for a simulation. Some packages simulate data in entirety, while others resample existing natural GWA datasets for use in simulation for study design.

Not all genetic simulation packages also simulate phenotypic data. HAPSIMU (Zhang et al. 2008) is one example of software that will simulate quantitative or qualitative traits for an additive genetic model with genetic parameters defined by the user.

There is also variation among simulation packages in the necessity for further statistical analysis after the simulation stage using other software. An additional consideration is the choice of a forward or backward model simulation, described further below. While all simulations will contain some simplifications of a data model, simulation is still one of the best ways to maximize the ability to assess important relationships between complex disease and the genome, through capturing how noise and data complexity will impact analyses.

The way to use a simulation approach for evaluating power is to first simulate the genetic data and phenotypic data. Each software package will have a range of variable parameters for simulation of the genetic data, likewise for phenotypic data if the software also simulates that kind of data. Then the researcher evaluates the resultant Type I and Type II error with the given parameters selected for simulation. Because the data are simulated, parameter sweeps can be made, adjusting factors such as effect size and sample size and allele frequency in multiple ways. In this way, a researcher can determine an optimal study design with desired power, with a Type I and Type II error rate considered reasonable for a given study.

Backward and Forward Model Simulations

The coalescent model, also frequently called a "backward model simulation," was developed first for population genetics and modeling evolutionary processes, such as haplotype divergence, mutation, and migration. Alleles in this approach are followed backward/retrospectively through a population, tracing back to a common ancestor. After defining the common ancestor, individuals are assigned to a coalescent tree. These simulations can be used to simulate sequence data and can be used to model complex population structures.

Forward simulations evolve a population, forward in time, generation by generation, similar to what can be observed in actual populations over time. These kinds of simulations allow the user to specify different scenarios including any genetic or environmental factors and allow a user to observe populations at different time points. Forward simulations are less efficient than coalescent approaches; however, with the increase in computational power as well as more efficient algorithms for forward simulations, these are now a more generally accessible simulation choice.

Coalescent Model Simulation – Short Genetic Sequences

ms (Hudson 2002) This software generates chromosomal segment samples for a population evolving under a Wright–Fischer neutral model.

http://home.uchicago.edu/~rhudson1/source/mksamples.html

SimCoal (Laval and Excoffier2004) Varying lengths of DNA can be simulated.

http://cmpg.unibe.ch/software/simcoal

CoaSIM (Mailund et al. 2005) It can be used to simulate samples of SNP, micro-satellite, and other haplotypes/genotype data, with further use as a way to construct cases/controls for a simulated association study.

https://github.com/mailund/CoaSim

Larger Coalescent Simulated Models

SNPSim (Posada and Wiuf 2003) This is for the generation of SNP haplotypes and diploid biallelic genotypes.

https://popmodels.cancercontrol.cancer.gov/gsr/packages/snpsim GENOME (Liang et al. 2007)

This program is for generating sequences on a whole genome scale where the resultant simulated data can be used with genome-wide study approaches.

http://csg.sph.umich.edu/liang/genome

MaCS: Markovian Coalescent Simulator (Chen et al. 2009) This software was developed to efficiently computationally generate simulated data by a modification of the coalescent model to enable simulating extremely long stretches of DNA.

http://github.com/gchen98/macs

Forward Model Simulations – Short Genetic Sequences

FreGene (Chadeau-Hyam et al. 2008) This is a forward selection model to simulate a single chromosome.

www.ebi.ac.uk/sites/ebi.ac.uk/files/groups/kersey_service_team/FREGEN/documentation_html.html

SimCoal2 (Laval and Excoffier 2004) This tool is for simulating complex recombination patterns under various demographic models.

http://cmpg.unibe.ch/software/simcoal2

Forward Model Simulations – Large Genetic Sequences

EASYPOP (Balloux 2001) This is a computer program for population genetics simulation, where output includes the multilocus genotypes after a given number of generations.

https://www.unil.ch/dee/en/home/menuinst/open-positions-and-public-resources/softwares-dataset/softwares/easypop.html

SIMUPOP (Peng and Amos 2008; Peng and Kimmel 2005) Non-random mating populations can be simulated with this software, where a large number of SNP markers can be generated.
http://simupop.sourceforge.net

ForSim (Lambert et al. 2008) This is used to do a forward simulation to develop a population, with user-specified number of pedigrees.

GenomeSimla (Edwards et al. 2008) This is a forward-time simulation software, where patterns of LD in family-based or case–control data can be generated.
https://github.com/RitchieLab/genomeSIMLA

Nemo 2.2 (Guillaume and Rougemont 2006) This is a forward-time simulation software that can be used for studying population genetics. This software can model up to two correlated quantitative traits, with an additive model for traits and the loci are pleiotropic.
http://nemo2.sourceforge.net

Resampling Simulation Tools

Another way to estimate power while designing a study is through resampling existing biological datasets. This allows for maintaining some of the complexity inherent in a natural dataset, while introducing specific genetic effect sizes and sample sizes to evaluate the power of analysis methods.

HAP-SAMPLE (Wright et al. 2007) A web-based tool for simulating case–control and affected-child trio studies by resampling Phase I/II HapMap SNP data. The user provides a list of SNPs and the disease model for the simulation. This tool can be used for candidate regions as well as genome-wide simulations.
https://sites.google.com/a/umich.edu/leeshawn/software

HAPSIMU (Zhang et al. 2008) This uses real data from the HapMap project and can simulate datasets for qualitative and quantitative traits.
http://l.web.umkc.edu/liujian

GWAsimulator (Li and Li 2008) A whole-genome simulation program for the simulation of case–control data from genomic SNP chips includes the option of supplying HapMap data for simulation purposes.
http://biostat.mc.vanderbilt.edu/wiki/Main/GWAsimulator

HAPGEN2 (Spencer et al. 2009; Su et al. 2011) This software uses HAPMAP3 and the 1000 genomes haplotype data as reference panels, and the software can simulate data for many populations. An additional R package is available for simulating interactions between multiple SNPs.
https://mathgen.stats.ox.ac.uk/genetics_software/hapgen/hapgen2.html

Software for Simulation of Phenotypic Data

Phenosim (Günther et al. 2011) This software simulated phenotypic data for genotypic data generated in coalescent simulations.
http://evoplant.uni-hohenheim.de/downloads

Power Simulations for Linkage Analysis

In common and genetically complex disorders, given the suspected underlying heterogeneity of most complex diseases (e.g. genetic and/or environmental contributions), estimating the underlying genetic model in advance is virtually impossible. Power simulations are a way to investigate possible scenarios for effectively approaching linkage studies.

Definitions for Power Assessments for Linkage Analysis

The **average logarithm of odds (LOD) score** is calculated as the sum of the LOD scores at a selected value of the recombination fraction divided by the number of replicates. This result is most usefully interpreted in conjunction with its standard deviation, which provides an assessment of the variability of the LOD score. This value provides the investigator with an estimate of how much information, on average, should be expected from the available pedigree information at a particular recombination fraction.

The **maximum attainable LOD score** is the highest LOD score obtained at any value of the recombination fraction in the entire set of replicates. This value allows the investigator to gain insight into the full information content of the available pedigree material, since the true maximum attainable LOD score is the highest LOD score when all meioses are fully informative under the assumed model and allele frequencies. Note however, the true maximum attainable LOD score may not be attained even in a series of multiple replicates if, by chance, not all the meioses are informative. For review of linkage analysis, see Chapter 6.

In considering the significance of a particular LOD score, it is useful to know whether all the available linkage information has been captured. The observed maximum LOD score of a pedigree may not match the maximum attainable LOD score for a pedigree for several reasons, including lack of informativeness of key individuals and differences in allele frequencies from simulated results and true results, differences between true and simulated penetrance, among others.

The **average maximum LOD score** is the sum of the highest LOD scores obtained in each replicate, regardless of where those scores occur on the LOD score curve, divided by the number of replicates. Note that the highest LOD score for any replicate need not, and usually will not, occur at the true recombination fraction. There are many caveats associated with the use of this particular measure of information content of a pedigree, the most compelling of which is that it tends to paint too rosy a picture of the true expected outcomes. Note that while the average LOD score is additive across pedigrees, the average maximum LOD score is not, which is an additional disadvantage to using the average maximum LOD score as a measure of pedigree information within a study.

Available pedigree simulation packages will also directly compute the Type I and Type II error rates and the power of the available material for specific constant values of the LOD score. For instance, the investigator may need to know the probability of obtaining a false-positive score (a LOD score of 3.0 or greater when the data were simulated assuming free recombination between the disease and marker locus), the probability of obtaining a false-negative score (a LOD score of −2.0 or less or some other specified constant value when the disease and marker loci were fixed as linked to one another), and the probability of correctly obtaining a LOD score of 3.0 or more when the disease and marker loci are set as linked to one another. In most cases, the investigator will choose to simulate hundreds or even thousands of replicates to obtain estimates of the necessary values (and their standard deviations when appropriate) that are as precise as possible.

Another useful application of computer simulation methods is in determining the empiric *p*-value associated with a study outcome. By generating genetic marker genotypes under the assumption of free recombination between the disease and marker locus for many replicates of the pedigree structures, the investigator can generate a series of LOD score results that could be obtained from a random genomic screen. By counting all simulated LOD scores exceeding the LOD score from the observed data and dividing by the total number of replicates, the investigator can find the empiric *p*-value for the observed data (Weeks et al. 1990). For this application, since the observation of the outcome is so rare, the investigator will typically perform tens of thousands of replicates to obtain a reasonable estimate of the empiric *p*-value.

Yet another application of computer simulation involves assessing the effect of maximization over models, which may be a more powerful approach than assuming a genetic model when the true model is unknown (Hodge et al. 1997). An investigator can choose to analyze several different genetic models (e.g. different degrees of dominance, disease allele penetrance, and frequency) with the same marker data to see which model gives the highest LOD score (Elston 1989; Greenberg 1989). As a general rule of thumb, the critical value of the test statistic should be increased by a LOD value of 0.3 when one of the foregoing parameters is maximized and by 0.6 when both penetrance and dominance/recessiveness are maximized.

Computer Simulation Methods for Linkage Analysis of Mendelian Disease

Available software packages that all allow for simulation of a marker loci linked to a single disease locus include SIMLINK, SLINK, and SIMULATE. SLINK and SIMLINK allow the simulation of genetic marker data for a locus linked to a disease gene and these two programs are very similar. One major algorithmic difference between the two programs is in the way they assign genetic marker genotypes given information on the trait phenotypes of members in the pedigree ("conditional" on disease status): SLINK uses a series of risk calculations and SIMLINK first traverses the entire pedigree, assigning disease genotypes conditional on disease phenotypes. The most critical differences from a practical perspective are that SIMLINK allows a thorough evaluation of the power and Type I error rates in one analysis, while multiple analyses must be performed to generate the same information with SLINK. However, SLINK allows the simulated data to be analyzed under a different model than that used to generate the data, which SIMLINK does not. All three programs allow simulation of a marker loci unlinked to a disease locus. Listed below are these three programs as well as five other software packages available for simulation of linkage study data for investigating the power a study would have with a range of parameters.

SIMLINK (Boehnke and Ploughman 1997)

The user collects family history data and then can use this software to determine if the number of families collected will be large enough to sufficiently detect linkage. The trait of interest can be autosomal or X-linked.

http://csg.sph.umich.edu/boehnke/simlink.php

SLINK: Simulation Program for Linkage Analysis (Ott 1989)

SLINK generates a series of simulated multilocus genotypes for a pedigree linked to a single-trait locus. The LINKAGE package, or additional packages such as MSIM, ISIM, and LSIM, can then be used to generate summary statistics and allow further analysis of the simulated data.

http://www.jurgott.org/linkage/SLINK.htm

A faster version rewritten in C is available:
https://www.ncbi.nlm.nih.gov/CBBresearch/Schaffer/fastlink.html

SUP: Slink Utility Program (Lemire 2006; Schäffer et al. 2011)

The number of markers that can be used with SLINK is technically not limited; however using more than five markers may take considerable computational time. Thus, a wrapper for SLINK named SUP was developed to make using a much larger number of markers more computationally feasible.

ALLEGRO (Gudbjartsson et al. 2000, 2005)

Allegro can be used to simulate multi-locus marker data either under linkage or no linkage. Allegro allows for analysis of extremely large pedigrees.
https://www.decode.com/software/allegro

MERLIN: Multipoint Engine for Rapid Likelihood Inference (Cook Jr 2002)

This is for the simulation of markers unlinked to a trait of interest for determining false-positive rates.
http://csg.sph.umich.edu/abecasis/Merlin/download

SimPED (Leal et al. 2005)

This is for simulating data with dense maps of SNP markers, generating haplotypes and/or genotypes for a large number of loci regardless of the pedigree structure.
http://bioinformatics.org/simped/doc.html

Power Studies for Linkage Analysis – Complex Disease

In complex diseases, power studies are not as straightforward as in Mendelian diseases but can help determine whether a particular disorder is amenable to gene mapping. Simulation can help approximate estimates of the sample size required to detect linkage making various assumptions about the underlying genetic model. Guidelines for sample size depend on how closely related the sampled individuals (usually affected) are, how tightly linked the disease and marker loci/locus are, how polymorphic the marker system is, and the mode of inheritance of the trait. These issues are frequently interrelated, and reliable estimates for these parameters are often unknown; consequently, developing standard guidelines is difficult and pursuing parameter sweeps through simulation can be useful.

Two of the primary assumptions that must be made about an underlying genetic model involve the frequency of the trait allele and the mode of inheritance of the disease-predisposing locus. The frequency of the trait allele affects power and sample size calculations as well: if the disease allele is rare, then the probability that two affected individuals within a pedigree are identical by descent (IBD) at the trait locus is high, whereas if the trait allele is common, the IBD probability decreases. Considered heuristically, the contribution of a distantly related affected relative pair (ARP) such as an affected cousin pair should be greater than the contribution of an ASP when the number and frequency of contributing loci are small. It is a priori more striking if a pair of affected cousins (who share only one-eighth of their genetic material in common by chance alone) share alleles in common than if a pair of affected siblings (who share half of their genetic material in common by chance alone) share alleles in common.

Inclusion of Unaffected Siblings

Much of the work in assessing power and sample size in complex disease for discrete traits has focused on the use of ASPs. The decision about whether to include unaffected siblings in a study is usually based on availability and cost of recruitment and the costs associated with the additional genotyping of these individuals. The advantages of collecting unaffected siblings are many. For instance, the inclusion of the genotypic marker data on unaffected siblings may increase the ability to determine the IBD status of genetic markers in ASPs, which is especially important in late-onset disorders where parents are unavailable for sampling. Additionally, as pointed out by Elston et al. (1996), performing a linkage study utilizing only ASPs is akin to performing a study without controls. Without the unaffected individuals, it is virtually impossible to determine whether a significant result is due to a true linkage or to some other factor, such as meiotic drive at a marker locus. The collection and availability of unaffected siblings also allows for the eventual assessment of penetrance and expression and for considering genes that may potentially modify the phenotype. Although some available methods allow for the inclusion of unaffected relatives, they may make the assumption that unaffected relatives are non-gene carriers. If the underlying disease gene has low penetrance, as is generally hypothesized for common variants influencing risk of common complex diseases, such an assumption may be unrealistic.

Affected Relative Pairs of Other Types

In many complex diseases, other types of ARPs will be identified; indeed, trends for different types in a disease may shed some light on the underlying genetic models. For instance, consider a trait in which two different loci are involved. If the disease alleles at the loci are rare and act epistatically, it is unlikely to be observed in ARPs who are more distantly related than siblings. However, if the disease alleles at the loci are relatively common and have low penetrance, the pedigrees may demonstrate affected avuncular or cousin pairs. Because it is unusual for a whole study to be designed around ARPs other than siblings, estimates of sample size for studies specifically geared to other types of ARPs are rare. When other types of ARPs are frequent, the investigator will frequently collect all affected patients and their connecting relatives and utilize an analysis approach other than one dependent on a sibpair design.

Other Considerations

Some difficulties in the estimates of sample size or power of available pedigree material in the investigation of complex traits warrant consideration. For instance, the presence of phenocopies or genetic heterogeneity will increase the amount of material necessary to correctly identify a linkage (Bishop and Williamson 1990). Additionally, the lower the heterozygosity of the marker to be tested, the more the material will be needed on average to detect a significant effect.

Genomic Screening Strategies: One-Stage versus Two-Stage Designs

Interestingly, although much literature is focused on the use of ASPs to study complex disease, it frequently is not the most powerful approach. Elston et al. (1996) point out that for a wide range of scenarios in which the genomic screening is done via a two-stage process (sometimes called a grid-tightening approach), the optimal sampling strategy for ARPs affected grandparent–grandchild pairs because of the limited opportunity for recombination. A two-stage genomic screen involves genotyping a certain number of genetic markers and then following up regions of interest defined by reaching or exceeding a certain predefined significance level. This is done by genotyping markers on either side of the "interesting" markers. Grandparent–grandchild pairs are difficult to find, particularly if the disorder has onset in adulthood. After grandparent–grandchild pairs, the optimal sampling strategies in terms of study cost usually follow the sequence of half-siblings followed by avuncular pairs followed by sibpairs. Affected cousin pairs have varying degrees of utility depending on several considerations. In general, as the relative risk ratio (λ) increases, so does the power of affected cousin pairs. Ease of sampling is almost the reverse, however, as sibpairs are generally the easiest to collect.

The computing algorithm DESPAIR is available through the Statistical Analysis for Genetic Epidemiology (S.A.G.E., http://darwin.cwru.edu/sage) package. DESPAIR is valuable for designing the optimal two-stage genomic screening strategy and also for estimating the number of pairs of affected relatives required to attain a certain power.

From a practical viewpoint, decisions regarding whether to genotype parents will depend on the relative costs of genotyping in the first stage or second stage and on the investigator's knowledge about underlying sample homogeneity (and how important the availability of complete IBD information will be relative to the required genotyping – in other words, whether allele frequencies will make a big difference in the outcome). As always, decisions about strategies will depend on the interplay between correctly detecting a linkage and following up false positives.

Software for Designing Linkage Analysis Studies of Complex Disease

SIMLA (Bass et al. 2004; Schmidt et al. 2005)

Multiple options are available with this software package, including specifying the number and location of disease markers and loci. The degree of marker disequilibrium with a disease locus and other markers can also be specified for this package, as well as pedigree size and rules for family ascertainment. Gene–gene and Gene–environment interactions can also be simulated.

http://hihg.med.miami.edu/software-download/simla-simulation-software

Quantitative Traits

Methods of mapping human QTLs can be applied to phenotypes ranging from height and weight to obesity, blood pressure, and any other trait that is measurable on a continuous scale. Assessing power and sample size is dependent on the particular analysis strategy that will be used. Some alternatives are discussed in the following sections.

Extreme Discordant Pairs

The sampling of extremely discordant relative pairs is a potentially powerful approach to identifying genes that predispose to human quantitative traits. Specifically, such relative pairs (usually siblings) should share very few genes IBD at the trait locus (Risch and Zhang 1995, 1996). Risch and Zhang (1996) have shown that the smallest genetic effect that can be identified using this approach with a reasonable sample size (i.e. 1500) is one that accounts for as little as 10% of the variance. Suppose, for instance, a researcher wants to detect a gene that accounts for 10% of the underlying phenotypic variance, has a trait allele frequency of 0.30, and has no residual correlation by employing a sample scheme that includes pairs of relatives in the top and bottom 10% of the overall distribution. Approximately 1482 sibpairs would be required to detect linkage allowing for a Type I error rate of 0.0001 and power of 0.80. In general, the presence of residual correlation between siblings, in which siblings appear to be more phenotypically similar than expected by the presence of a major genetic locus, will tend to decrease the necessary sample size, especially as the heritability decreases. As explained by Risch and Zhang (1996), this reduction in estimated sample size is because phenotypically discordant sibpairs will be more likely to be genetically discordant at the locus of interest. This is important because there is likely to be residual sibling correlation due to other genetic effects for loci with low heritability. With high heritability, most sibling correlation is probably due to that locus, and hence there is unlikely to be a large residual correlation.

Although in retrospect the collection of extremely discordant or extremely concordant relative pairs rather than pairs with less extreme phenotypes seems intuitive, this idea has served to revolutionize the consideration of human QTLs. There are some potential drawbacks with either approach. To ascertain the extremes of the distribution, hundreds or thousands of families may need to be screened. If the screening tests are expensive, time-consuming, or logistically difficult, such an approach may not be feasible. However, if such screening can be undertaken with relative ease, a much smaller dataset can be subjected to the rigors of a genomic screen and ultimate fine-mapping of susceptibility genes.

Sampling Consideration for the Variance Component Method

The power of the variance component approach in general varies according to several factors, such as genetic and environmental effects and the resemblance between pedigree members. However, the heritability of the QTL is a key component of the power calculation. The greater the QTL effect is, the greater the statistical power is for a given sample size. A QTL with small effect may require considerably larger sample sizes (either larger family size or larger family numbers) to be detectable with comparable power (Williams and Blangero 1999). The details of sample size and power for the variance component method were discussed in Williams and Blangero (1999). The residual additive heritability (a^2, the proportion of the total genetic variance that is not explained by the QTL) and the QTL heritability (q^2) are the main parameters for the sample size and power estimations. An example of sample size prediction for 80% power at a LOD score of 3.0 is listed in Table 12.3. For instance, a QTL with 40% heritability under 0.3 residual additive heritability requires 713 sibpairs, or 559 nuclear families of family size 4 (parents and two siblings), to reach

Table 12.3 Number of sibships and nuclear families in variance component linkage analysis.

	$a^2 = 0.0$		$a^2 = 0.3$	
n_{sibs}	$q^2 = 0.1$	$q^2 = 0.4$	$q^2 = 0.1$	$q^2 = 0.1$
Sibships				
2	16 500 (33 000)	921 (1 842)	14 732 (29 463)	713 (1 426)
3	5 476 (16 429)	294 (881)	4 697 (14 090)	217 (652)
4	2 728 (10 910)	142 (567)	2 270 (9 079)	102 (408)
Nuclear families				
2	16 358 (65 431)	836 (3 342)	13 369 (53 474)	559 (2 235)
3	5 433 (27 166)	272 (1 361)	4 354 (21 771)	182 (908)
4	2 708 (16 246)	134 (801)	2 136 (12 818)	89 (533)

Note: Number of sampling units and number of individuals (in parentheses) required for 80% power at a LOD score of 3.0/Entried have been rounded to the nearest integer.
Source: Williams and Blangero (1999).

80% power to detect linkage. Overall, additional QTL heritability can reduce the sample size more rapidly than additional residual additive genetic heritability. Collecting parental phenotypes concurrently with sibling data is more efficient than ascertaining additional sibships.

Software for Designing Linkage Analysis Studies for Quantitative Traits

SOLAR: Sequential Oligogenic Linkage Analysis Routines (Almasy and Blangero 1998)

This software can be used for quantitative trait simulation. An extension of the original SOLAR package called SOLAR-Eclipse is maintained at the University of Maryland at Baltimore:

http://solar-eclipse-genetics.org/brief-overview.html

MERLIN: Multipoint Engine for Rapid Likelihood Inference (Cook Jr 2002)

This software is used for the simulation of markers unlinked to a trait of interest for determining false-positive rates.

http://csg.sph.umich.edu/abecasis/merlin/index.html

SimuPOP (Peng and Amos 2008; Peng and Kimmel 2005)

Here, populations can be generated, and pedigrees can be sampled from these populations.

http://simupop.sourceforge.net/Main/Features

Summary

This chapter has detailed several motivations for estimating the power a study will have under various parameters. Investigating a priori the considerations for study design is recommended as studies take time and resources. Many of these methods can be used as well in an ad hoc fashion, to investigate hypotheses formed after the analysis of results and further explore and/or investigate questions that arise after a study has been completed.

Obtaining estimates of power and sample size to design a robust study can improve investigating the complex architecture that exists between the genome and phenotypic traits and outcomes. Many approaches are available for calculating power for various scenarios, from analytic power calculations to data simulation, which can even include resampling natural data. Thus, even with a novel analysis methodology for genomic data or a very complex data structure, approaches are available for aiding study design. These power calculations can provide a sense for whether statistical methodologies are performing as expected, or whether modifications to the study design, sample size, or analysis approach need to be made.

References

Almasy, L. and Blangero, J. (1998). Multipoint quantitative-trait linkage analysis in general pedigrees. *Am. J. Hum. Genet.* 62: 1198–1211.

Balloux, F. (2001). EASYPOP (version 1.7): a computer program for population genetics simulations. *J. Hered.* 92: 301–302.

Bass, M.P., Martin, E.R., and Hauser, E.R. (2004). Pedigree generation for analysis of genetic linkage and association. *Pac. Symp. Biocomput. Pac. Symp. Biocomput.* 2004: 93–103.

Bishop, D.T. and Williamson, J.A. (1990). The power of identity-by-state methods for linkage analysis. *Am. J. Hum. Genet.* 46: 254–265.

Boehnke, M., and Ploughman, L.M. (1997). SIMLINK: a program for estimating the power of a proposed linkage study by computer simulation. Version 4.12, 2 April 1997. https://csg.sph.umich.edu/boehnke/simlink.php.

Brown, B.W. (2004). Power calculations for the transmission/disequilibrium and affected sib pair tests using elementary probability methods. *Genet. Res.* 83: 133–141.

Chadeau-Hyam, M., Hoggart, C.J., O'Reilly, P.F. et al. (2008). Fregene: simulation of realistic sequence-level data in populations and ascertained samples. *BMC Bioinform* 9: 364.

Chen, W.M. and Deng, H.W. (2001). A general and accurate approach for computing the statistical power of the transmission disequilibrium test for complex disease genes. *Genet. Epidemiol.* 21: 53–67.

Chen, G.K., Marjoram, P., and Wall, J.D. (2009). Fast and flexible simulation of DNA sequence data. *Genome Res.* 19: 136–142.

Cook, E.H. Jr. (2002). Merlin: faster linkage analysis with improved genotyping error detection. *Pharmacogen. J.* 2: 139–140.

Cortes, A. and Brown, M.A. (2011). Promise and pitfalls of the Immunochip. *Arthritis Res. Ther.* 13: 101.

Edwards, T.L., Bush, W.S., Turner, S.D. et al. (2008). Generating Linkage Disequilibrium Patterns in Data Simulations Using genomeSIMLA. In: *Evolutionary Computation, Machine Learning and Data Mining in Bioinformatics. EvoBIO 2008*, Lecture Notes in Computer Science, vol. 4973 (eds. E. Marchiori and J.H. Moore). Berlin, Heidelberg: Springer https://doi.org/10.1007/978-3-540-78757-0_3.

Edwards, T.L., Lewis, K., Velez, D.R. et al. (2009). Exploring the performance of Multifactor Dimensionality Reduction in large scale SNP studies and in the presence of genetic heterogeneity among epistatic disease models. *Hum. Hered.* 67: 183–192.

Elston, R.C. (1989). Man bites dog? The validity of maximizing LOD scores to determine mode of inheritance. *Am. J. Med. Genet.* 34: 487–488.

Elston, R.C., Guo, X., and Williams, L.V. (1996). Two-stage global search designs for linkage analysis using pairs of affected relatives. *Genet. Epidemiol.* 13: 535–558.

Ferreira, M.A.R., Sham, P., Daly, M.J., and Purcell, S. (2007). Ascertainment through family history of disease often decreases the power of family-based association studies. *Behav. Genet.* 37: 631–636.

Gauderman, W.J. (2002). Sample size requirements for association studies of gene-gene interaction. *Am. J. Epidemiol.* 155: 478–484.

Gauderman, W.J. (2003). Candidate gene association analysis for a quantitative trait, using parent-offspring trios. *Genet. Epidemiol.* 25: 327–338.

Gordon, D., Finch, S.J., Nothnagel, M., and Ott, J. (2002). Power and sample size calculations for case-control genetic association tests when errors are present: application to single nucleotide polymorphisms. *Hum. Hered.* 54: 22–33.

Gordon, D., Levenstien, M.A., Finch, S.J., and Ott, J. (2003). Errors and linkage disequilibrium interact multiplicatively when computing sample sizes for genetic case-control association studies. *Pac. Symp. Biocomput. Pac. Symp. Biocomput.* 2003: 490–501.

Gordon, D., Haynes, C., Blumenfeld, J., and Finch, S.J. (2005). PAWE-3D: visualizing power for association with error in case–control genetic studies of complex traits. *Bioinformatics* 21: 3935–3937.

Greenberg, D.A. (1989). Inferring mode of inheritance by comparison of LOD scores. *Am. J. Med. Genet.* 34: 480–486.

Gudbjartsson, D.F., Jonasson, K., Frigge, M.L., and Kong, A. (2000). Allegro, a new computer program for multipoint linkage analysis. *Nat. Genet.* 25: 12–13.

Gudbjartsson, D.F., Thorvaldsson, T., Kong, A. et al. (2005). Allegro version 2. *Nat. Genet.* 37: 1015–1016.

Guillaume, F. and Rougemont, J. (2006). Nemo: an evolutionary and population genetics programming framework. *Bioinforma. Oxf. Engl.* 22: 2556–2557.

Günther, T., Gawenda, I., and Schmid, K.J. (2011). Phenosim – a software to simulate phenotypes for testing in genome-wide association studies. *BMC Bioinform* 12: 265.

Hodge, S.E., Abreu, P.C., and Greenberg, D.A. (1997). Magnitude of type I error when single-locus linkage analysis is maximized over models: a simulation study. *Am. J. Hum. Genet.* 60: 217–227.

Hudson, R.R. (2002). Generating samples under a Wright-Fisher neutral model of genetic variation. *Bioinforma. Oxf. Engl.* 18: 337–338.

Lambert, B.W., Terwilliger, J.D., and Weiss, K.M. (2008). ForSim: a tool for exploring the genetic architecture of complex traits with controlled truth. *Bioinforma. Oxf. Engl.* 24: 1821–1822.

Laval, G. and Excoffier, L. (2004). SIMCOAL 2.0: a program to simulate genomic diversity over large recombining regions in a subdivided population with a complex history. *Bioinforma. Oxf. Engl.* 20: 2485–2487.

Leal, S.M., Yan, K., and Müller-Myhsok, B. (2005). SimPed: a simulation program to generate haplotype and genotype data for pedigree structures. *Hum. Hered.* 60: 119–122.

Lemire, M.S.U.P. (2006). An extension to SLINK to allow a larger number of marker loci to be simulated in pedigrees conditional on trait values. *BMC Genet.* 7: 40.

Li, C. and Li, M. (2008). GWAsimulator: a rapid whole-genome simulation program. *Bioinforma. Oxf. Engl.* 24: 140–142.

Liang, L., Zöllner, S., and Abecasis, G.R. (2007). GENOME: a rapid coalescent-based whole genome simulator. *Bioinforma. Oxf. Engl.* 23: 1565–1567.

Mailund, T., Schierup, M.H., Pedersen, C.N.S. et al. (2005). CoaSim: a flexible environment for simulating genetic data under coalescent models. *BMC Bioinform* 6: 252. https://doi.org/10.1186/1471-2105-6-252.

Menashe, I., Rosenberg, P.S., and Chen, B.E. (2008). PGA: power calculator for case-control genetic association analyses. *BMC Genet.* 9: 36.

Ott, J. (1989). Computer-simulation methods in human linkage analysis. *Proc. Natl. Acad. Sci. U. S. A.* 86: 4175–4178.

Park, J.-H., Wacholder, S., Gail, M.H. et al. (2010). Estimation of effect size distribution from genome-wide association studies and implications for future discoveries. *Nat. Genet.* 42: 570–575. https://doi.org/10.1038/ng.610. Epub 2010 Jun 20.

Peng, B. and Amos, C.I. (2008). Forward-time simulations of non-random mating populations using simuPOP. *Bioinforma. Oxf. Engl.* 24: 1408–1409.

Peng, B. and Kimmel, M. (2005). simuPOP: a forward-time population genetics simulation environment. *Bioinforma. Oxf. Engl.* 21: 3686–3687.

Posada, D. and Wiuf, C. (2003). Simulating haplotype blocks in the human genome. *Bioinforma. Oxf. Engl.* 19: 289–290.

Purcell, S., Cherny, S.S., and Sham, P.C. (2003). Genetic power calculator: design of linkage and association genetic mapping studies of complex traits. *Bioinforma. Oxf. Engl.* 19: 149–150.

Risch, N. and Zhang, H. (1995). Extreme discordant sib pairs for mapping quantitative trait loci in humans. *Science* 268: 1584–1589.

Risch, N.J. and Zhang, H. (1996). Mapping quantitative trait loci with extreme discordant sib pairs: sampling considerations. *Am. J. Hum. Genet.* 58: 836–843.

Schäffer, A.A., Lemire, M., Ott, J. et al. (2011). Coordinated conditional simulation with SLINK and SUP of many markers linked or associated to a trait in large pedigrees. *Hum. Hered.* 71: 126–134.

Schmidt, M., Hauser, E.R., Martin, E.R., and Schmidt, S. (2005). Extension of the SIMLA package for generating pedigrees with complex inheritance patterns: environmental covariates, gene-gene and gene-environment interaction. *Stat. Appl. Genet. Mol. Biol.* 4: Article15.

Skol, A.D., Scott, L.J., Abecasis, G.R., and Boehnke, M. (2006). Joint analysis is more efficient than replication-based analysis for two-stage genome-wide association studies. *Nat. Genet.* 38: 209–213.

Spencer, C.C.A., Su, Z., Donnelly, P., and Marchini, J. (2009). Designing genome-wide association studies: sample size, power, imputation, and the choice of genotyping chip. *PLoS Genet.* 5: e1000477.

Su, Z., Marchini, J., and Donnelly, P. (2011). HAPGEN2: simulation of multiple disease SNPs. *Bioinforma. Oxf. Engl.* 27: 2304–2305.

Thornton-Wells, T.A., Moore, J.H., and Haines, J.L. (2004). Genetics, statistics and human disease: analytical retooling for complexity. *Trends Genet. TIG* 20: 640–647.

Udler, M.S., Tyrer, J., and Easton, D.F. (2010). Evaluating the power to discriminate between highly correlated SNPs in genetic association studies. *Genet. Epidemiol.* 34: 463–468.

Weeks, D.E., Lehner, T., Squires-Wheeler, E. et al. (1990). Measuring the inflation of the LOD score due to its maximization over model parameter values in human linkage analysis. *Genet. Epidemiol.* 7: 237–243.

Williams, J.T. and Blangero, J. (1999). Power of variance component linkage analysis to detect quantitative trait loci. *Ann. Hum. Genet.* 63: 545–563.

Wright, F.A., Huang, H., Guan, X. et al. (2007). Simulating association studies: a data-based resampling method for candidate regions or whole genome scans. *Bioinforma. Oxf. Engl.* 23 (19): 2581–2588.

Zhang, F., Liu, J., Chen, J., and Deng, H.-W. (2008). HAPSIMU: a genetic simulation platform for population-based association studies. *BMC Bioinform* 9: 331.

Index

Page numbers in *italics* denote figures and numbers in **bold** denote tables.

Genetic Analysis of Complex Diseases, Third Edition. Edited by William K. Scott and Marylyn D. Ritchie.

h

j

k

l

m